Resilient Cities and Land Take Effective Management Through Sustainable Urban Planning Tools

Resilient Cities and Land Take Effective Management Through Sustainable Urban Planning Tools

Guest Editors

Debora Anelli
Pierluigi Morano
Marco Locurcio
Francesco Tajani

MDPI

Basel • Beijing • Wuhan • Barcelona • Belgrade • Novi Sad • Cluj • Manchester

Guest Editors
Debora Anelli
DICATECh
Polytechnic University of Bari
Bari
Italy

Pierluigi Morano
Department of Civil
Engineering and Architecture
Polytechnic University of Bari
Bari
Italy

Marco Locurcio
DICATECh
Polytechnic University of Bari
Bari
Italy

Francesco Tajani
Department of Architecture
and Design
Sapienza University of Rome
Rome
Italy

Editorial Office
MDPI AG
Grosspeteranlage 5
4052 Basel, Switzerland

This is a reprint of the Special Issue, published open access by the journal *Land* (ISSN 2073-445X), freely accessible at: https://www.mdpi.com/journal/land/special_issues/O8S2U6887Z.

For citation purposes, cite each article independently as indicated on the article page online and as indicated below:

Lastname, A.A.; Lastname, B.B. Article Title. *Journal Name* **Year**, *Volume Number*, Page Range.

ISBN 978-3-7258-6348-8 (Hbk)
ISBN 978-3-7258-6349-5 (PDF)
https://doi.org/10.3390/books978-3-7258-6349-5

Contents

land

MDPI

Editorial

Resilient Cities and Land Take Effective Management Through Sustainable Urban Planning Tools

Debora Anelli [1,*], Pierluigi Morano [2], Marco Locurcio [2] and Francesco Tajani [1]

[1] Department of Architecture and Design, Sapienza University of Rome, Via Flaminia 359, 00196 Rome, Italy; francesco.tajani@uniroma1.it

[2] Department of Civil, Environmental, Land, Building Engineering and Chemistry, Polytechnic University of Bari, Via Orabona 4, 70125 Bari, Italy; pierluigi.morano@poliba.it (P.M.); marco.locurcio@poliba.it (M.L.)

* Correspondence: debora.anelli@uniroma1.it

1. Introduction

In the contemporary context of cities' increasing exposure to climate risks and the fragility of urban infrastructure and soil productivity, the concept of urban resilience is central to sustainable planning processes [1–6]. Cities are increasingly becoming subject to extreme events and complex transformations; therefore, they require effective tools to adapt to ongoing changes and manage territorial resources prudently and equitably. It is of principal importance to optimize the use of these resources and avoid their unsustainable exploitation [7,8].

This new edition of the present Special Issue builds on the previous one [9], broadening the scope of analysis and strengthening the focus on the connection between urban resilience and limiting land use, through the use of practical tools and novel evaluation models [10–14]. The main international frameworks for sustainable development remain benchmarks, including the UN 2030 Agenda, the European Green Deal, and the latest Recovery and Resilience Plans. Additionally, the New Urban Agenda (2016), which aims to reorient urban planning and management towards inclusive, safe, resilient, and sustainable cities, and the OECD Principles on Urban Policy (2019), which promote integrated, inclusive, intelligent, and sustainable urban policies, are considered. Finally, the Global Sustainable Development Report (2023) calls for a move beyond sectoral approaches and promotes transformative governance [15–17]. It is precisely with this multisectoral perspective that the present Special Issue positions itself [18–21].

This collection explores the development of resilient cities from a multidisciplinary and multiscale perspective. The covered topics include urban planning tools, spatial modeling approaches, composite indicators, digital technologies, and economic and financial support for integrated strategies. Particular focus is given to urban regeneration processes, risk management, adaptive planning, and the integration of environmental, social, and infrastructural components.

The ten articles published in this Special Issue demonstrate a variety of adopted approaches and the relevance of challenges addressed in different geographical contexts. The methodologies employed span multi-criteria assessment models, spatial and predictive analysis techniques, and field surveys and empirical case studies. The contributions can be divided into four macro-areas (Figure 1).

Figure 1. Macro-areas of the contributions in this Special Issue.

These macro-areas comprise the following:

- The assessment and management of urban and infrastructural resilience, with methodological proposals for measuring the interaction between risks and the adaptive capacity of cities;
- Planning tools and policies for land take-over mitigation, including analysis of land use in rapidly urbanizing areas and strategies for rational land use;
- Innovation and urban governance through smart cities and digitalization, with experimental evidence on the benefits of digital transformation for urban innovation;
- Social inclusion, equity, and participation in urban regeneration, taking into account the impacts of public policies and public–private partnerships.

Overall, the research included in this edition provides valuable theoretical, methodological, and practical contributions, helping to advance the scientific landscape regarding the potential and challenges of planning resilient urban areas. It is hoped that these reflections will inspire new insights and support public and private decision-makers in developing comprehensively sustainable and inclusive urban strategies.

2. Authors' Institutions

This Special Issue showcases 10 papers published by authors based at 18 different international universities and research institutes, as well as graduate schools and colleges, as shown in Table 1.

Table 1. Overview of authors' affiliations and institutions.

Typology	Number of Institutions	Institution
Public universities	11	• University of Illinois at Urbana-Champaign, USA • Nanning Normal University, China • City University of Hong Kong • Swinburne University of Technology, Australia • Victoria University, Australia • Zhejiang University, China • Texas A&M University, USA • Sichuan University, China • Chengdu University, China • Instituto Superior Técnico, University of Lisbon, Portugal • Northwest University, China
Private universities	4	• University of Nottingham Ningbo (China) • Kyonggi University (Republic of Korea) • Lusófona University (Portugal) • Universidad de las Américas Puebla (UDLAP) (Mexico)
Academic research centers (public or private)	6	• Key Laboratory of Environment Change and Resources Use in Beibu Gulf (Ministry of Education, China) • Centre for Innovation in Territory, Urbanism and Architecture (CiTUA), Instituto Superior Técnico, University of Lisbon (Portugal) • RCM2+ (Research Centre in Asset Management and System Engineering), Lusófona University (Portugal) • Institute of Silk Road Studies, Northwest University (China) • Collaborative Research Centre for Archaeology of the Silk Roads, Northwest University (China) • InLAB (Territory and Urban Artifact Research Group), Universidad de las Américas Puebla (Mexico)
Government research centers (technical or professional)	1	• Zhengzhou Urban Planning Institute
Government agencies	1	• Sichuan Provincial Water Resources Department

The organizations involved in this Special Issue are geographically and institutionally diverse, which confirms the international and multidisciplinary nature of research on urban resilience and sustainable land management. Public universities represent the largest component, with 11 institutions spread across the USA, China, Australia, Portugal, and Hong Kong. This prevalence can be explained by the fact that resilience, land use, and urban innovation are strongly debated and focused on in public universities in developed countries, thanks to the following: national policies regarding ecological transition and climate neutrality; investment in urban applied research; and close collaboration between cities, local authorities, and communities.

Meanwhile, the influence of private universities and academic research centers in different countries highlights growing attention to emerging or developing contexts where rapid urbanization and land management present challenges that require innovative, adapted solutions. Finally, contributions from governmental and technical institutions confirm the operational interest in the topics addressed and the need for effective tools in planning practice.

3. Main Topics Addressed

The articles in this Special Issue, selected for their scientific and applicative relevance, are organized into four thematic macro-areas that reflect the main research strands that emerge: tools for the assessment of urban resilience; land take-over management and spatial transformations; urban innovation and smart cities; and public policies, participation, and inclusion.

1. *Tools and models for the assessment of urban resilience*

This category comprises contributions that propose methodologies to assess urban or infrastructural resilience through quantitative models, complex networks, or geodesign approaches (Figure 2).

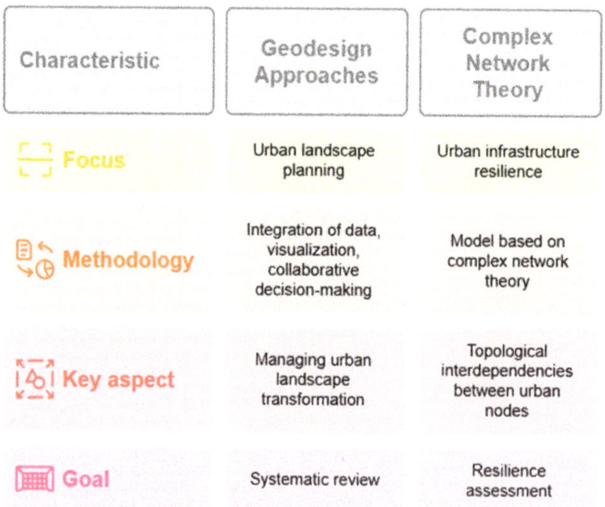

Figure 2. Main features of papers related to tools and models for assessment of urban resilience.

- *A Review on the Use of Geodesign Processes in Managing Urban Landscape Transformation di Bo Pang and Brian Deal* outlines a systematic review of geodesign approaches applied to urban landscape planning and transformation, with a focus on the integration of data, visualization and collaborative decision-making.
- *Evaluation of Urban Infrastructure Resilience Based on Complex Network Theory* di Dong, W., Zhou, Y., Guo, D., Chen, Z., & Wang, J. proposes an innovative model based on complex network theory to assess the resilience of urban infrastructures, considering topological interdependencies between urban nodes.

2. *Land take-over management and spatial transformations*

These articles analyze the dynamics of urban expansion, land consumption, the growth of cities in changing contexts, and fiscal and financial policies related to territorial government (Figure 3).

Characteristic	Urban Expansion Analysis	Land Finance Analysis	Silk Road Network Analysis
Focus	Urban sprawl in Africa	Land finance, local debt	Cultural networks, urban evolution
Methodology	Remote sensing data, spatial analysis	Economic data analysis	Interactive spatial model
Geographic Region	Africa	China	Silk Road

Figure 3. Main features of papers related to land take-over management and spatial transformations.

- *Analysis of the Spatial Pattern of Urban Expansion in African Countries from 1990 to 2018* by Liu, B., Xie, S., Chen, M., Yao, N., & Liu, W. presents a large-scale study in which remote sensing data and spatial analysis techniques were used to describe urban sprawl in Africa and its implications for land management.
- In *Land Finance, Local Government Debt and Economic Growth: Evidence from China*, Zhao, Y., Peng, S., Zhang, Q., Wang, Y., Gong, C., & Lu, X. analyze the link between land finance, local debt, and urban development, highlighting risks and opportunities in tax policies related to urbanization.
- *Reconstructing the Silk Road Network: Insights from a Large-Scale Spatial Interaction Model* by Shen, Y., Liu, J., Han, J., & Wan, X. details how an interactive spatial model was used to reconstruct the historical and geographical structure of the Silk Road network, offering insights into how cultural networks influence urban evolution.

3. *Urban innovation and smart cities*

Additionally, this Special Issue includes studies exploring the role of digital technologies, smart indicators, and urban innovation strategies for sustainable and resilient development (Figure 4).

Feature	Chinese Smart Cities	Urban Modeling in Global South
Focus	Performance comparison	Systematic Review
Metrics	Innovation, digitalization, sustainability, governance	Methodological gaps, sustainable digitalization

Figure 4. Main features of papers related to urban innovation and smart cities.

- In *Promoting Urban Innovation through Smart Cities: Evidence from the Top 20 Smart Cities in China*, Ji, M., Jin, M., Chen, L., Liu, Y., & Tian, Y. compare the performance of 20 Chinese smart cities through indicators of innovation, digitalization, sustainability, and urban governance.
- *Urban Modeling in the Global South and Sustainable Development: A Systematic Review* by Anne K. Kurjenoja provides a systematic review of urban models in the context of the Global South, highlighting methodological gaps and potential for the sustainable digitalization of planning practices.

4. *Public policies, participation, and inclusion*

This category features contributions that deepen insights into the interaction between policies, local actors, and decision-making processes, with a focus on social justice, territorial equity, and stakeholder involvement (Figure 5).

Feature	Study of Karachi	Study of Houston	Study of Rio de Janeiro
Focus	Barriers to active transport	Post-disaster recovery decisions	PPP in urban regeneration
Key factor	Regulatory, cultural and infrastructural barriers	Social, political and economic influences	Risks of social exclusion, benefit distribution

Figure 5. Main features of papers related to public policies, participation, and inclusion.

- In *Barriers Affecting Promotion of Active Transportation: A Study in Karachi, Pakistan*, Olayode, I.O., Chau, H.W., & Jamei, E. carry out an analysis of the regulatory, cultural, and infrastructural obstacles to the promotion of active mobility in a complex and high-density urban context.
- *Home Elevation Decisions in Post-Disaster Recovery: Social and Policy Influences in Houston, TX* by García, I., Tao, Z., Orduña, J., Martínez-Román, L., & Welideniya, W. is a qualitative study that explores the social, political, and economic factors that influence post-disaster reconstruction and adaptation decisions.
- In *Public–Private Partnerships in Urban Regeneration Projects: The Case of Porto Maravilha, Rio de Janeiro*, Vale de Paula, P., Cunha Marques, R., & Gonçalves, J.M. critically assess the effectiveness of PPP in urban regeneration processes, analyzing the risks of social exclusion and imbalances in the distribution of benefits.

The figure below (Figure 6) graphically summarizes the network of conceptual relationships that connects the ten articles published in this Special Issue. Each major node represents a paper, identified by a unique color and a shortened title label. The minor nodes, connected to the first ones by arrows of the same color, represent the main topics addressed. The circular organization allows the immediate visualization of the connections both within and across the studies.

Figure 6. Network graph of mutual connections between published papers and main topics.

This visualization shows how some topics emerge as recurring, acting as conceptual bridges between multiple contributions. This is the case, for example, for the following:

- *Urban Planning* links to at least four articles, including those on urban resilience in Zhengzhou, flood risk management through geodesign, and the analysis of urban expansion models in Africa. This highlights the centrality of planning as a cognitive infrastructure capable of integrating diverse (technological, social, and environmental) and transversal approaches.
- *Smart Governance and Urban Innovation* are addressed in numerous articles, including those focused on areas that are apparently distant from one another, such as urban regeneration through PPP or the promotion of active mobility.
- *Green Economy, Socio-Territorial Planning, and Neoliberal Urbanism* trace a common thread between studies focused on varying national and international contexts (China, Brazil, the Global South), but they are united via reflection on the relationship between economic models and urban realities.

Further, the links between the papers—mediated by common topics—also suggest the presence of significant conceptual clusters:

An initial nucleus develops around the themes of urban resilience and vulnerability, connected through keywords such as *flood adaptation*, *housing vulnerability*, and *risk-resilience index*.

A second cluster revolves around urban innovation, in which the roles of digital governance, smart city advancement, sustainable mobility, and participatory urban design are intertwined.

A third axis links studies with geopolitical and cultural relevance, such as those on the Silk Road and planning in the Global South, which share references to UNESCO, *heritage, spatial analysis*, and *the Global South*.

Overall, the graph not only visually represents the thematic richness of this Special Issue but also demonstrates the scientific cohesion and interconnection that characterize it. More than a simple collection, it is thus configured as a comprehensive ecosystem, where concepts, methods, and territories intertwine, generating new trajectories of interdisciplinary research.

4. Conclusions

The ten articles collected in this Special Issue offer a rich and detailed cross-section of the main challenges that contemporary cities face in their progress towards sustainability, inclusiveness, and resilience. While these studies address different geographical and thematic contexts—from the Global South to advanced economies and from territorial planning to urban finance—it is possible to identify some common trajectories and shared reflections.

First of all, the structural role of territorial inequality significantly emerges. Whether it involves urban evictions in Mexico due to gentrification driven by neoliberal models or post-disaster housing vulnerability in the United States, such inequality becomes a mirror of economic and social imbalances, often exacerbated by political decisions and planning tools. A second transversal thematic axis concerns the conflict between urban growth and environmental sustainability. The analysis of urban sprawl in Africa and the use of shared socioeconomic pathways (SSPs) show that development projections often ignore the fragility of local ecosystems. Meanwhile, one article on land finance in China shows how financial logic can hinder the green transition, if not accompanied by adequate regulatory tools.

The circular visualization presented in this Editorial reinforces these conclusions, highlighting the density of thematic connections between the papers and the ways in which concepts such as "urban planning", "green economy", or "digitalization" act as hubs around which multiple contributions are structured. These interconnections signal not only theoretical convergence but also the significance of a holistic approach to managing cities, capable of simultaneously focusing on design, governance, technology, and social justice.

Finally, many of the presented contributions suggest that the future of sustainable cities will depend on the ability to integrate multidisciplinary and multi-scalar perspectives, from digital technologies for geodesign and risk management to collaborative governance in urban regeneration processes. The theme that transverses this entire collection of research is that urban complexity is not only a technical challenge but a significant political issue.

In this sense, this Special Issue represents an important theoretical and applied contribution for scholars, administrators, and planners, offering analytical tools and operational insights to build fairer, more adaptive, and intelligent cities [22–26].

Author Contributions: This Editorial is the result of the joint work of all authors. All authors have read and agreed to the published version of the manuscript.

Funding: This research received no external funding.

Acknowledgments: We thank the Guest Editor for their constant support, the reviewers for their productive and insightful suggestions, all the authors for submitting their high-quality scientific contributions to this Special Issue, and the MDPI team for making the publication process efficient and successful.

Conflicts of Interest: The authors declare no conflicts of interest.

References

1. Mendizabal, M.; Heidrich, O.; Feliu, E.; García-Blanco, G.; Mendizabal, A. Stimulating urban transition and transformation to achieve sustainable and resilient cities. *Renew. Sustain. Energy Rev.* **2018**, *94*, 410–418. [CrossRef]
2. Mitchell, D.; Enemark, S.; Van der Molen, P. Climate resilient urban development: Why responsible land governance is important. *Land Use Policy* **2015**, *48*, 190–198. [CrossRef]
3. Datola, G. Implementing urban resilience in urban planning: A comprehensive framework for urban resilience evaluation. *Sustain. Cities Soc.* **2023**, *98*, 104821. [CrossRef]
4. Palla, A.; Pezzagno, M.; Spadaro, I.; Ermini, R. Participatory Approach to Planning Urban Resilience to Climate Change: Brescia, Genoa, and Matera—Three Case Studies from Italy Compared. *Sustainability* **2024**, *16*, 2170. [CrossRef]
5. Guo, N.; Wu, F.; Sun, D.; Shi, C.; Gao, X. Mechanisms of resilience in cities at different development phases: A system dynamics approach. *Urban Clim.* **2024**, *53*, 101793. [CrossRef]
6. Acuti, D.; Bellucci, M. Resilient cities and regions: Planning, initiatives, and perspectives. In *Climate Action*; Springer: Berlin/Heidelberg, Germany, 2020; pp. 763–774.
7. Barbarossa, L.; Pappalardo, V.; Martinico, F. Building the resilient city. Strategies and tools for the city masterplan. *UPLanD-J. Urban Plan. Landsc. Environ. Des.* **2018**, *3*, 15–24.
8. Moraci, F.; Errigo, M.F.; Fazia, C.; Burgio, G.; Foresta, S. Making less vulnerable cities: Resilience as a new paradigm of smart planning. *Sustainability* **2018**, *10*, 755. [CrossRef]
9. Locurcio, M.; Tajani, F.; Anelli, D. Sustainable urban planning models for new smart cities and effective management of land take dynamics. *Land* **2023**, *12*, 621. [CrossRef]
10. Jha, A.K.; Miner, T.W.; Stanton-Geddes, Z. (Eds.) *Building Urban Resilience: Principles, Tools, and Practice*; World Bank Publications: Washington, DC, USA, 2013.
11. Hao, H.; Wang, Y.; Chen, J. Empowering Scenario Planning with Artificial Intelligence: A Perspective on Building Smart and Resilient Cities. *Engineering* **2024**, *43*, 272–283. [CrossRef]
12. Anwar, M.R.; Sakti, L.D. Integrating artificial intelligence and environmental science for sustainable urban planning. *IAIC Trans. Sustain. Digit. Innov. (ITSDI)* **2024**, *5*, 179–191. [CrossRef]
13. Bocca, A. Sustainable development and proximity city. The environmental role of new public spaces. *TeMA-J. Land Use Mobil. Environ.* **2024**, *17*, 71–87.
14. Karimi, F.; Sultana, S. Urban expansion prediction and land use/land cover change modeling for sustainable urban development. *Sustainability* **2024**, *16*, 2285. [CrossRef]
15. Sicuaio, T.; Zhao, P.; Pilesjo, P.; Shindyapin, A.; Mansourian, A. Sustainable and resilient land use planning: A multi-objective optimization approach. *ISPRS Int. J. Geo-Inf.* **2024**, *13*, 99. [CrossRef]
16. Mehryar, S.; Sasson, I.; Surminski, S. Supporting urban adaptation to climate change: What role can resilience measurement tools play? *Urban Clim.* **2022**, *41*, 101047. [CrossRef]
17. Kalfas, D.; Kalogiannidis, S.; Chatzitheodoridis, F.; Toska, E. Urbanization and land use planning for achieving the sustainable development goals (SDGs): A case study of Greece. *Urban Sci.* **2023**, *7*, 43. [CrossRef]
18. Asadzadeh, A.; Fekete, A.; Khazai, B.; Moghadas, M.; Zebardast, E.; Basirat, M.; Kötter, T. Capacitating urban governance and planning systems to drive transformative resilience. *Sustain. Cities Soc.* **2023**, *96*, 104637. [CrossRef]
19. Sharifi, A.; Yamagata, Y. Resilience-oriented urban planning. In *Resilience-Oriented Urban Planning: Theoretical and Empirical Insights*; Springer: Berlin/Heidelberg, Germany, 2018; pp. 3–27.
20. Suleimany, M.; Gonbad, M.R.S.; Naghibizadeh, S.; Niri, S.D. Artificial intelligence as a tool for building more resilient cities in the climate change era: A systematic literature review. In *Artificial Intelligence and Machine Learning Applications for Sustainable Development*; CRC Press: Boca Raton, FL, USA, 2025; pp. 60–81.
21. Almulhim, A.I. Building Urban Resilience Through Smart City Planning: A Systematic Literature Review. *Smart Cities* **2025**, *8*, 22. [CrossRef]
22. Morano, P.; Tajani, F.; Guarnaccia, C.; Anelli, D. An optimization decision support model for sustainable urban regeneration investments. *WSEAS Trans. Environ. Dev.* **2021**, *17*, 1245–1251. [CrossRef]
23. Locurcio, M.; Tajani, F.; Anelli, D.; Ranieri, R. A multi-criteria composite indicator to support sustainable investment choices in the built environment. *Valori E Valutazioni* **2022**, *30*, 85–100. [CrossRef]
24. Locurcio, M.; Tajani, F.; Morano, P.; Anelli, D.; Manganelli, B. Credit risk management of property investments through multi-criteria indicators. *Risks* **2021**, *9*, 106. [CrossRef]

25. Calbick, K.S.; Day, J.C.; Gunton, T.I. Land use planning implementation: A 'best practices' assessment. *Environ. A J. Interdiscip. Stud.* **2003**, *31*, 69–82.
26. Jones, C.; Baker, M.; Carter, J.; Jay, S.; Short, M.; Wood, C. *Strategic Environmental Assessment and Land Use Planning: An International Evaluation*; Taylor & Francis: Abingdon, UK, 2013.

land

MDPI

Article

Public-Private Partnerships in Urban Regeneration Projects: The Brazilian Context and the Case of "*Porto Maravilha*" in Rio de Janeiro

Paula Vale de Paula [1,*], Rui Cunha Marques [2] and Jorge Manuel Gonçalves [1]

[1] Centre for Innovation in Territory, Urbanism and Architecture (CiTUA), Instituto Superior Tecnico, University of Lisbon, 1049-001 Lisboa, Portugal; jorgemgoncalves@tecnico.ulisboa.pt

[2] Research Centre in Asset Management and System Engineering (RCM2+), Department of Industrial Engineering and Management, Lusofona University, 1749-024 Lisboa, Portugal; rui.marques@ulusofona.pt

* Correspondence: paulavaledepaula@tecnico.ulisboa.pt

Abstract: Urban regeneration projects have been part of urban planning in different cities to recover strategic areas and combat new land consumption. Often, Public-Private Partnerships (PPPs) agreements are signed to meet the demands of these projects. In the Brazilian context, the general PPP law classifies concessions into two types: sponsored concessions or administrative concessions. Urban regeneration projects in the Brazilian context can be carried out through different policies and instruments, such as those in the City Statute. Despite this, there are several obstacles in how these practices are developed, which require greater attention to the enabling environment and the Critical Success Factors (CSFs) that can improve the performance of these public procurement practices. Based on this, this study aims to analyze the Brazilian context of PPP arrangements in urban regeneration projects and the specific case of the "*Porto Maravilha*" in Rio de Janeiro. To this scope, the study adopts a previously established CSF framework. The data analyzed are derived from secondary sources obtained through documentation collection and primary sources obtained through interviews. Regarding the PPP process, the results indicate financial and execution problems. Concerning the project, which remains unfinished, the results point to a transformation of the port area, but with significant social impacts.

Keywords: Public-Private Partnerships (PPPs); urban regeneration; Critical Success Factors (CSFs); Brazil; *Porto Maravilha*

1. Introduction

In different urban contexts, the intense urbanization currently being experienced is exerting strong pressure on cities. Today, more than half of the world's population lives in cities. The highest urban growth rates are found in developing countries, leading to extensive suburbanization of peripheral areas or agglomeration in slums [1].

In the face of growing and intense urbanization, measures must be considered to achieve greater sustainability in urban contexts, which is one of the Sustainable Development Goals (SDGs) proposed by the United Nations—SDG 11, Sustainable Cities and Communities.

Urban regeneration practices are presented as a path to sustainability. They respond to decline urban degradation and decline scenarios, giving new functions to obsolete areas and offering an alternative to the urbanization of new areas [2].

Nevertheless, urban regeneration practices are often complex, involving different risks and requiring high investments. In addition, the public sector, as the provider of various types of infrastructure and public services, often does not have the resources to meet all the current urban demands. In this sense, private sector action through Public-Private Partnership (PPP) agreements can be strategic in urban regeneration and developing urban regeneration projects.

Urban regeneration projects carried out through PPP agreements are used to develop projects in different urban contexts. Among the project typologies, waterfront projects stand out. In many cases, former ports and central areas in the city are considered strategic for the city from an urban point of view and, therefore, interesting to regenerate. The regeneration of these areas through PPP arrangements is often related to spaces for culture, leisure, and recreation, seeking to give the city a new image. In some cases, the regeneration processes of these areas may also be motivated by mega-events such as the World Cups, the Olympic Games, or World Exhibitions. Within neoliberalist logic, they can be associated with processes of exclusion and gentrification [2].

On the other hand, the use of PPP arrangements presents both positive and negative points. The advantages of using PPP arrangements include incorporating private sector capacity, knowledge, and innovation, sharing risk with the private sector, reducing the overall cost of the project through greater private sector efficiency, and increasing the probability of meeting project deadlines [3,4].

Some critical points can also be observed in the use of PPP arrangements. Some of the essential points are additional costs that, if not managed properly, can compromise economic viability; loss of responsibility; mitigation of unexpected risks that are eventually assumed by the public sector; and management difficulties over long periods of time [3,4].

Therefore, urban regeneration practices carried out through PPP arrangements can be controversial. They fail to achieve the initial objectives proposed and result in negative consequences from an urban, social, and financial point of view.

In Brazil, PPP arrangements are carried out through two modalities: concessionary PPP or sponsored PPP [5]. On the other hand, urban regeneration practices are carried out mainly in large and medium-sized cities seeking to recover their central areas. Urban regeneration practices can be carried out through different policies and instruments, such as those found in the City Statute [6].

The urban regeneration of Rio de Janeiro's port area through the "Porto Maravilha" project is an interesting and complex example of a Public-Private Partnership. The ambitious project, whose PPP arrangement was formalized in 2009, envisioned the urban regeneration of 500 ha of the central and port area [7]. Despite developing important parts of the project, such as museums and open spaces, the project was not completed due to problems that compromised the PPP arrangement.

Considering this scenario, this study aims to analyze the Brazilian context regarding PPP arrangements in urban regeneration projects and the specific case of "Porto Maravilha". The study focuses on PPP arrangements in urban regeneration projects from a perspective that is not restricted to urban planning and includes important contractual and management aspects of these practices.

The study identifies best practices through Critical Success Factors (CSFs). CSFs refer to elements, conditions, events, and activities that must be considered for an organization or a project to achieve its goals and ensure its success [8,9].

The analysis aims to identify which critical success factors are observed in the Brazilian context and in the specific case study, and which factors can be better worked on. Meeting the proposed objective, the study also aims to answer the following research questions:

- How are PPP arrangements understood and applied in urban regeneration projects in the Brazilian context?

- Which CSFs are most relevant in the Brazilian context? Which factors could be better addressed?
- Which CFSs are most relevant in the *"Porto Maravilha"* case study? Which factors could be better addressed?

To this scope, the study uses a previously established framework, which considers clusters related to the external environment related to the PPP projects—a country's enabling environment that facilitates or does not facilitate the implementation of partnerships and exogenous factors found in that country—and the internal environment of the PPP projects—referring to the partnership itself and its different phases [10].

The study consists of five sections: this brief introduction, a second section on the research methodology, a third section analyzing the Brazilian context and presenting the case study, a fourth section discussing the results obtained, and a final section drawing conclusions.

2. Materials and Methods

The study aims to analyze the Brazilian context of PPP agreements in urban regeneration projects, also delving into a specific case of a PPP agreement for urban regeneration of a waterfront area—the case of the Port of Rio de Janeiro and the *Porto Maravilha* partnership. The analysis focuses on the CSFs observed in the Brazilian context and the PPP analyzed, and the issues that could be improved.

To this scope, the study uses a previously established framework of CSFs in PPP agreements for urban regeneration projects [10]. The framework considers five distinct clusters, two relate to the external environment, and three to the internal environment of the partnership. The external environment includes the enabling environment and exogenous factors. The internal environment includes the three phases of the partnership: the preparation phase, the procurement phase, and the contract management phase (Figure 1).

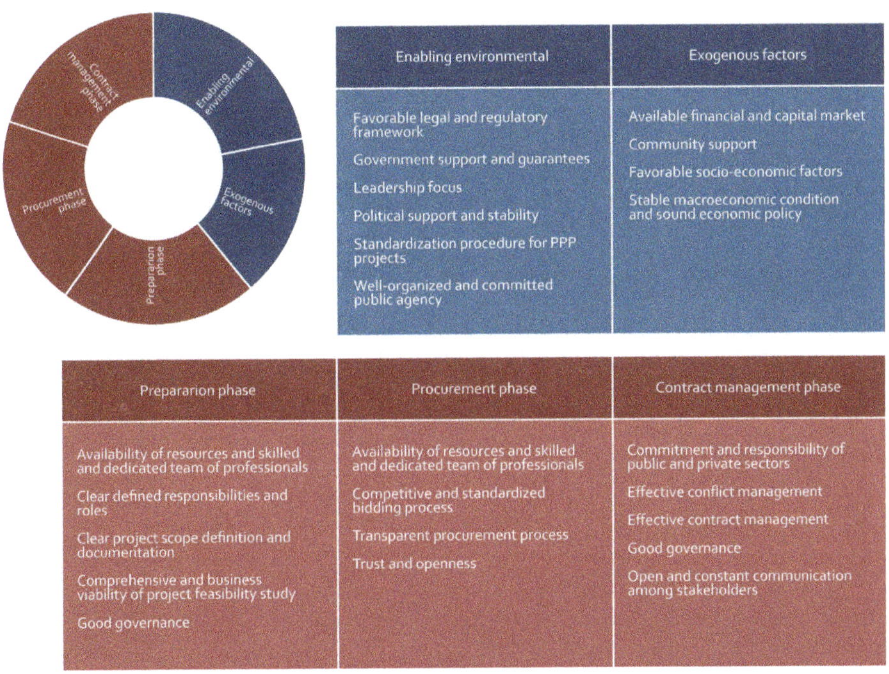

Figure 1. Framework of PPP CSFs for Urban Regeneration Projects. Source: [10]—adapted.

The enabling environment consists of the entire environment external to the PPP projects, which can be controlled in such a way as to facilitate the development of partnerships. Concerning the enabling environment, six factors were identified:

- Favorable legal and regulatory framework: laws and regulations that follow the country's legislation at its different levels and that facilitate the development of partnerships, protecting the rights and interests of all parties.
- Government support and guarantees: Government support in the form of guarantees ensures the project's financial viability and creditworthiness.
- Leadership focus: the capacity of the public sector to exercise its leadership role in implementing the partnership.
- Political support and stability: political stability at different levels beyond the period of a political mandate that can guarantee the continuity of the partnership.
- Standardization procedure for PPP projects: standardization of the process to assure the public sector that the partnership is carried out in accordance with governmental and local objectives and to facilitate the private sector to prepare for the partnership.
- Well-organized and committed public agency: organized and committed public body, with technical capacity and experience in partnerships.

Exogenous factors include factors that cannot be controlled, but which also influence the development of PPP projects. The following factors were identified:

- Available financial and capital market: available financial and capital markets allow for low financing costs and offer a diversified range of financial products.
- Community support: The community's acceptance, understanding, and participation are important to ensure the smooth running of the PPP, avoid delays, facilitate the hiring of local labor, provide projects valued by citizens, manage expectations, and mitigate disagreements.
- Favorable socio-economic factors: a favorable socio-economic scenario in which the public sector and the population have a robust payment capacity, creating an environment that encourages partnerships.
- Stable macroeconomic conditions and sound economic policy: sound economic policy capable of creating a stable and growing economic environment, allowing for a low cost of capital and greater certainty in the market, thus reducing risks.

For the preparation phase, the following factors were identified:

- Availability of resources and a skilled and dedicated team of professionals: because these are long processes, it is necessary to guarantee resources for the PPP during the transition; furthermore, it is essential to have a dedicated and competent team of professionals to accompany the entire PPP.
- Clearly defined responsibilities and roles: the different roles and functions of the actors involved have been clearly defined already in the preparation phase.
- Clear project scope definition and documentation: clear definition, already in the preparation phase, of the project documentation, including objective, scope, geographical area, technical project, functional responsibilities, and previous experience in similar projects.
- Comprehensive and business viability of project feasibility study: a comprehensive business feasibility study of the project, including technical feasibility studies, financial and socio-economic analyses, and realistic assessments of costs and benefits.
- Good governance: managing the relationships between the actors involved is essential at all stages of the PPP, including the preparatory phase.

For the procurement phase, the following factors were identified:

- Availability of resources and a skilled and dedicated team of professionals: just as in the preparatory phase, the availability of resources and a team to accompany the PPP is fundamental during the contracting phase.
- Competitive and standardized bidding process: a standardized process must define the rules a priori and present them to bidders, facilitating the preparation of bids and allowing them to be fairly compared; the competitive process, in turn, guarantees the best cost-benefit ratios.
- Transparent procurement process: a transparent procurement process that reduces transaction costs and shortens negotiation and contracting times.
- Trust and openness: openness between the parties to create trust and a solid relationship, increasing their willingness to cooperate.

Finally, regarding the contract management phase, the following factors were identified:

- Commitment and responsibility of public and private sectors: the commitment of both parties to the objectives of the partnership and the project, including allocating their best resources.
- Effective conflict management: effective and efficient conflict management, as early as possible, and involving those responsible in both sectors, to avoid aggravating the conflict and compromising the partnership.
- Effective contract management: effective and efficient contract management by controlling costs, time, security, materials, techniques, and skills to avoid renegotiations and compromising the partnership.
- Good governance: managing the relationships between the different players involved is important at all stages of the PPP arrangements, including the contract management phase.
- Open and constant communication among stakeholders: openness and communication between all the actors involved—not just the public and private sectors—ensures that the interests and concerns of all parties are managed, increasing trust and commitment to project management.

The external environment will be used to analyze the Brazilian context in general, while the PPP's internal environment will guide the analysis of the specific case study.

The study was carried out through desk-based research on the Brazilian context and the specific case study. The document analysis includes secondary data referring to laws, regulations, official documents, reports, scientific articles, and promotional information about the Porto Maravilha project.

To complement the case study analysis, primary data were extracted from interviews conducted with stakeholders involved in the case study. The use of interviews as a primary data collection method was extremely important since it allowed us to hear different stakeholders involved in the process, with different points of view that can often be contradictory.

In this sense, five interviews were conducted with stakeholders or experts in the Porto Maravilha partnership and project in March and April 2025 (Table 1). The interviews were conducted online in a semi-structured format based on open-ended questions (Table A1—Appendix A).

Table 1. Interviews—interviewee profile.

Interviewee 1	Former president of the public company responsible for Porto Maravilha
Interviewee 2	Former employee of the Private Consortium
Interviewee 3	Employee of the Port Authority of Rio de Janeiro

Table 1. *Cont.*

Interviewee 4	Representative of the local community (Morro da Providencia)
Interviewee 5	Academic and researcher on the Porto Maravilha project

3. Results

3.1. Brazilian Context

To analyze the Brazilian context based on the clusters identified previously, the study focused first on the enabling environment and, subsequently, on the exogenous factors that influence the Brazilian scenario on PPP arrangements in urban regeneration projects.

3.1.1. Enabling Environment in Brazil

In Brazil, the legal framework for PPP projects is based mainly on Law No. 11,079/2004—"*Normas gerais para licitação e contratação de parceria público-privada no âmbito da administração pública*" (General rules for bidding and contracting public-private partnerships within the public administration). This law establishes general rules for bidding and contracting PPP projects within the sphere of powers of the Federal Government, the States, the Federal District, and the Municipalities. The law defines the PPP arrangement as "the administrative concession contract, in the sponsored or administrative modality" [5].

The sponsored concession corresponds to the concession of public services and goods when it "involves, in addition to the tariff charged to users, pecuniary consideration from the public partner to the private partner" [5]. In addition to Law No. 11,079/2004, the sponsored concession is also governed by Law No. 8987/1995, which provides for the concession and permission regime for the provision of public services provided for in art. 175 of the Federal Constitution and makes other provisions [11].

The administrative concession, in turn, corresponds to "a contract for the provision of services of which the Public Administration is the direct or indirect user, even if it involves the execution of infrastructure or the supply and installation of goods" [5]. In addition to Law No. 11,079/2004, the administrative concession is also governed by Law No. 8987/1995 and Law No. 9074/1995, which establish rules for granting and extending concessions and permissions for public services and make other provisions [12].

Law No. 11,097/2004 also refers to important issues related to PPP arrangements. The document defines that it is forbidden to enter into partnerships in the following circumstances: where the value of the contract is less than BRL 10,000,000.00 (ten million reais—1.6 million USD); where the period for providing the service is less than five years; or where the sole purpose is to supply labor, supply and install equipment, or carry out public works [5].

The document also defines some guidelines to be observed when signing PPP projects, namely: efficiency in fulfilling state missions and in the use of society's resources; respect for the interests and rights of the recipients of the services and the private entities entrusted with their execution; non-excludability of regulatory and jurisdictional functions, the exercise of police power and other activities exclusive to the state; fiscal responsibility when signing and executing partnerships; transparency of procedures and decisions; objective distribution of risks between the parties; financial sustainability and socio-economic advantages of partnership projects [5].

Concerning the contracts to be established, they must follow the elements identified in art. 23 of Law No. 8987/1995 and art. 5 of Law No. 11,097/2004. The contracts must refer to the object, area, and term of the concession; the manner, form, and conditions of providing the service; the criteria, indicators, formulas, and parameters defining the quality of the service; the price of the service and the criteria and procedures for adjusting and reviewing tariffs; the rights, guarantees, and obligations of the granting authority and the concession holder, including those related to the foreseeable need for future alteration and expansion of the service and consequent modernization, improvement and expansion of equipment and facilities; the rights and duties of users to obtain and use the service; the form of supervision of the installations, equipment, methods, and practices of execution of the service, as well as the indication of the competent bodies to exercise it; the contractual and administrative penalties to which the concession holder is subject and their form of application; the cases of termination of the concession; the reversible assets; the criteria for calculating and paying the indemnities due to the concession holder, when applicable; the conditions for extending the contract; the obligation, form, and frequency of the concession holder's rendering of accounts to the granting authority; the requirement to publish the concession holder's periodic financial statements; and the forum and amicable way of resolving contractual differences [11].

Concerning guarantees, the pecuniary obligations contracted by the Public Administration in a PPP contract may be guaranteed by: earmarking revenues, in compliance with the provisions of item IV of article 167 of the Federal Constitution; establishing or using special funds provided for by law; contracting insurance guarantees with insurance companies that the Public Administration does not control; guarantees provided by international organizations or financial institutions; guarantees provided by a guarantee fund or state-owned company created for this purpose; other mechanisms allowed by law [5].

Finally, Law No. 11,079/2004 also deals with bidding processes, defining that contracting in PPP processes will be preceded by bidding in the form of a tender or competitive dialogue [5]. The document also refers to other laws that, in part, regulate the bidding process: Law No. 8666, of 21 June 1993 [13], currently repealed by Law No. 14,133 of 1 April 2021—the Administrative Bidding and Contracts Law [14]; Law No. 8987, of 13 February 1995, which provides for the concession and permission regime for the provision of public services provided for in art. 175 of the Federal Constitution [11]; and Law No. 9307, of 23 September 1996, which provides for arbitration [15].

Another important aspect of the Brazilian scenario regarding PPP arrangements is the presence of PPP Units, which are responsible for assisting PPP processes. These units, however, are present on a state level in some states such as Bahia, Minas Gerais, Pernambuco, São Paulo, Rio Grande do Sul, Espirito Santo, Porto Alegre, and Piauí.

Urban regeneration practices, on the other hand, are mainly carried out in the central areas of large and medium-sized cities. In this context, some documents stand out: the *Política de Desenvolvimento Urbano* (Urban Development Policy) defined in articles 182 and 183 of the Federal Constitution, the *Estatuto da Cidade* (City Statute), the *Política Nacional de Desenvolvimento Urbano* (National Urban Development Policy), the *Política Nacional de Habitação* (National Housing Policy), the *Plano Nacional de Habitação* (National Housing Plan), and the *Programa de Reabilitação de Áreas Urbanas Centrais* (Rehabilitation Program for Central Urban Areas).

The City Statute, approved in 2001 through Law No. 10,257, is the ordinary law that regulates Brazilian urban policy, and articles 182 and 183 of the Federal Constitution [6]. The Statute provides a series of instruments that municipalities can incorporate into their Municipal Master Plans and subsequently apply. The instruments provided by the City Statute can be of three types [16]: instruments to induce urban development, instruments to democratize urban management, and instruments to regularize land ownership.

Among the instruments for inducing urban development, the City Statute provides for an instrument called *Operação Urbana Consorciada—OUC* (Consortium Urban Operation). An *OUC* is defined as "the set of interventions and measures coordinated by the municipal authority, with the participation of property owners, residents, permanent users, and private investors, to achieve structural urban transformations, social improvements, and environmental enhancement in an area" [6]. Therefore, it is an instrument aimed at the structural transformation of an urban area through a combination of public and private investment [17].

Among the incentives, the City Statute defines that an *OUC* can provide for: the modification of indices and characteristics of subdivision, use, and occupation of land and subsoil, and changes to building regulations, considering the environmental impact resulting from them; and the regularization of constructions, renovations or extensions carried out in disagreement with current legislation [6].

The approval of an *OUC* should be carried out at the municipal level, and the coordination of the *OUC* is the responsibility of the City Council. The City Statute states that a "specific municipal law, based on the master plan, may delimit an area for the application of consortium operations" [6]. In other words, a specific municipal law should be adopted for each *OUC*.

This law must include a definition of the area to be affected; a basic program for occupying the area; an economic and social assistance program for the population directly affected by the operation; the purposes of the operation; a preliminary neighborhood impact study; the compensation to be demanded from owners, permanent users, and private investors; and the form of control over the operation, which must be shared with representatives of civil society [6].

Furthermore, the City Statute also states that "the specific law approving the consortium urban operation may provide for the municipality to emit a certain number of additional construction potential certificates, which will be sold at auction or used directly to pay for the work necessary for the operation itself" [6].

As stated in the City Statute, the municipality is responsible for controlling partnerships, assuming leadership of operations, and ensuring that partnerships are aligned with the municipal master plan and, therefore, with the city's urban planning and collective interest. The specific law also guarantees a standardization of the information provided about the partnership and clearly defines the objectives of each of the parties and the incentives.

3.1.2. Exogenous Factors in Brazil

The exogenous factors analyzed include the availability of a financial and capital market, community support, favorable socio-economic factors, and a stable macroeconomic condition and sound economic policy.

The Brazilian capital market has developed and expanded considerably since the early 2000s regarding the availability of financial and capital markets. This development and expansion were the result of constant and significant reforms in the regulatory framework, modernization of the financial infrastructure, and increased participation of national and international investors [18,19].

The Brazilian equity capital markets are mainly regulated by the Brazilian Securities and Exchange Commission (CVM), which is jointly regulated by the National Monetary Council (CMN), the Brazilian Association of Financial and Capital Market Entities (ANBIMA), and B3 S.A.—*Brasil, Bolsa, Balcão (B3)*, the Brazilian stock exchange [19].

Brazil's capital markets are the largest and most dynamic in Latin America, known for their robust trading volumes and diverse investment options. Investors in Brazil have access to a wide range of financial products, from traditional stocks and bonds to sophisticated instruments such as Exchange Traded Funds (ETFs), structured products, and Environmental, Social, and Governance (ESG) focused opportunities. This diversity allows investors to adapt their strategies to different risk profiles and market conditions [18].

Regarding community support, there is a strong polarization between those who defend a strong State with greater power of intervention and those who defend greater decentralization of the State and free action of private agents [20], thus reflecting a polarization regarding support for PPPs. In general, opposition to PPP arrangements is related to the transparency of these processes [21] and corruption issues [22].

Concerning socio-economic factors, the country presents a controversial scenario. Despite being the 10th largest economy in the world and the largest economy in Latin America, Brazil has historically been one of the most unequal countries in the world [23]. With 216 million inhabitants living in different realities across 8.5 million km^2, the differences are found to manifest in many "Brazils" [24]. The country's regions present relevant socio-economic disparities: even before COVID-19, some regions in the north and northeast of the country had a Human Capital Index (HCI) of around 40%, while some regions in the richer southeast of Brazil had HCI levels of around 70% [24].

In 2020, with the beginning of the COVID-19 pandemic, the country was about to face historic socio-economic disruptions. The pandemic took a huge human toll, causing the deaths of thousands of people (194,949 people just in 2020 and 714,736 people so far); the country's economy suffered its worst contraction in recorded history; the closure of businesses led to an unprecedented exit of workers [25]. The crisis generated by COVID-19 was the second in the country's recent history. After the global financial crisis of 2008 and the response to macroeconomic stimulus, the country entered a technical recession in 2014. The economic recession was also accompanied by a political crisis, marked by intense polarization. The crisis that began in 2014 and the COVID pandemic have increased already serious socio-economic disparities [25].

Finally, to understand the country's macroeconomic scenario, some indicators relating to annual Gross Domestic Product (GDP) per capita growth in percentual and inflation were analyzed, based on the data obtained from the World Bank over the last twenty years.

Concerning annual GDP per capita, the World Bank data shows negative peaks in four years—2009 (−1.1%), 2015 (−4.4%), 2016 (−4.1%), and 2020 (−3.9%). Despite the growth in 2021 (+4.2%), the growth in 2023 was only 2.4%. In terms of GDP per capita growth, the macroeconomic scenario does not indicate a robust economy, but rather an economic slowdown.

Regarding inflation, the country has had considerable inflation peaks in recent years, particularly in 2022, with a 9.3% inflation rate. In 2023, inflation stood at 4.6%, representing a high and unfavorable inflation rate for the country's macroeconomic scenario.

3.2. *"Porto Maravilha" Case Study*

After a general overview of the Brazilian context, the study focused on the case of *"Porto Maravilha"*. It begins with an initial contextualization of Rio de Janeiro and its port. It then presents the proposed urban regeneration project. Finally, it focuses on the PPP responsible for developing the project.

3.2.1. Context of Rio de Janeiro and the Port

Rio de Janeiro is the capital of the state with the same name, located in the southeastern region of Brazil (Figure 2). The city of Rio de Janeiro has a population of 6.2 million inhabitants, while its metropolitan area has 13.4 million inhabitants. In economic terms, Rio de Janeiro is the second-largest metropolitan area and one of the country's most important centers for services, culture, and tourism [26]. As Brazil's main gateway, Rio de Janeiro has good accessibility at regional, national, and international levels, whether in terms of airports or the road network [26]. Despite this, the city is also characterized by enormous contrasts, with differences in the quality of life of different groups of citizens, poverty, and inequality. It is also well known for its favelas.

Figure 2. The context of Rio de Janeiro. Source: Prepared by the authors.

In the context of Rio de Janeiro, the port area is also strategically located, close to *Santos Dumont* Airport, *Tom Jobim* International Airport, and important areas and points of interest in the city, such as *Maracanã* Stadium, *Corcovado*, *Copacabana*, and *Pão de Açucar* (Figure 3).

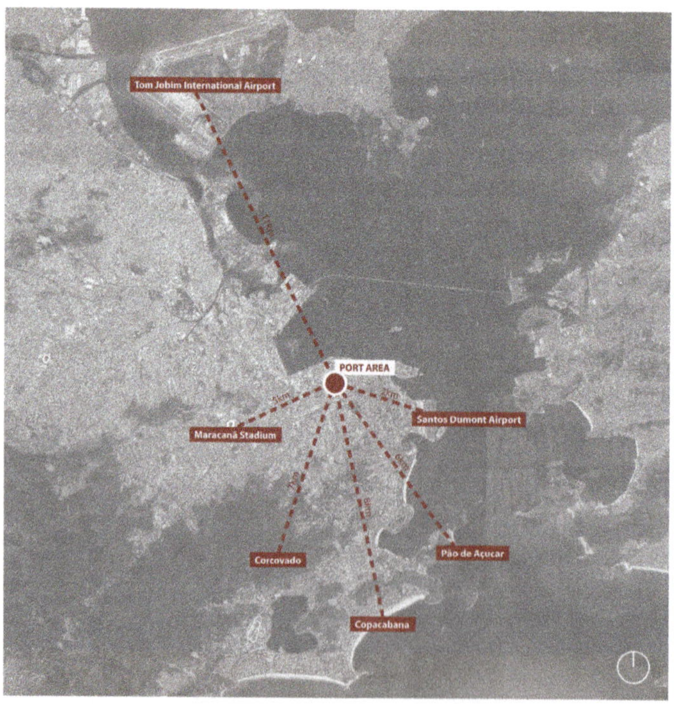

Figure 3. The context of the Port of Rio de Janeiro. Source: Prepared by the authors.

The Port of Rio de Janeiro emerged in the 18th century, in an area close to the first Portuguese settlement. Still, in the middle of the 18th century, the port expanded considerably due to the coffee boom in Brazil. For a long time, the Port of Rio de Janeiro was closely linked to Brazil's economic, social, and cultural development [26].

However, with the changes in the Brazilian economy and in maritime transport technologies, the Port of Rio de Janeiro went through a phase of decline and abandonment. Although part of the port infrastructure has been maintained, with the modernization of port activities, logistical difficulties related to the city's growth, and changes in the city's economic activities, the extension of the port region that was originally present has lost its primary function [27].

In this sense, since the 1970s, the port area has been continuously degraded, and proposals for renovation have failed. The lack of alignment between the different players—the state, the municipality, private entities, and federal agencies—was pointed out as one of the main barriers that explain the failure of projects [27]. The port area of Rio de Janeiro is complex, large, and has land features. It is close to the metropolis's historical and financial center.

3.2.2. *Porto Maravilha* Proposed Project

The *Porto Maravilha* project seeks to regenerate Rio de Janeiro's port area through recreation, tourism, cultural complexes, and housing, like what can be seen in other port cities. The project is part of the context of the major mega-events that took place in Rio de Janeiro: the World Cup in 2014 and the Olympic Games in 2016. The mega-events were thus essential motivations for the urban regeneration of the port area [28].

The project covers an area of approximately 500 ha and foresees the construction of around 4.5 million square meters over a period of 30 years. The delimited perimeter contains three integral neighborhoods—*Saúde, Gamboa,* and *Santo Cristo,* four partial neighborhoods—*Centro da Cidade, Cidade Nova, São Cristóvão,* and *Caju,* and historic favelas in the region—*Morro da Providência, Moreira Pinto, São Diogo, Pedra Lisa, Livramento,* and *Conceição* (Figure 4) [29].

As a point of arrival for the enslaved population and later home to freed slaves, the area is rich in cultural aspects related to the African population. Part of the port area, known as *Pequena África* (Small Africa), is the historical home of the Afro-Brazilian population in Rio de Janeiro [30].

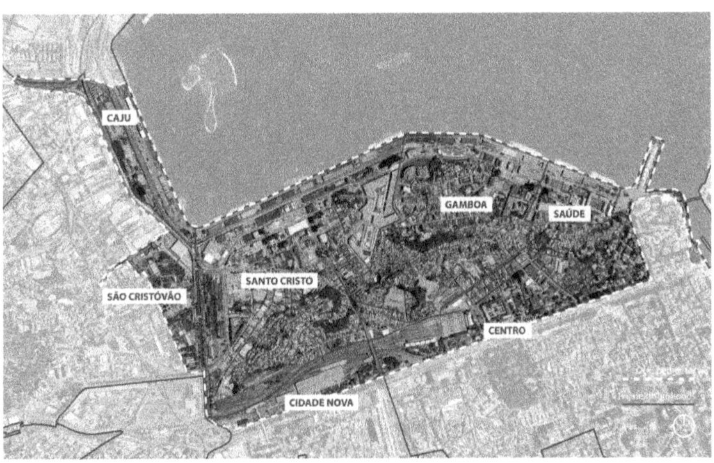

Figure 4. Delimitation of the intervention area with the subdivision of neighborhoods. Source: Prepared by the authors based on [31].

Before the project started, the Porto Maravilha area had a population of around 28,000 inhabitants, mainly located on the hills in precarious settlements. The area was also crossed by a large viaduct, called the *Elevado da Perimetral* (Perimetral Elevated), with an intense flow of vehicles. Because of this, the area was considered a transit zone, with many abandoned and degraded buildings.

The project was designed to restore urban infrastructure, transport, the environment, and historical and cultural heritage. It was also intended to improve housing conditions and attract new residents to the area. In addition to improving housing, the project also expected the arrival of large companies, new tax incentives, and the provision of public services [32].

Regarding mobility, the project aims to completely change the concept of urban mobility in the port and central area, creating a system that favors public transport, values the idea of living close to work, creates new spaces for pedestrians, creates new cycle paths, and integrates means of transport [32].

The project includes major works and the provision of various services. Among the works are: the demolition of the *Elevado da Perimetral* (Perimetral Elevated), the construction of the *Museu de Arte do Rio—MAR* (Rio Art Museum) and the *Museu do Amanhã* (Museum of Tomorrow), the construction of the Via *Binário do Porto e Túnel Rio 450* (Port Binary Road and Rio 450 Tunnel), the construction of the Via *Expressa e Túnel Prefeito Marcello Alencar* (Express road and Mayor Marcello Alencar Tunnel), the *Nova Orla Conde* (New Conde waterfront), the *Veículo Leve Sobre Trilhos* (Light Rail Vehicles), 70 km of roads re-urbanized and 650,000 m^2 of sidewalks renovated, 700 km of urban infrastructure networks rebuilt (water, sewage, drainage), 17 km of new cycle paths, 15,000 trees [32].

The services provided include: conservation and maintenance of the road system; conservation and maintenance of green areas and squares; maintenance and repair of public lighting and sidewalks; execution of urban cleaning services; selective garbage collection; maintenance of the drainage network and collective galleries; maintenance of traffic signs; installation and maintenance of bicycle parks; maintenance and conservation of tourist, historical, and geographical points and monuments; and citizen service [32].

The planned works and services are expected to have a significant impact on the area: an increase in population from 32,000 to 100,000 inhabitants in 10 years; an increase in the green area; a 50% increase in traffic flow capacity in the region; a reduction in air and noise pollution, with the removal of the Perimetral and a reduction in heavy transport in the region; an increase in soil permeability; an increase and improvement in the quality of public services provided; the transformation of the region into a reference point for the city [32].

In terms of social commitments, the project was based on the assumption that the residents of the area should remain there and proposed a series of measures: the creation of housing of social interest; installation of nurseries, emergency care units, and schools to meet the expected population density; integration between the various modes of public transport, facilitating accessibility and communication with other areas; restoration of the area's environmental quality; generation of direct and permanent employment in the region; regularization and formalization of economic activities; professional training; creation of the *Porto Maravilha* Cultural and *Porto Maravilha* Citizen Programmes; support for community development initiatives [32].

3.2.3. *OUC Porto Maravilha*

The *OUC Porto Maravilha* was established in 2009 through Complementary Law No. 101/2009 [33]. However, the procedures that led to the implementation of the *OUC* began in 2006, through Decree No. 26,852, of 8 August 2006, which creates and delimits the *Área de Especial Interesse Urbanistico (AEIU) da Região Portuária do Rio de Janeiro*—Area of Special Urban Interest of the Port Region of Rio de Janeiro for the purposes it indicates, declares real estate to be of public utility, establishes conditions for partnerships with the private sector, and authorizes the creation of a Working Group [34].

After the creation and delimitation of the Area of Special Interest, a series of other documents related to the area were published, such as the different decrees to constitute and create the Working Group responsible for analyzing the studies for modeling a concession or PPP agreement for the implementation of the Recovery and Revitalization Plan for the Port Region, as well as decrees to create and alter the Working Group to supervise the implementation of the *Porto Maravilha* Project [35–41].

Complementary Law No. 101/2009 modified the Master Plan and authorized the Executive Power to institute the *OUC Porto Maravilha.* It is important to highlight that Law No. 101/2009 presents the *OUC* and its basic program in a very generic way, basically referring to a few major urban interventions and massive building densification—mainly because of the increase in density that would be allowed with the sale of *Certificado de Potencial Adicional de Construção—CEPACs* (Certificates of Additional Construction Potential).

Concerning the definition of the intervention plan in Law No. 101/2009, the basic program for the occupation of the area does not clearly define objectives or provide a clear idea of the type of occupation intended. Without any more detailed explanation, the program refers to a "new occupation pattern" to be implemented in the area and the subdivision into sectors, but without relating them to the different types of urban spaces within the perimeter of the operation [42]. The lack of clarity in the basic program and the occupation plan, the enormous dimension of the area, and the possibility of intense densification make the project very complex and make it difficult to understand its development over time.

Complementary Law No. 101/2009 was followed by Complementary Law No. 102/2009, which created the *Companhia de Desenvolvimento Urbano da Região do Porto do Rio de Janeiro—CDURP* (Urban Development Company of the Port Region of Rio de Janeiro), responsible for the development of the *AEIU* of the Port Region of Rio de Janeiro [43], and Complementary Law No. 105/2009, which instituted the Municipal Program of Public-Private Partnerships—*Propar-Rio* [44]. Also, in 2009, Decree No. 31,620 established the *CDURP* and approved its statutes [45].

CDURP is a mixed-capital company, independent financially and administratively from the municipality. It is responsible for promoting the development of the Port through concessions and managing the assets transferred to it [46]. In the first months of 2010, *CDURP* granted a 24-month preliminary license for the AEIU Neighborhood Impact Study and a preliminary license for the *OUC*.

Two decrees were published in July 2010. The first is Decree No. 32,575, of 28 July 2010, which includes the Port of Rio Revitalization Project in the Municipal Plan for Public-Private Partnerships [47]. The second is Decree No. 32,576, of 28 July 2010, which provides for the delegation of the management of services of local interest and public services of municipal competence to be provided within the boundaries of the AEIU of the Port of Rio Region [48].

Also, in July 2010, Public Tender No. 001 was opened [49]. The public tender notice defined that the bidding process would be carried out and judged in two phases, with the first phase referring to the analysis of the qualification documents and the second phase referring to the analysis of the technical proposal and the economic proposal. The tender document stipulated that the execution of the contract would be divided into fifteen successive phases, following the general terms of the General Urbanization Plan for the Port AEIU [49].

The public tender notice also defines the terms relating to the counterpart and guarantee of payment by the public authority, the amount of which would be the subject of dispute in the competition. In general terms, the maximum estimated value of the contract was eight billion reais (4.5 billion USD) [49]. The notice defined the conditions for participation and included models for the documents to be submitted by bidders, including documents relating to qualification, legal standing, tax regularity, technical qualification, economic-financial qualification, technical proposal, and economic-financial proposal [49].

In the following month, Decree No. 32,666, of 11 August 2010, determined the issuance of *CEPAC* [50]. The Municipality of Rio de Janeiro has issued 6,436,722 *CEPACs*, in book-entry form, for BRL 545.00 (five hundred and forty-five reais—310.00 USD) each, to be used as additional construction potential in the Rio de Janeiro Port Region AEIU.

The amounts obtained from the sale of these *CEPACs*, as well as other assets transferred to *CDURP*, such as 408,000 m^2 of land in the Porto region, capitalize this state-owned company to promote the *Porto Maravilha* PPP [46].

In October 2010, the Qualification Trial, the Technical Proposal Trial, and the Final Score were carried out. The winner was the *Porto Novo* Consortium, made up of Construction companies *OAS* Ltd.a, *Norberto Odebrecht* Brasil S/A, and *Carioca Christiani-Nielsen Engenharia S/A*. In November 2010, the announcement of the signing of the administrative concession contract was published. Figure 5 shows the main steps that led to the implementation of the PPP arrangement.

After the contract was signed, the Public Tender No. 002/2010 was also opened to hire specialized technical services for the management and supervision of the Administrative Concession Contract for the revitalization, operation, and maintenance of the AEIU in the Port region of Rio de Janeiro. Qualification was judged in February 2011, and the result was announced in April 2012, with the *Ecoplan/JHP* Consortium winning. Approval and award took place in May 2012 [51].

In June 2011, the Real Estate Investment Fund Porto Maravilha, owned by *Caixa Econômica Federal*, bought a single lot of *CEPACs* in Rio de Janeiro's Port Zone for BRL 3.5 billion (2.2 billion USD), during an auction organized by the Rio de Janeiro Municipal Government [51].

The PPP Porto Maravilha contract was managed through the service orders that mark the start of the project phases. The order for the first phase was placed in June 2011. In addition, several amendments to the contract were published [52], modifying the original agreement.

New decrees and laws were also published during contract management. Municipal Decree No. 36,665 of 1 January 2013, provides for the review of the economic and financial balance of administrative contracts [53]. Complementary Law No. 147/2014 amends the wording of art. 24 of Complementary Law No. 105/2009 to change the percentage limit for contracting PPP agreements from 3% (three percent) to 5% (five percent) of the net current revenue for the year and the projected net current revenue [53]. Municipal Decree No. 46,787 of 7 November 2019, creates a Working Group to analyze documentation and determine possible distortions in executing the Administrative Concession Contract PPP-01 [54].

In June 2018, it was reported that *Caixa Econômica Federal* declared the illiquidity of the Real Estate Investment Fund *Porto Maravilha*, responsible for *OUC* payments, leaving *CDURP* to maintain the flow of transfers under the PPP contract. In January 2023, *CDURP* announced that the project had been halted due to the illiquidity of the Real Estate Investment Fund.

To date, approximately 85% of the project has been completed. The main works include the demolition of the Elevado da Perimetral, the construction of tunnels, urban infrastructure such as water supply and wastewater collection and treatment, and cultural buildings such as museums.

Figure 5. Timeline of the constitution of Porto Maravilha PPP. Source: prepared by the authors based on [51].

3.2.4. *OUC* and Project *Porto Maravilha*

Despite the interruption of the partnership responsible for the Porto Maravilha project, some aspects should be highlighted, both in a positive and negative view.

In terms of urban aspects, the project brought improvements to the infrastructure. The demolition of the Elevado da Perimetral, a viaduct that cut through the area, was a crucial element in the greater integration of the area with the rest of the city (Figure 6). The construction of the tunnels, as an alternative to the viaduct, was also positive for the area. In addition, infrastructure works such as new roads, sidewalks, street lighting, and wastewater collection and treatment sewage have benefited the area (Figure 7). The expansion of the Light Rail Vehicle has also improved transportation, increasing the area's connectivity.

Figure 6. Demolition of the *Elevado da Perimetral* (Perimetral Elevated). Source: [32].

Figure 7. New Waterfront. Source: [32].

Regarding the urban planning aspects, interviewee 1 states that "the region had an extremely poor and inefficient urban infrastructure. It has all been redone—I'm talking about water, wastewater, drainage, energy, telecommunications, gas... All of this has been redone in this region. In fact, 85% of the planned construction program has already been carried out. All this was done in the period 2011–2016. Not to mention the urban aspects: there has been an increase in the number of trees and improvements to the sidewalks. This area was often subject to flooding because the drainage system was unable to drain the water. In addition to considering the usual technical parameters for making a drainage network—which is rainfall recurrence and, in the case of the vicinity of the Port, tidal recurrence—the new drainage network also took into account studies by the city council estimating the impact of rising sea levels in that region" (translated from Portuguese).

He complements this by saying that "because of this whole process, the VLT was implemented, changing urban mobility patterns in the region. Clean energy has a low impact on noise, air pollution, and all that. The demolition of the Perimetral and the construction of the tunnels—unfortunately, I don't know if this is being monitored or not—have certainly had a positive impact on reducing air pollution, the impact on traffic itself, and several listed buildings in the region" (translated from Portuguese).

Some anchor architectural projects are worth highlighting. The construction of the Rio Art Museum (MAR) and the Museum of Tomorrow have created new points of interest for the city, motivating tourism and culture (Figures 8 and 9). New public spaces, such as squares, have also made a place for tourism.

Figure 8. *Museu do Amanhã* (Museum of Tomorrow). Source: [32].

Figure 9. *Museu de Arte do Rio* (Rio Art Museum). Source: [32].

Concerning these elements, interviewee 5 states that "the bet that was made on the region becoming part of Rio de Janeiro's tourist circuit has intensified. At first, there were those anchor projects—the Rio Art Museum, the Museum of Tomorrow, some restoration of buildings of historical importance, with some archaeological discoveries" (translated from Portuguese).

On the other hand, interviewee 4 points out that "the municipality took 114 million reais from the infrastructure of Morro do Pinto and Morro da Providência and reallocated it to more than 300 million reais for the construction, designed by Calatrava, of the Museum of Tomorrow. So, they took the money from a place that needed it to build the Museum of Tomorrow" (translated from Portuguese).

Historical and cultural aspects were reinforced in the area with the discovery of spaces such as *Cais do Valongo* (Valongo Wharf), an old pier located in the port area related to the arrival of Africans enslaved in America (Figure 10). African history related to the area was further reinforced with the African Heritage Circuit and the Museum of Afro-Brazilian Cultural History—MUHCAB (Figure 11).

Figure 10. *Cais do Valongo* (Valongo Wharf). Source: [28].

Figure 11. Ethnicity Murals. Source: [28].

Regarding the reinforcement of African heritage in the region, interviewee 4 points out that "if today there is a great deal of visitation in the port area, if today the African heritage circuit has become well established, it's because of the rediscovery *of Cais do Valongo*, but it's worth remembering that they weren't looking for the pier, they were just doing some work and found it" (translated from Portuguese).

Despite the work and improvements carried out, the project did not achieve its initial vision of real estate development, mainly related to commercial uses. Due to the change in the economic context, the private market reduced its interest in the area, leading to a period of low demand. After the period of crisis, the private market began to return to the area, but commercial use lost its focus, giving way to more high- and middle-income residential use, as well as short-term residential use.

About this change in the project's vision, interviewee 5 states that "we're seeing a change in the profile of what was initially envisioned. What was envisioned when the project was conceived? Basically, the area would have a strong demand for corporate spaces... and not only. The area was always thought of as having potential for leisure and entertainment activities. It would also have an incentive for residential use, but the project's main focus was corporate real estate, especially driven by the gas and oil industry... at the time, it was discovered with pre-salt. Petrobras expanded a lot, so the demand for corporate space in the region was thought to come from that. Then came the economic crisis, a whole series of things... and that bet didn't materialize. So, what we're seeing there today is basically a massive construction of small-scale residential real estate... studios, one-bedroom apartments, and at most two-bedroom apartments... and a lot of this new segment is being driven by short-term rentals, Airbnb, and other platforms" (translated from Portuguese).

From a social perspective, despite the improvements to public and cultural spaces, the project has been strongly criticized for its negative impacts on the most vulnerable population who previously lived there. The criticisms refer, above all, to processes of gentrification and expulsion of the most vulnerable population. The idea of presenting the area as an initial urban emptiness reinforces the social criticism since it reduces the importance of the existing residents.

Regarding this idea of the area being an urban emptiness, interviewee 5 states that "we have this discourse of urban emptiness, of characterizing these areas as urban emptiness... It's partly pertinent. But, in part, it also invisibles a lot of things. The port region of Rio, in fact, is a region that has a series of warehouses that have already been used industrially in ports and are underused. They could have been used for other purposes and for much more intensive use. On the other hand, the region is not exactly a population vacuum. When the project was launched, the area had a significant number of people, almost 30,000 people living in the port area, mainly in the slums of the port area" (translated from Portuguese).

Concerning the displacement of the low-income population, interviewee 4 added that "they invited the poorest residents to leave. They didn't invite them, they expelled them... some people were bruised, wounded, and psychologically threatened... To obey the order of development and progress" (translated from Portuguese).

Another relevant criticism from a social perspective concerns the social housing that was initially planned for the area but was never built. Interviewee 4 reinforces this criticism, commenting that "the question is, who are or would be these almost 70,000 new inhabitants? Since the Ministry of the City, at the start of the Porto Maravilha project, demanded 5000 low-income housing units—5000 houses or apartments—for people earning up to 3 minimum salaries, and to this day, not a single house or apartment has been built for people earning up to 3 minimum salaries... quite the opposite, the city has removed the poorest people, giving them no chance to stay and no chance to build new housing for low-income people here in the port region" (translated from Portuguese).

Finally, concerning the PPP arrangement and the process, some points are worth highlighting.

Using a PPP arrangement as an integrated model for building infrastructure and providing services stands out as a differential. Interviewee 1 points out that "there was a complexity to the works that required them to be coordinated, so there had to be a beginning-middle-end to plan. And then, when an urban operation is created, the law approves a program of interventions: I'm going to increase the potential and sell CEPAC to pay for this program of works here. What did the city council do? It included some services such as traffic management, urban cleaning, and maintenance of public lighting, in other words, some urban services, and turned this into a package, an object, for a public concession in the form of a PPP. Why was this done? Because then you contract everything at once, you have a company responsible for everything. This makes it easier for the municipality to manage the whole process. Because even if the concessionaire is going to subcontract the services, it must respond to an intervention strategy that covers everything that needs to happen in the area in the medium term". He adds that "from the point of view of managing the PPP contract, this direct and unique interaction allowed the municipality to manage the process more efficiently. In the end, for a few years, it served as an experiment in truly integrated management of public works and services" (translated from Portuguese).

On the other hand, doubts were raised about the real necessity of a PPP arrangement to develop the project. Interviewee 5 points out that "when we look at the number of resources that have been mobilized there, from the various levels of government, it's not true that it wouldn't be feasible to do a project without relying on the private sector as a last-resort financier. Because deep down, it wasn't. The volume of public resources mobilized there to get this system up and running was very high. So, there were public resources. The city council was able to carry out this project—when I say the city council, I mean not just the city council, but all the public bodies involved" (translated from Portuguese).

Also, regarding the entities involved, the process received criticism concerning the benefits generated for the entities and how they were balanced. Interviewee 3 believes that the port authority has not benefited as it should from the process, stating that "the port authority had properties, and the municipality used these properties, and they were not fully indemnified" (translated from Portuguese).

The speed at which the process was intended—with 85% of the work carried out in 5 years—is also a factor for discussion. On this issue, interviewee 2 states that "you take a quick shot, as was done—a project of 5 to 6 years of massive investment, intense transformation—but then it lost traction, and there was that immense infrastructure, well built, well made, but without use, without continuity. The other view is, let's do it step by step, more slowly... make a plot, invest, go slowly, build gradually, get a feel for the market, balance demand with supply" (translated from Portuguese).

The financial dependence related to the sale of CEPACs is another weakness in the process. Interviewee 2 reinforces this point, stating that "firstly, with the issue of the value of CEPACs... the tender that took place, the value that was attributed, the potential, the level of risk that Caixa took on as manager and the scenario it envisioned for capitalizing CEPACs, was a practically impossible scenario... that nothing would happen, that Rio de Janeiro would develop, that the country's economy would grow constantly, there would be no crisis, income would rise. And economic factors that threatened this scenario were not considered" (translated from Portuguese).

The overly optimistic expectations related to the country's economic context at the time of the PPP arrangement also demonstrate a fragility, since scenarios that could affect the planning carried out—and which were confirmed—were not considered, generating problems in the financial viability and progress of the project. Interviewee 2 points out that "the assumptions that were adopted for the economic and financial bases of the project were ambitious from the point of view of leverage, that is, how much it was imagined that it would generate in terms of resources and the speed at which it would generate them, the speed of investment at which things would happen, the economic aspects of the country... so no consideration was given to possible economic downturns. The economy of Brazil and Rio de Janeiro collapsed in 2016, so this was a fundamental aspect" (translated from Portuguese).

4. Discussion

The analysis considered the external and internal environments of the PPP project. The clusters and factors related to the external environment—enabling environment and exogenous factors—were used to discuss the Brazilian context concerning PPP arrangements. The clusters and factors related to the internal environment—preparatory, procurement, and contract management phases—were used to analyze the specific urban regeneration project and the OUC Porto Maravilha case.

From the analysis, it was possible to identify the CSFs in the Brazilian context and in the case of the *Porto Maravilha* PPP project, as well as the critical points that could be better developed to facilitate and improve the implementation and development of PPP arrangements and provide better value for money.

4.1. Brazilian Context

In the Brazilian context, PPP arrangements can be carried out through sponsored concession or administrative concession. Law No. 11,079/2004—General rules for bidding and contracting PPPs within the scope of public administration are the primary document referring to PPP arrangements in the country, followed by other laws—Law No. 8987/1995, Law No. 9074/1995, Law No. 8666/1993, Law No. 14,133/2021, and Law No. 9307/1996. These documents are responsible for defining PPP projects, the bidding process, contract elements, and guarantees.

The existence of PPP units–agencies explicitly created to support the development of PPP arrangements–also reinforces the public sector's leadership. Despite this, PPP units are defined at the state level, and only a few states have established such units.

Urban regeneration in the Brazilian context is more intense in medium and large urban centers. The City Statute, Law No. 10,257/2001, establishes instruments at the federal level to be implemented at the municipal level. Among the main instruments, the Consortium Urban Operation is responsible for the structural transformation of an urban area through a combination of public and private investment. According to the City Statute, OUC refers to a set of interventions and measures coordinated by the public authorities, but which can involve several other actors—property owners, residents, permanent users, and private investors. The interventions are aimed at structural urban transformations and social and environmental improvements in a specific area.

The law also defines the types of incentives that can be made available and the possibility of issuing additional construction certificates, known as *CEPACs*. Finally, the City Statute defines the municipality as one that must implement an OUC through a specific municipal law for each operation.

Regarding exogenous factors, the availability of financial and capital markets stands out, with a large trading volume, diverse investment options, and a wide range of financial products. However, other exogenous factors are not favorable to the implementation of PPPs.

Community support is characterized by great polarization: part of the population believes that public power should remain more centralized. In contrast, part of the population defends the greater power of the private sector. Socio-economic factors do not create a favorable environment either, since there are great disparities between the country's various states, especially between the north/northeast and the south/southeast. Finally, the macroeconomic scenario, analyzed through GDP per capita and inflation, also does not present good conditions for implementing PPPs.

4.2. "Porto Maravilha" Case Study

Some aspects of the OUC Porto Maravilha and its different phases should be highlighted.

During the preparation phase of the contract, some aspects are favorable for the implementation of the PPP project, such as the creation of a specific public company for this purpose. The public tender for the private partner selection also represented a positive aspect, since it clearly defined all the elements to be presented, including models for a standardized process. The well-designed tender notice allowed for a competitive bidding process.

On the other hand, it is worth noting that Complementary Law No. 101/2009, which established the *OUC Porto Maravilha*, is unclear regarding the intended intervention plan for the area. The lack of clarity in the plan—which only defined sectors and new construction amounts, leaving many aspects to be discussed, a posteriori of the law and the contract itself—combined with the considerable dimension of the area and the possibility of intense densification, increased the complexity of the project. Furthermore, the sale of CEPACs to financially enable the project creates fragility from a financial perspective, since it depends on an external payment for the project to progress.

Following the bidding process's results, CDURP and the Porto Novo Consortium signed a partnership contract. As established by the tender notice, the contract was to be carried out in phases.

Some important events during the contract's management are important to highlight. The first was the hiring, also through a public tender, of the *Ecoplan/JHP* Consortium, which would be responsible for managing and supervising the Administrative Concession Contract. Another important element was the purchase of all the *CEPACs* available by the Real Estate Investment Fund Porto Maravilha, owned by *Caixa Econômica Federal,* since the amounts obtained from the sale of *CEPACs* would guarantee the development of the OUC.

The contract management phase was generally marked by a series of additional terms to the original contract, new decrees, and complementary laws. Finally, the illiquidity of the Real Estate Investment Fund *Porto Maravilha* halted the project's progress, aggravating the situation. So far, approximately 85% of the work has been completed.

In general, the PPP arrangement and the Porto Maravilha project are complex and ambitious, hindering their full development.

The PPP arrangement was carried out in a very favorable economic context and, as a result, did not take into account different future scenarios that could hinder the project's progress. The sale of CEPACs to pay for the PPP creates a certain fragility from a financial point of view since it depends on an external agent and the market's interest in buying CEPACs from that agent.

Furthermore, the idea of making the partnership and the project viable through the sale of additional constructive potential is based, above all, on increasing the value of these titles—which is only possible with a priori improvements and, eventually, with the expulsion of less favored residents, reinforcing a gentrifying character.

As far as the project is concerned, despite the numerous improvements seen in the area—improvements to mobility, the construction of squares, and cultural facilities—there are significant social impacts, especially for the low-income population.

5. Conclusions

Focusing on PPP arrangements in urban regeneration projects, this study analyzed the specific Brazilian context on the topic and the specific case of the *"Porto Maravilha"* partnership in Rio de Janeiro. Finally, the study aimed to understand the CSFs in the Brazilian context and the case study, as well as the issues that could be improved.

To this purpose, the study was based on a pre-established framework of CSFs for PPPs in urban regeneration projects. The framework divides the external environment of the PPP arrangement and the internal environment, considering the different phases of the PPP project. The framework thus facilitates the analysis of the Brazilian context concerning PPP arrangements and the specific case study concerning the *Porto Maravilha* partnership. To carry out the study, data were collected from official documents, laws, and decrees, available academic literature, and promotional information. In addition, interviews were conducted with actors involved in the *Porto Maravilha* PPP project.

The analysis of the external environment indicated that Brazil's enabling environment presents important factors for the development of PPP arrangements. The most relevant factor refers to the legal and regulatory framework, i.e., the legislation that defines the types of partnerships and the means for these partnerships to be carried out. The presence of PPP units, although only in a few states, is also a factor that demonstrates commitment to the development of PPP arrangements.

Concerning exogenous factors, the most relevant factor is the wide availability of financial and capital markets. On the other hand, other critical factors need to be better addressed: socio-economic conditions capable of creating the best scenario for PPP agreements, stable and growing macroeconomic conditions, and the support and acceptance of the local community.

Regarding the Porto Maravilha case study, a few points deserve attention. Firstly, the complexity of the proposed project. This is a large area in a state of considerable disrepair and abandonment. In addition, other aspects should be considered, such as historical aspects—given the historical importance of the area—and especially social aspects—given the presence of the hills and the low-income population. The sale of construction potential would considerably increase the density of the area, and the lack of a priori definition of the basic design elements also made the project more complex.

In terms of the PPP arrangement, flexibility in different aspects stands out. Firstly, it allows for the integration of different urban services, large-scale infrastructures, and interventions in public spaces, providing an integrated approach. From a financial point of view, it allows financial tools to be adapted to local real estate conditions and public finances. Another important flexibility is the possibility of changing important elements of the process according to the socio-economic shifts that are observed, such as the initial vision being altered according to changes in the market.

However, the wide range of works and services made the arrangement quite complex. In addition, the dependence on the sale of CEPACs to make the project financially viable reveals a weakness and jeopardizes the smooth running of the project. Finally, overly optimistic expectations regarding the economic scenario also compromised the project.

In this sense, despite the relevance of some critical factors in the different phases of the PPP arrangement—such as the availability of resources and a team of qualified and dedicated professionals in the preparation phase and procurement phase; and a competitive and standardized process in the procurement phase—the commitment of other factors was crucial for the development of the PPP arrangement and the project—in the preparation phase, a clear definition of the scope of the project; in the contract management phase, effective contract management and conflict management.

Despite this, 85% of the project has been completed to date, completely changing the Porto Maravilha area. Improvements in urban mobility and infrastructure, new public spaces, and important cultural facilities such as museums have been made. On the other hand, social impacts have affected low-income populations.

The case of Porto Maravilha represents a scenario of urban regeneration of port areas and waterfronts that are often present in different urban contexts. Despite creating a new image for the city, mainly based on tourism and leisure, it brings with it social impacts, especially related to gentrification.

The analysis of the Brazilian context and especially of the Porto Maravilha OUC raises important points that should be considered in PPP arrangements in urban regeneration projects. The first point concerns the financial sustainability of a PPP agreement, which must be solid enough to complete the partnership and the proposed project. The choice of solid and committed stakeholders is also essential to avoid different types of conflicts.

The public sector's leadership is crucial, especially as a representative of common interests. The engagement of all the stakeholders affected by the intervention, including the local population, is essential to ensure that the partnership and the project meet their needs and expectations, avoiding conflicts and negative impacts.

Regarding the project to be developed, it is important to clearly define the perimeter and plan for the area, including at a phased level. Defining the project from the beginning avoids major uncertainties and conflicts during implementation. It is important to consider the size of the area, the amount to be built, the specifics of the area, and historical and social aspects. An overly ambitious project can result in unmet expectations and areas falling into disuse again.

PPP arrangements in urban regeneration projects are complex, including urban, architectural, social, contractual, management, and political aspects. Awareness of the different areas related to these practices is necessary for a greater chance of success.

By analyzing a specific context and case, this study contributes to a better understanding of the good practices associated with PPPs in urban regeneration projects, especially in the case of Brazil.

Despite this, the study has limitations. The first limitation concerns the analysis of some factors. Some factors are subjective, such as good governance and openness between parties, which make objective analysis difficult. A second limitation is the analysis of only one urban regeneration operation, which does not represent the whole context of the analyzed country.

In this sense, future studies could apply the proposed framework to explore other operations or contexts. Analyzing other contexts and specific cases will further expand knowledge about the implementation and management of this type of partnership, contributing to better results from both a financial and social point of view.

Author Contributions: Conceptualization, P.V.d.P. and R.C.M.; methodology, P.V.d.P. and R.C.M.; data curation, P.V.d.P.; formal analysis, P.V.d.P. and R.C.M.; writing—original draft preparation, P.V.d.P.; writing—review and editing, R.C.M. and J.M.G.; visualization, P.V.d.P. All authors have read and agreed to the published version of the manuscript.

Funding: The author is grateful for the Foundation for Science and Technology's support through funding UIDB/05703 from the research unit CiTUA.

Data Availability Statement: The original contributions presented in this study are included in the article. Further inquiries can be directed to the corresponding author.

Conflicts of Interest: The authors declare no conflicts of interest.

Appendix A

Table A1. Interviews—open-ended questions.

How was Rio de Janeiro's port area before the project?
What was the context of the Porto Maravilha OUC and PPP?
Which external entities were involved?
How do you evaluate the urban results obtained through the area's regeneration? What aspects do you see as positive, and what could be addressed better?
How do you evaluate the partnership between the public sector and the Porto Novo Consortium? What aspects do you see as positive, and what could be addressed better?
Were there any conflicts observed during the development of the Porto Maravilha partnership and project? How were these conflicts managed?
How did the local population receive the Porto Maravilha partnership and project? Was there support or opposition?
If the process was done again, what could be done differently?

References

1. Amirtahmasebi, R.; Orloff, M.; Wahba, S.; Altman, A. *Regenerating Urban Land: A Practitioner's Guide to Leveraging Private Investment*; World Bank Publications: Washington, DC, USA, 2016.
2. de Paula, P.V.; Marques, R.C.; Gonçalves, J.M. Public–Private Partnerships in Urban Regeneration Projects: A Review. *J. Urban Plan. Dev.* **2023**, *149*, 04022056. [CrossRef]
3. Berrone, P.; Ricart, J.E.; Duch, A.I.; Bernardo, V.; Salvador, J.; Piedra Peña, J.; Rodríguez Planas, M. EASIER: An Evaluation Model for Public–Private Partnerships Contributing to the Sustainable Development Goals. *Sustainability* **2019**, *11*, 2339. [CrossRef]
4. Habitat, U.N. *Public-Private Partnerships in Housing and Urban Developments*; UN-HABITAT: Nairobi, Kenya, 2011.
5. Brasil. Lei No. 11.079, de 30 de Dezembro de 2004—Institui Normas Gerais Para Licitação e Contratação de Parceria Público Privada No Âmbito Da Administração Pública. 2004. Available online: https://www.planalto.gov.br/ccivil_03/_ato2004-2006/2004/lei/l11079.htm (accessed on 1 September 2024).
6. Brasil. Lei No. 10.257, de 10 de Julho de 2001—Regulamenta Os Arts. 182 e 183 da Constituição Federal, Estabelece Diretrizes Gerais da Política Urbana e dá Outras Providências. 2001. Available online: https://www.planalto.gov.br/ccivil_03/leis/leis_2001/l10257.htm (accessed on 1 September 2024).
7. World Bank. *Porto Maravilha—Case Study. Urban Regeneration KSB*; World Bank: Washington, DC, USA, 2020.
8. Fortune, J.; White, D. Framing of Project Critical Success Factors by a Systems Model. *Int. J. Proj. Manag.* **2006**, *24*, 53–65. [CrossRef]
9. Leidecker, J.K.; Bruno, A.V. Identifying and Using Critical Success Factors. *Long Range Plan.* **1984**, *17*, 23–32. [CrossRef]
10. Vale de Paula, P.; Cunha Marques, R.; Gonçalves, J.M. Critical Success Factors for Public–Private Partnerships in Urban Regeneration Projects. *Infrastructures* **2024**, *9*, 195. [CrossRef]
11. Brasil. Lei No. 8.987, de 13 de Fevereiro de 1995—Dispõe Sobre o Regime de Concessão e Permissão da Prestação de Serviços Públicos Previsto No Art. 175 da Constituição Federal, e dá Outras Providências. 1995. Available online: https://www.planalto.gov.br/ccivil_03/leis/l8987compilada.htm (accessed on 1 September 2024).
12. Brasil. Lei No. 9.074, de 7 de Julho de 1995—Estabelece Normas Para Outorga e Prorrogações das Concessões e Permissões de Serviços Públicos e dá Outras Providências. 1995. Available online: https://www.planalto.gov.br/ccivil_03/leis/l9074cons.htm (accessed on 1 September 2024).
13. Brasil. Lei n. 8.666, de 21 de Junho de 1993—Regulamenta o Art. 37, Inciso XXI, da Constituição Federal, Institui Normas Para Licitações e Contratos da Administração Pública e dá Outras Providências. 1993. Available online: https://www.gov.br/saude/pt-br/acesso-a-informacao/banco-de-precos/legislacao/lei-no-8-666-de-21-de-junho-de-1993.pdf/view (accessed on 1 September 2024).
14. Brasil. Lei n. 14.133 de 1 de Abril de 2021—Lei de Licitações e Contratos Administrativos. 2021. Available online: https://www.planalto.gov.br/ccivil_03/_ato2019-2022/2021/lei/l14133.htm (accessed on 1 September 2024).
15. Brasil. Lei n. 9.307, de 23 de Setembro de 1996—Dispõe Sobre a Arbitragem. 1996. Available online: https://www.planalto.gov.br/ccivil_03/leis/l9307.htm (accessed on 1 September 2024).
16. Rolnik, R.; Saule Júnior, N. *Estatuto da Cidade: Guia Para Implementação Pelos Municípios e Cidadãos*; Polis: Brasília, Brazil, 2001.
17. Guilhen, A.J. *A Parceria Público-Privada Como Instrumento Urbanístico: O Caso do Município de São Paulo-SP*; UFSCAR: São Carlos, Brazil, 2014.
18. ANBIMA. Capital Markets in Brazil. 2024. Available online: https://international.anbima.com.br/reports/brazilian-industry (accessed on 1 September 2024).
19. Chambers and Partners. BRAZIL: An Introduction to Capital Markets Law. 2024. Available online: https://chambers.com/content/item/5398 (accessed on 1 September 2024).
20. Thamer, R.; Lazzarini, S.G. Projetos de Parceria Público-Privada: Fatores Que Influenciam o Avanço Dessas Iniciativas. *Rev. Adm. Pública* **2015**, *49*, 819–846. [CrossRef]
21. Itaparica, A.L.M. *Transparência Em Parcerias Público-Privadas No Brasil: Uma Análise dos Desafios e Perspectivas da Produção Técnico-Científica*; Instituto Serzedello Corrêa, Escola Superior do Tribunal de Contas da União: Brasílila, Brazil, 2023.
22. Melo-Silva, G.; Lourenço, R.L.; Angotti, M. Parcerias Público-Privadas: Modernização Administrativa e Relacionamentos Econômicos Imersos Em Conflitos de Interesse e Corrupção. *Rev. Adm. Pública* **2021**, *55*, 538–558. [CrossRef]
23. Balakrishnan, R.; ToSCANI, F.; Vargas, M. Poverty and Inequality in Brazil and Latin America. In *Brazil: Boom, Bust, and the Road to Recovery*; World Bank: Washington, DC, USA, 2018; pp. 139–169.
24. World Bank. The World Bank in Brazil. 2024. Available online: https://www.worldbank.org/en/country/brazil/overview (accessed on 1 September 2024).
25. World Bank. *Brazil Poverty and Equity Assessment: Looking Ahead of Two Crises*; World Bank: Washington, DC, USA, 2022.
26. Raymundo, G. *Towards Urban Waterfront Redevelopment. The Case of Rio de Janeiro*; Cogitatio Press: Lisbon, Portugal, 2006.
27. Sampaio, M.M. *Operação Urbana Consorciada Porto Maravilha: Estudo de Caso de Uma Parceria Público-Privada*; Universidade Federal do Rio de Janeiro, Instituto de Economia: Rio de Janeiro, Brazil, 2014.

28. Davies, M.; Pontes Nogueira, J. A Wonderful Global City? Resisting Urban Regeneration in Olympic Rio. *J. Int. Relat. Dev. (Ljubl)* **2023**, *26*, 530–556. [CrossRef]
29. Rossi, B.S.d.C. *Dependência e Gestão Urbana: Analisando a Presença do Capital Estrangeiro No Porto Maravilha*; Universidade Federal do Rio de Janeiro, Instituto de Pesquisa e Planejamento Urbano e Regional: Rio de Janeiro, Brazil, 2022.
30. Rodrigues, J.V. *Camadas de Memória Entre o Mar e o Morro: Da Pequena África Ao Porto Maravilha*; Universidade de Brasília: Brasília, Brazil, 2013.
31. Werneck, M. *Porto Maravilha: Agentes, Coalizões de Poder e Neoliberalização No Rio de Janeiro*; Universidade Federal do Rio de Janeiro: Rio de Janeiro, Brazil, 2016.
32. Rio de Janeiro. 2023. Available online: https://www.ccpar.rio/projeto/porto-maravilha/operacao-urbana/ (accessed on 1 September 2024).
33. Rio de Janeiro. *Lei Complementar No. 101, de 23/11/2009, Que Modifica o Plano Diretor e Autoriza o Poder Executivo a Instituir a Operação Urbana Consorciada da Região do Porto do Rio, Na Área de Especial Interesse Urbanístico—AEIU*; Câmara Municipal do Rio de Janeiro: Rio de Janeiro, Brazil, 2009.
34. Rio de Janeiro. *Decreto Municipal No. 26852, de 08/08/2006, Que Cria e Delimita a Área de Especial Interesse da Região Portuária do Rio de Janeiro Para Que Indica, Declara de Utilidade Pública Imóveis, Estabelece Condições Para Parceria Com o Setor Privado e Autoriza a Cons-Tituição de Grupo de Trabalho*; Câmara Municipal do Rio de Janeiro: Rio de Janeiro, Brazil, 2006.
35. Rio de Janeiro. *Decreto Municipal No. 26866, de 09/08/2006, Que Constitui o Grupo de Trabalho Previsto No Decreto n.o 26852, de 08/08/2006, Esta-Belecendo Que a Sua Composição Terá Representantes das Secretarias Municipais de Urbanismo, Fazenda e do Instituto Municipal de Urbanismo Pereira Passo—IPP*; Câmara Municipal do Rio de Janeiro: Rio de Janeiro, Brazil, 2006.
36. Rio de Janeiro. *Decreto Municipal No. 27665, de 12/03/2007, Que Cria o Grupo de Trabalho Para Acompanhar as Disposições Constantes No Decreto n.o 26852, de 08/08/2006. Fica Estabelecida a Data de 10/09/2007 Como Limite Para Entrega dos Estudos Relacionados Ao Projeto*; Câmara Municipal do Rio de Janeiro: Rio de Janeiro, Brazil, 2007.
37. Rio de Janeiro. *Decreto Municipal No. 30355, de 01/01/2009, Que Cria Grupo de Trabalho Para Supervisionar a Implantação do Projeto Porto Mara-Vilha*; Câmara Municipal do Rio de Janeiro: Rio de Janeiro, Brazil, 2009.
38. Rio de Janeiro. *Decreto Municipal No. 30397, de 08/01/2009, Que Altera o Decreto Municipal n.o 30355, de 01/01/2009, Para Incluir a Secretaria Mu-Nicipal de Cultura Na Composição do Grupo de Trabalho Que Supervisionará a Implantação do Projeto Porto Maravilha*; Câmara Municipal do Rio de Janeiro: Rio de Janeiro, Brazil, 2009.
39. Rio de Janeiro. *Decreto Municipal No. 30406, de 12/01/2009, Que Altera o Decreto Municipal n.o 30355, de 01/01/2009, Para Incluir a Secretaria Mu-Nicipal de Transportes e a Secretaria Municipal de Meio Ambiente No Grupo de Trabalho Que Cuida Da Implantação do Projeto Porto Maravilha*; Câmara Municipal do Rio de Janeiro: Rio de Janeiro, Brazil, 2009.
40. Rio de Janeiro. *Decreto Municipal No. 30448, de 04/02/2009, Que Altera o Decreto Municipal n.o 30355, de 01/01/2009, Para Incluir a Procuradoria Geral Do Município do Rio de Janeiro Na Composição Do Grupo de Trabalho Que Supervisionará a Implantação do Projeto Porto Maravilha*; Câmara Municipal do Rio de Janeiro: Rio de Janeiro, Brazil, 2009.
41. Rio de Janeiro. *Decreto Municipal No. 30475, de 17/02/2009, Que Dispõe Sobre a Criação de Grupo Executivo Para Implementação das Deliberações Do Grupo de Trabalho, Criado Pelo Decreto Municipal n.o 30355, de 01/01/2009, e Implantação do Projeto Porto Maravilha*; Câmara Municipal do Rio de Janeiro: Rio de Janeiro, Brazil, 2009.
42. Barandier, H.; Machado, D.P. "Porto Maravilha" e o Destino Da Área Portuária No Rio de Janeiro. In Proceedings of the III Encontro da Associação Nacional de Pesquisa e Pós-Graduação em Arquitetura e Urbanismo, São Paulo, Brazil, 20–24 October 2024; Volume 3.
43. Rio de Janeiro. *Lei Complementar No. 102, de 23/11/2009, Que Cria a Companhia de Desenvolvimento Urbano Da Região Do Porto Do Rio de Janeiro—CDURP, Responsável Pelo Desenvolvimento da AEIU Da Região do Porto do Rio de Janeiro*; Câmara Municipal do Rio de Janeiro: Rio de Janeiro, Brazil, 2009.
44. Rio de Janeiro. *Lei Complementar No. 105, de 22/12/2009, Que Institui o Programa Municipal de Parcerias Público-Privadas—Propar-Rio*; Câmara Municipal do Rio de Janeiro: Rio de Janeiro, Brazil, 2009.
45. Rio de Janeiro. *Decreto No. 31620, de 21/12/2009, Que Institui a Companhia de Desenvolvimento Urbano da Região do Porto do Rio de Janeiro S/A—CDURP e Aprova Seu Estatuto Social*; Câmara Municipal do Rio de Janeiro: Rio de Janeiro, Brazil, 2009.
46. Amaro, G.O. *O Poder Público Como Stakeholder Das Empresas: A Iniciativaprivada e o Estado No Projeto Porto Maravilha*; Fundação Getúlio Vargas: Rio de Janeiro, Brazil, 2017.
47. Rio de Janeiro. *Decreto No. 32575, de 28/07/2010, Que Dispõe Sobre a Inclusão do Projeto de Revitalização da Região do Porto do Rio No Plano Municipal de Parcerias Público-Privadas*; Câmara Municipal do Rio de Janeiro: Rio de Janeiro, Brazil, 2010.
48. Rio de Janeiro. *Decreto No. 32576, de 28/07/2010, Que Dispõe Sobre a Delegação Da Gestão Dos Serviços de Interesse Local e Serviços Públicos de Competência Municipal a Serem Prestados Nos Limites da Área de Especial Interesse Urbanístico—AEIU Da Região do Porto Do Rio*; Câmara Municipal do Rio de Janeiro: Rio de Janeiro, Brazil, 2010.

49. Rio de Janeiro. *Edital de Licitação—Concorrência Pública No. 001/2010 Para Contratação, Em Regime de Parceria Público Privada, Modalidade Concessão Administrativa, da Revitalização, Operação e Manutenção da AEIU da Região Portuária do Rio de Janeiro*; Câmara Municipal do Rio de Janeiro: Rio de Janeiro, Brazil, 2010.
50. Rio de Janeiro. *Decreto No. 32666, de 11/08/2010, Que Determina a Emissão dos Certificados de Potencial Adicional de Construção—CEPAC*; Câmara Municipal do Rio de Janeiro: Rio de Janeiro, Brazil, 2010.
51. Radar PPP. Porto Maravilha (Município do Rio de Janeiro)—Radar PPP. 2010. Available online: https://radarppp.com/resumo-de-contratos-de-ppps/porto-maravilha-municipio-do-rio-de-janeiro/ (accessed on 1 September 2024).
52. Rio de Janeiro. *1o. Termo Aditivo Ao Contrato da Parceria Público-Privada—PPP Na Modalidade de Concessão Administrativa, Celebrado Entre a Companhia de Desenvolvimento Urbano da Região do Porto do Rio, Como Contratante, e a Concessionária Porto Novo S/A, Como Contratada*; Câmara Municipal do Rio de Janeiro: Rio de Janeiro, Brazil, 2012.
53. Rio de Janeiro. *Lei Complementar No. 147/2014, de 24/10/2014, Que Altera a Redação Do Art. 24 da Lei Complementar n.º 105, de 22/12/2009, Para Alterar o Limite Percentual Para Contratação de Parceria Público-Privada, de 3% (Por Cento) Para 5% (Cinco Por Cento) da Receita Corrente Líquida do Exercício e da Receita Corrente Líquida Projetada; Cria o Parágrafo Único Para Vincular a Alteração dos Limites à Modificação da Lei Federal n.º 11.079, de 30/12/2004*; Câmara Municipal do Rio de Janeiro: Rio de Janeiro, Brazil, 2014.
54. Rio de Janeiro. *Decreto No. 46787, de 07/11/2019, Que Cria Grupo de Trabalho Com a Finalidade de Analisar Documentação e Apurar Possíveis Dis-Torções Na Execução do Contrato de Concessão Administrativa Contrato PPP-01*; Câmara Municipal do Rio de Janeiro: Rio de Janeiro, Brazil, 2019.

land

MDPI

Article

Home Elevation Decisions in Post-Disaster Recovery: Social Vulnerability, Policy Gaps, and Lessons from Houston

Ivis García [1,*], Zhihan Tao [2], Julia Orduña [3], Leslie Martínez-Román [1] and Windya Welideniya [1]

1 Landscape Architecture & Urban Planning, Texas A&M University, 789 Ross Street, College Station, TX 77840, USA; lmartinezroman@tamu.edu (L.M.-R.); windyawelideniya@email.tamu.edu (W.W.)
2 Department of Landscape Architecture, University of Urbana Champaign, Cunningham, 611 Lorado Taft Dr Suite 101, Champaign, IL 61820, USA; zhihan@illinois.edu
3 Texas Housers, 20 N. Sampson St., Houston, TX 77003, USA; julia@texashousing.org
* Correspondence: ivis.garcia@tamu.edu

Abstract: This study examines the factors influencing home elevation decisions among participants in Houston's Homeowner Assistance Program (HoAP) and the Texas General Land Office's Homeowner Assistance Program (HAP) in the aftermath of Hurricane Harvey and other flood events. Using a mixed-methods approach, we conducted surveys and semi-structured interviews with 50 homeowners, supplemented by secondary data analyses of program records and GIS-based flood risk assessments. Additionally, 25 undergraduate students engaged in a structured field trip, conducting site observations, interacting with residents, and discussing home elevation with experts. The findings reveal disparities in home elevation outcomes, with lower completion rates in socially vulnerable neighborhoods despite program eligibility. The study also identifies key factors influencing elevation decisions, including mobility concerns, financial constraints, neighborhood esthetics, and perceptions of long-term flood risk. Homeowners aged 60–79 were more likely to elevate their homes, while individuals with disabilities faced additional barriers. This research highlights the need for targeted policy interventions to improve program equity and ensure that vulnerable populations receive adequate support. Beyond its case study implications, this research contributes to broader discussions on disaster recovery, climate adaptation, and urban resilience. It also serves as a model for integrating student learning into community-based participatory research. While this study is limited in scope, it offers insights into the intersection of social vulnerability and housing adaptation, informing future policy efforts to enhance flood resilience in historically marginalized communities.

Keywords: flooding; housing elevation; housing retrofit; mitigation; housing assistance programs

1. Introduction

Hurricane Harvey made landfall on 25 August 2017, bringing unprecedented rainfall and catastrophic flooding to Houston and the surrounding region. As one of the costliest natural disasters in U.S. history, Harvey caused widespread damage, displacing thousands of residents and affecting over 300,000 structures. Houston's vulnerability to flooding is largely due to its low-lying topography, rapid urban development, and proximity to the Gulf of Mexico/America. The city's extensive bayou system, which historically managed stormwater, has been strained by increasing impermeable surfaces and inadequate drainage infrastructure, exacerbating flood risks. These geographic and environmental factors make Houston an essential case study for understanding flood resilience and home elevation policies.

Disasters do not impact all communities equally, and Hurricane Harvey exposed stark disparities in recovery outcomes. Low-income and historically marginalized communities often lack the financial resources, political influence, and infrastructure necessary for effective disaster mitigation. These communities are more likely to be located in flood-prone areas and to experience barriers in accessing government assistance. Addressing these inequalities requires targeted policies that not only provide financial aid but also ensure equitable access to home elevation and reconstruction programs.

In response to Harvey, federal, state, and local agencies implemented home elevation policies to reduce future flood risk. Two major programs—the City of Houston's Homeowner Assistance Program (HoAP) and the Texas General Land Office's (GLO) Homeowner Assistance Program (HAP)—were established to support home repair, reconstruction, and elevation. These programs were intended to increase flood resilience by elevating homes in flood-prone areas, yet disparities in program participation and completion rates persist. This raises critical questions about the factors influencing homeowners' decisions to elevate their homes and the extent to which these programs are effectively reaching vulnerable populations. This research addresses four key research questions:

- How does the demographic makeup of Houston correlate with disaster vulnerability?
- Which social, personal, and demographic factors influence home elevation among participants in home repair programs?
- What are the practical considerations and perceptions of flood risks that influence homeowners' decisions to elevate their homes?

Despite growing research on climate adaptation and disaster mitigation, relatively little attention has been paid to structured home elevation programs within government-funded post-disaster recovery efforts. While studies have examined large-scale flood mitigation strategies and individual homeowner resilience, there is a gap in understanding regarding how elevation programs function in practice, particularly in urban environments like Houston. This research fills that gap by analyzing how homeowners navigate these programs, assess risks, and make elevation decisions.

To explore these issues, we employed a mixed-methods approach, combining surveys and semi-structured interviews with 50 homeowners who participated in home repair programs. Additionally, 25 undergraduate students from Texas A&M University visited affected neighborhoods, engaging with residents, and asked questions of two elevation experts in a recorded lecture discussion.

The findings indicate that home repair completion rates were lower in socially vulnerable neighborhoods, highlighting disparities in program accessibility and implementation. Among the surveyed homeowners, 70.27% of those aged 60–79 opted for home elevation, particularly in frequently flooded areas. Key factors influencing elevation decisions included accessibility concerns, neighborhood esthetics, and perceptions of elevation effectiveness. While some residents viewed elevation as a crucial mitigation strategy, others were hesitant due to financial constraints, skepticism about structural integrity, or concerns over the visual impact on their neighborhood.

This study contributes to the broader discourse on climate adaptation, disaster recovery, and housing resilience by identifying policy gaps in home elevation assistance programs. It underscores the need for more tailored outreach and equitable distribution of resources to flood-prone communities. While limited in scope, this research serves as a pilot study to support future investigations into the intersection of social vulnerability and housing adaptation. These findings provide valuable insights for policymakers, planners, and scholars working to improve disaster mitigation strategies. Additionally, this study demonstrates the value of integrating student learning into community-based research, offering a model for applied, transdisciplinary engagement in urban planning and disaster resilience studies.

1.1. Literature Review

Houston's demographic composition is highly diverse, with significant Hispanic, African American, and immigrant populations. Many of these communities reside in economically disadvantaged neighborhoods that lack adequate infrastructure and resources, making them more vulnerable to disasters. Socioeconomic disparities, language barriers, and limited access to healthcare and transportation contribute to unequal disaster preparedness, responses, and recovery efforts [1].

Research shows that low-income neighborhoods in Houston tend to be located at lower elevations, making them more susceptible to flooding. Even when accounting for latitude, longitude, and population density, these communities experience a higher flood risk due to historical patterns of development and disinvestment [2]. Using Google Maps and American Community Survey data, Lu (2017) found that lower median household incomes, higher poverty rates, and a greater percentage of ethnic minorities and non-citizens were correlated with lower neighborhood elevations [2]. This spatial inequality is not unique to Houston; similar trends are observed in other major U.S. metropolitan areas, revealing a broader pattern of environmental injustice that disproportionately exposes marginalized communities to flood hazards.

Zahran et al. (2008) further established that socially vulnerable populations in Eastern Texas suffered higher casualty rates during floods [3]. Their study highlights that while mitigation efforts can reduce risk, socioeconomic factors significantly influence survival outcomes. Communities with greater resources and political capital tend to recover faster, while low-income residents often struggle to access disaster aid and rebuilding programs.

International studies have examined individual decision-making in flood risk mitigation, particularly in relation to home elevation. Grothmann and Reusswig (2006) found that in Germany, personal risk perception and belief in the effectiveness of mitigation measures were key drivers behind decisions to elevate homes or implement flood-resistant modifications [4]. In France, Poussin et al. (2014) combined survey and preference data to analyze flood mitigation adoption rates, revealing that households were more likely to take action when they perceived a direct benefit [5].

Botzen et al. (2013) studied homeowners in the Netherlands and found that 52% were willing to invest USD 10,000 in elevation measures to protect their properties [6]. Their findings also suggest that homeowners prefer physical flood protections over financial mechanisms such as insurance, emphasizing a willingness to pay for certainty in risk reduction.

While these studies provide insights into flood mitigation behaviors globally, they do not address homeowners already enrolled in a structured reconstruction program with elevation options. Understanding these decision-making processes within government-backed programs is crucial in designing effective disaster resilience policies.

Previous research has highlighted the importance of federal and local mitigation grants in making home elevation financially feasible [7]. Programs such as FEMA's Hazard Mitigation Grant Program (HMGP), Flood Mitigation Assistance Program (FMA), and Pre-Disaster Mitigation Grant Program (PDM) provide funding to help homeowners elevate their properties. However, access to these funds is often inequitable, with low-income homeowners facing bureaucratic barriers, a lack of information, and financial limitations that prevent them from fully utilizing the available resources [8,9].

The Homeowner Assistance Program (HAP) in Houston is an example of a large-scale government initiative aimed at flood resilience. However, disparities in program outcomes suggest the inequitable distribution of resources, with some communities benefiting more than others. This study investigates how program participants perceive and navigate home elevation decisions, contributing to a deeper understanding of the barriers and incentives within structured recovery programs.

Flooding in Houston is a combined result of natural vulnerability and urban development patterns. The city's flat coastal geography and proximity to the Gulf of Mexico inherently expose it to high flood risks [10]. However, the rapid urban expansion since the mid-20th century has exacerbated these risks, with unchecked development reducing the natural drainage capacity and straining the existing infrastructure [11].

Hurricane Harvey (25 August–3 September 2017) serves as a landmark event in Houston's flood history. The storm dumped an estimated 27 trillion gallons of rain over Texas, leading to catastrophic flooding that damaged over 300,000 structures and displaced more than 30,000 residents in Houston alone [12]. The economic damages reached approximately USD 125 billion, making Harvey one of the costliest disasters in U.S. history [13].

The aftermath of Harvey exposed systemic weaknesses in Houston's flood management policies. Many affected residents struggled to access assistance programs, while outdated floodplain maps and inconsistent building regulations failed to prevent new construction in high-risk areas [14]. Although FEMA and the Texas General Land Office have since revised floodplain designations and elevation requirements, the effectiveness of these changes remains a subject of ongoing debate.

Houston's floodplains are mapped by FEMA, and these maps are updated regularly to reflect changes in topography, hydrology, and urban development. However, the fast-changing nature of land use, climate, and storm patterns means that these maps can quickly become outdated, making flood risk assessment and management challenging. The discrepancies in mapped versus actual flood risks create uncertainty for homeowners when deciding whether to elevate their homes or take other protective measures.

1.2. Research Gap

While numerous studies have explored flood risk perception and mitigation strategies, there is limited research on how homeowners make elevation decisions within structured reconstruction programs. By examining Houston's Homeowner Assistance Program (HoAP) and Texas General Land Office's Homeowner Assistance Program (HAP), this study contributes to the broader discourse on climate adaptation, social vulnerability, and disaster resilience planning. Understanding who chooses to elevate, why, and what challenges they face is essential in designing equitable and effective housing recovery programs in flood-prone communities.

2. Materials and Methods

2.1. Study Site and Access

Houston, Texas, exemplifies the tension between rapid urban expansion and vulnerability to natural hazards, particularly flooding. The city's geographic and topographic features, combined with rapid development and inadequate drainage infrastructure, have exacerbated its susceptibility to severe flood events. For this study, we purposefully selected Houston due to its frequent flooding, its diverse population, and the availability of structured home repair and elevation assistance programs following Hurricane Harvey.

Our research focused on participants in home repair programs targeting flood-prone neighborhoods. These communities are characterized by socioeconomic disparities, a mix of long-term residents and transient populations, and varying degrees of flood risk awareness. The diverse social and economic conditions in these neighborhoods influence residents' access to recovery programs and their decision-making processes regarding home elevation.

To facilitate community engagement and research participation, we partnered with Texas Housers, a nonprofit advocating for housing justice and disaster recovery for low-income households. Their established relationships within these neighborhoods helped build trust with residents, many of whom were initially hesitant to participate in research activities. Texas Housers also assisted in identifying and recruiting program participants, ensuring that the study reflected the perspectives of those most affected by flood-related displacement and home repair challenges.

2.2. Data Collection

2.2.1. Field Trip

As part of this study, 25 undergraduate students from Texas A&M University's Urban and Regional Studies Capstone Course, accompanied by two faculty members, conducted a structured field trip to one of the thirteen flood-affected neighborhoods. Neighborhoods were selected based on their proximity to flood-prone bayous and high concentrations of homes eligible for elevation programs. During this one day, 8 to 10 h field trip on 29 February 2024, students engaged in direct observations of flood-damaged homes and elevated structures. They also engaged in informal conversations (not recorded) with seven homeowners about their flood experiences and perspectives on home elevation. (Faculty members formally interviewed these participants). Faculty also involved students in a recorded lecture with two elevation specialists, who provided a technical overview of elevation methods and financial considerations for affected homeowners. Students were able to ask open-ended questions.

Students also met with representatives from Texas Housers, who facilitated access to four program participants and arranged site visits to homes undergoing elevation or reconstruction. The field trip provided valuable firsthand insights into the physical, financial, and emotional challenges related to home elevation and reinforced students' understanding of community-based research in disaster recovery planning.

2.2.2. Secondary Data Collection and Analysis

To assess the physical and social factors influencing home elevation decisions, we collected and analyzed in Geographic Information System (GIS) secondary data sources, including the following:

- Flooding frequency: FEMA's National Flood Hazard Layer (NFHL) (2024).
- Topographic elevation data: LiDAR Elevation Data (DEM) from the Texas Natural Resources Information System (TNRIS) and Texas Geographic Information Office (TxGIO) (2023).
- Social vulnerability indicators: CDC's Social Vulnerability Index (2020) to evaluate the intersection of flood risk and socioeconomic conditions.
- Home elevation program records: Publicly available data from the City of Houston's HoAP program (2024), detailing the number and locations of elevated homes.

Additionally, we submitted a public record request to the Texas General Land Office (GLO) for program data related to elevation requirements, demographics of applicants, and home foundation types (slab, pier, or beam). These data were not received in time for inclusion in this report but will be analyzed in future research.

2.2.3. Survey and Interviews

To explore homeowners' perceptions of home elevation, we conducted 50 surveys and in-depth interviews with participants in the City of Houston's and Texas GLO's home repair programs. All participants were either eligible for or had been offered home elevation assistance.

Participants were recruited from 13 neighborhoods across Houston, identified through flood impact maps and program participation records. The selected neighborhoods represented diverse geographic clusters across the north, south, east, and west of downtown Houston, capturing a range of flood experiences and recovery challenges. We visited 130 addresses; 4 people declined, and 76 people were not home or did not answer the door (we left a flyer). We were able to conduct 50 surveys/interviews. Even though the flyer included information about completing a survey online, no one did so. The participation rate was 39.7%. GLO records show that 1608 participated in the program. The sample represents approximately 3.1% of the total program participants.

Surveys were administered door-to-door, with homeowners choosing whether to complete the survey inside or outside their homes. Each visit lasted between 20 min and over an hour (most around 40 min), allowing time for in-depth discussions and site observations. Paper-based surveys measured demographic variables, flood history, decision-making factors, and perceptions of government assistance. A consent form was given to participants for them to keep for their records.

Interviews were recorded with participant consent, while detailed notes were taken for those who declined recording. A follow-up interview was conducted to validate survey responses and ensure data consistency. In addition to resident interviews, we conducted structured interviews with two elevation specialists during the field trip. These experts provided technical explanations about elevation processes and financial and regulatory challenges, offering additional context for interpreting homeowner decisions. We promised confidentiality to all participants.

2.3. Quantitative and Qualitative Data Analysis

Survey responses ($N = 50$) were analyzed using descriptive statistics (mean, median, and standard deviation). Chi-square tests were used to examine relationships between demographic factors (e.g., age and disability status) and home elevation decisions.

Interview transcripts were coded thematically using Dedoose web application for mixed methods research. Key themes included perceived flood risks, government assistance, mobility concerns, and neighborhood esthetics. Comparative analysis was conducted to identify common patterns across different demographic groups.

2.4. Ethical Considerations and Researcher Positionality

This study was conducted in accordance with ethical guidelines, as approved by the Institutional Review Board (IRB) (ID: STUDY2004-0011) at Texas A&M University.

Participants were informed about the study's purpose, risks, and confidentiality protections before agreeing to participate. Participants were given a consent form for them to read and kept. Homeowners received USD 20 cash compensation for their time and were promised access to the study results. We collected the contact information of participants but de-identified the information for the data collected. All participant names were replaced with pseudonyms. Interviews and survey responses were anonymized to protect identities. Findings will be shared with participants through focus groups and follow-up presentations (in-person and via conference calls). This was designed to ensure that residents benefited from the research and could use the results to advocate for better disaster recovery policies.

3. Results

3.1. Analysis of Secondary Data

3.1.1. Floodplain Map

Approximately 20% of Texas's land area, equating to roughly 56,000 square miles, falls within the 100-year floodplain, as identified by the Texas Water Development Board [15]. While specific percentages for Houston and Harris County are not detailed in the provided sources, it is notable that more than 7000 housing units have been constructed within Harris County's 100-year floodplain since 2010 [15]. This development trend underscores the region's ongoing challenges with urban expansion into flood-prone areas.

Given that Harris County encompasses a significant portion of Houston, Figure 1 highlights the city's substantial exposure to flood risks. The concentration of flood-prone areas along Houston's bayous and low-lying regions, such as neighborhoods near Buffalo Bayou, Brays Bayou, and Greens Bayou, further emphasizes the need for comprehensive flood management strategies.

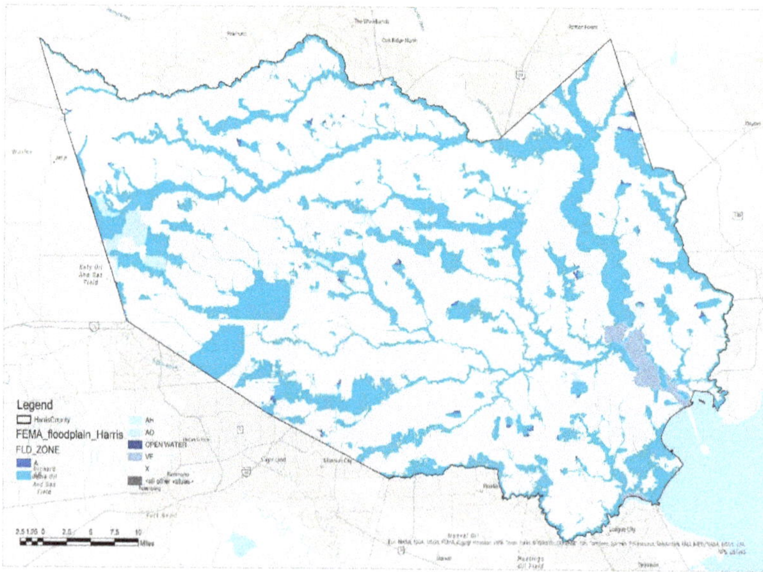

Figure 1. Federal Emergency Management Agency floodplain map of Harris County, Texas. *Note*: From *FEMA's National Flood Hazard Layer (NFHL) Viewer* [Dataset] by Federal Emergency Management Agency, n.d. Retrieved 1 October 2024 (https://www.fema.gov/flood-maps/national-flood-hazard-layer, accessed on 1 October 2024).

It is important to note that existing floodplain maps, last officially updated in 2007, may not fully reflect current risk levels due to factors such as urban development and climate change. Consequently, ongoing efforts to revise these maps are designed to provide a more accurate representation of flood-prone areas, ensuring that residents and policymakers have up-to-date information to guide future development and flood mitigation strategies.

3.1.2. Social Vulnerability Index

The CDC Social Vulnerability Index (SVI) is a quantitative tool designed to assess how susceptible communities are to environmental hazards and disasters [16]. It incorporates multiple socioeconomic, demographic, and housing factors to determine the relative vulnerability of different populations. The SVI has been widely used in disaster management and resilience planning to identify communities that may require additional resources and support during emergencies [17]. By integrating these variables, the Index provides a comprehensive framework for understanding social vulnerability at a granular level.

In Houston, Texas, social vulnerability is particularly pronounced due to significant socioeconomic disparities across neighborhoods. The city's diverse racial and ethnic composition, income inequality, and uneven access to essential services contribute to varying degrees of vulnerability. These disparities are particularly relevant in flood-prone areas, where historically marginalized communities face greater risks and challenges in recovery efforts. For example, Figure 2 illustrates the percentage of people living below 150% of the poverty line in high-risk flood zones across Harris County, emphasizing the intersection of flood exposure and economic disadvantage. Understanding these patterns is essential in distributing targeted resources and improving disaster resilience in vulnerable communities.

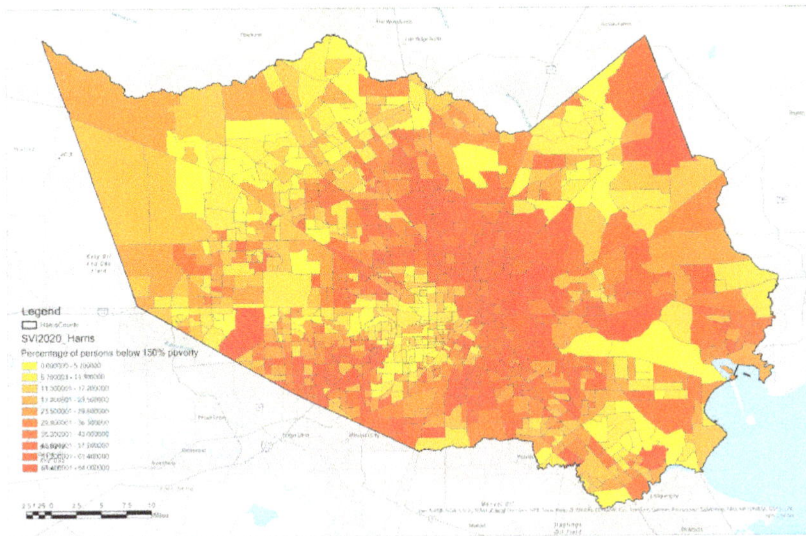

Figure 2. Percentage of persons living below the 150% poverty line in Harris County, Texas, by census tract. Note: From CDC/ATSDR Social Vulnerability Index 2020 Database, Harris County, Texas, by Centers for Disease Control and Prevention ([17], https://www.atsdr.cdc.gov/place-health/php/svi/svi-data-documentation-download.html?CDC_AAref_Val=https://www.atsdr.cdc.gov/placeandhealth/svi/data_documentation_download.html, accessed on 1 October 2024).

Figure 3 illustrates the percentage of the nonwhite population residing in each census tract within Harris County. The left map highlights census tracts with elevated poverty rates, with darker red shades indicating areas where a higher proportion of residents live below the poverty line. The right map uses the same red shading to represent areas with higher levels of social vulnerability, as measured by the CDC Social Vulnerability Index (SVI).

The relationship between poverty, racial disparities, and environmental vulnerability is particularly significant in Houston, where historically marginalized communities often face greater challenges in disaster preparedness, response, and recovery. Repeated flooding events in the city have disproportionately impacted these communities, further exacerbating socioeconomic inequalities. By mapping race, poverty, and social vulnerability together, Figure 3 underscores the systemic risks faced by nonwhite and low-income populations, reinforcing the need for equitable disaster mitigation and recovery efforts.

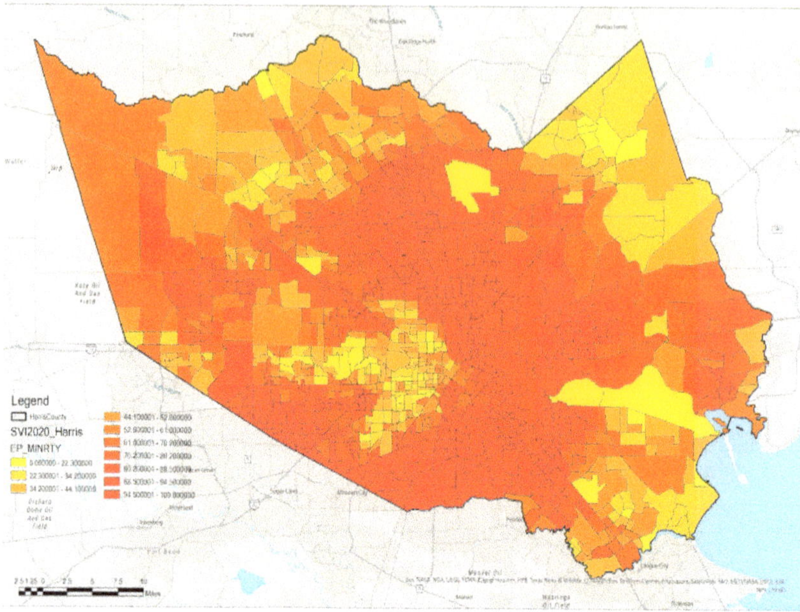

Figure 3. Percentage of nonwhite population in Harris County, Texas, by census tract. Note: From CDC/ATSDR Social Vulnerability Index 2020 Database, Harris County, Texas, by Centers for Disease Control and Prevention (https://www.atsdr.cdc.gov/placeandhealth/svi/data_documentation_download.html, accessed on 1 October 2024).

The Social Vulnerability Index (SVI) incorporates multiple dimensions, including socioeconomic status, household composition, racial and ethnic minority status, and housing and transportation factors. Figure 4 illustrates the spatial distribution of social vulnerability across Houston, revealing significant regional disparities.

According to the City of Houston, super neighborhoods were established to foster collaboration among residents of adjacent communities, allowing them to collectively identify, prioritize, and address local concerns. Many of these designated areas, particularly those located in eastern and southern Houston, exhibit higher levels of social vulnerability. In Figure 4, census blocks are shaded to represent varying degrees of vulnerability, with darker areas indicating higher social vulnerability percentiles. These patterns highlight the unequal distribution of risks and resources, emphasizing the need for targeted interventions to enhance resilience in the most vulnerable communities.

Interviews revealed distinct clusters of newly constructed and easily identifiable "Harvey homes" in high-risk, lower-income neighborhoods. However, new construction was noticeably absent in more affluent areas such as Meyerland and Brays Bayou, which also have lower Social Vulnerability Index (SoVI) scores. While some homes in these wealthier neighborhoods may have been elevated, they were generally not demolished and rebuilt, as observed in lower-income communities.

This disparity suggests that home reconstruction efforts vary based on income and location, with lower-income areas experiencing more extensive demolitions and replacements, whereas higher-income areas retained pre-existing housing structures with modifications. As a result, no survey respondents were identified in these wealthier neighborhoods, highlighting potential inequities in recovery efforts. This pattern underscores the geographic concentration of flood risk and social vulnerability, emphasizing the need for targeted interventions that address disparities in disaster recovery and housing resilience strategies.

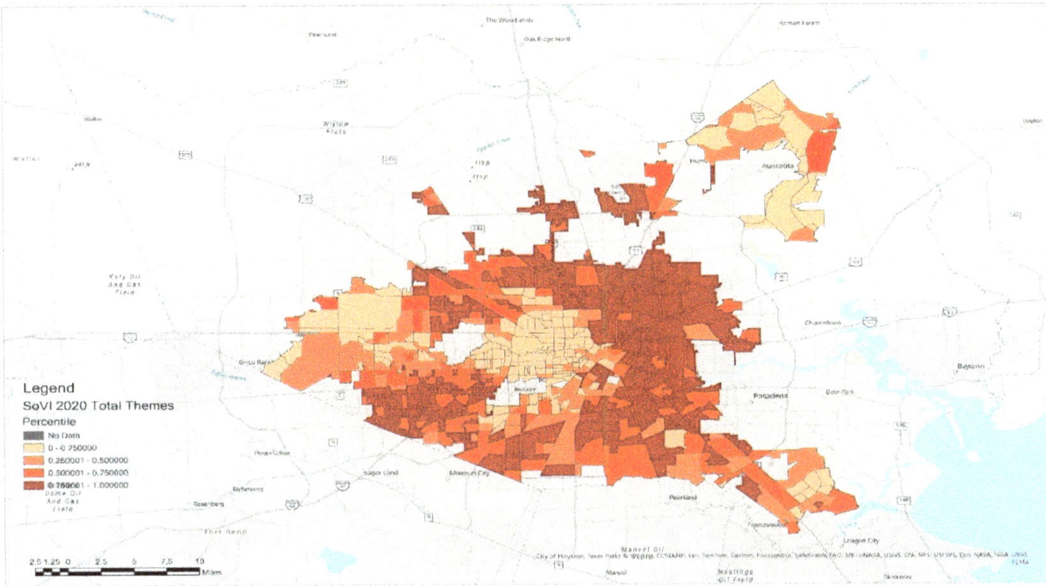

Figure 4. Total social vulnerability percentiles of census tracts in Harris County, Texas. Note: From CDC/ATSDR Social Vulnerability Index 2020 Database, Harris County, Texas, by Centers for Disease Control and Prevention (https://www.atsdr.cdc.gov/placeandhealth/svi/data_documentation_download.html, accessed on 1 October 2024).

3.1.3. Current Status of the Homeowner Assistance Program

As of 31 May 2024, the Homeowner Assistance Program (HoAP) had completed 158 home reconstructions, while the Texas General Land Office (GLO) had completed 1608 homes in Houston. However, completion rates varied significantly across different neighborhoods, with notable geographic disparities in project distribution.

Figure 5 presents the total number of completed home elevation projects within the study area. This map visually represents where the most home elevations have taken place, highlighting neighborhoods with higher absolute numbers of completed projects. However, this figure does not account for how many projects were initiated versus how many were completed, nor does it reflect the proportion of completed projects relative to total applications. The Kingwood, East Houston, and Memorial super neighborhoods recorded the highest number of completed projects, reflecting concentrated recovery efforts in these areas. Similarly, western and northern Houston demonstrated the highest completion rates, meaning that a greater proportion of announced projects were successfully completed.

Figure 6 focuses on completion rates, showing the percentage of projects completed out of the total number of approved or initiated projects in each neighborhood. This figure provides a clearer measure of program efficiency by illustrating where home elevation projects are more likely to be fully executed compared to areas with lower completion rates.

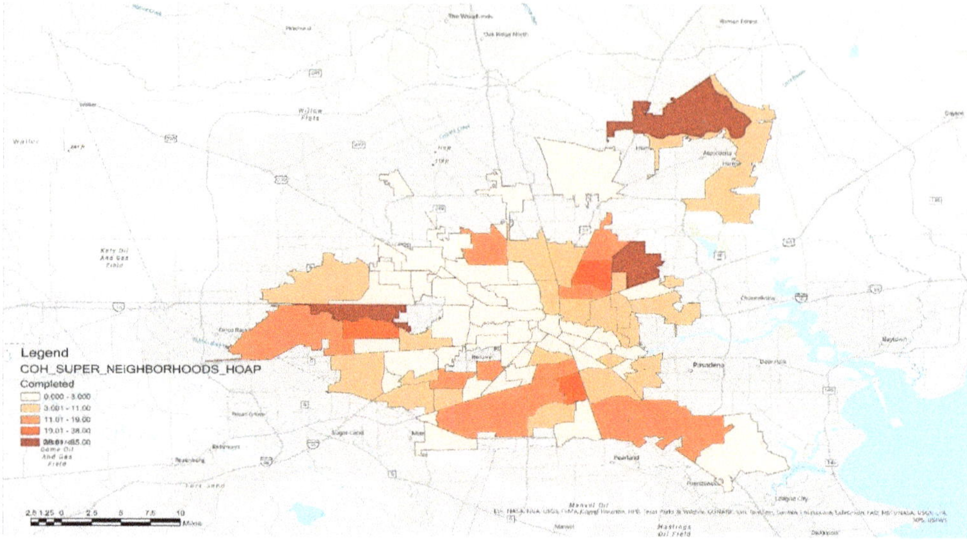

Figure 5. Total number of completed home elevation projects in the Homeowner Assistance Program projects by super neighborhoods in Houston, Texas (Data accessed on 3 May 2024). Note: Map constructed by the research team using data from Transparency and Reporting Portal (https://recovery.houstontx.gov/dr17/transparency/, accessed on 1 October 2024), City of Houston Recovery Programs, n.d. Retrieved 5 March 2024. (https://recovery.houstontx.gov/dr17/transparency/, accessed on 3 May 2024). According to the City of Houston, super neighborhoods were established to encourage collaboration among residents of neighboring communities, enabling them to collectively identify, prioritize, and address the needs and concerns of the larger community.

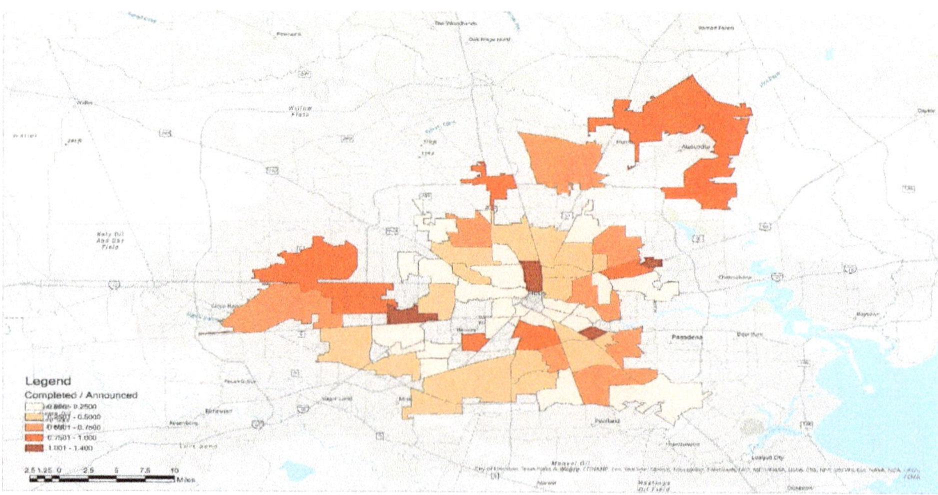

Figure 6. Completion rates of Homeowner Assistance Program projects in super neighborhoods in Houston, Texas. Note: Map constructed by the research team using data from Transparency and Reporting Portal (https://recovery.houstontx.gov/dr17/transparency.html, accessed on 3 May 2024), City of Houston Recovery Programs, n.d. Retrieved 5 March 2024. (https://recovery.houstontx.gov/dr17/transparency.html, accessed on 3 May 2024).

In contrast, southern neighborhoods of Houston, which are characterized by higher minority populations, greater poverty levels, and aging infrastructure, exhibited lower completion rates despite their heightened vulnerability. Figure 7 integrates social vulnerability data into the completion rates from Figure 6 to assess disparities in program outcomes. This map overlays program effectiveness with community-level social vulnerability indicators to identify whether more vulnerable populations are receiving the same level of support as less vulnerable ones. The figure uses a gradient scale from low vulnerability and high completion rates to high vulnerability and low completion rates, helping to pinpoint the communities most in need of targeted policy interventions.

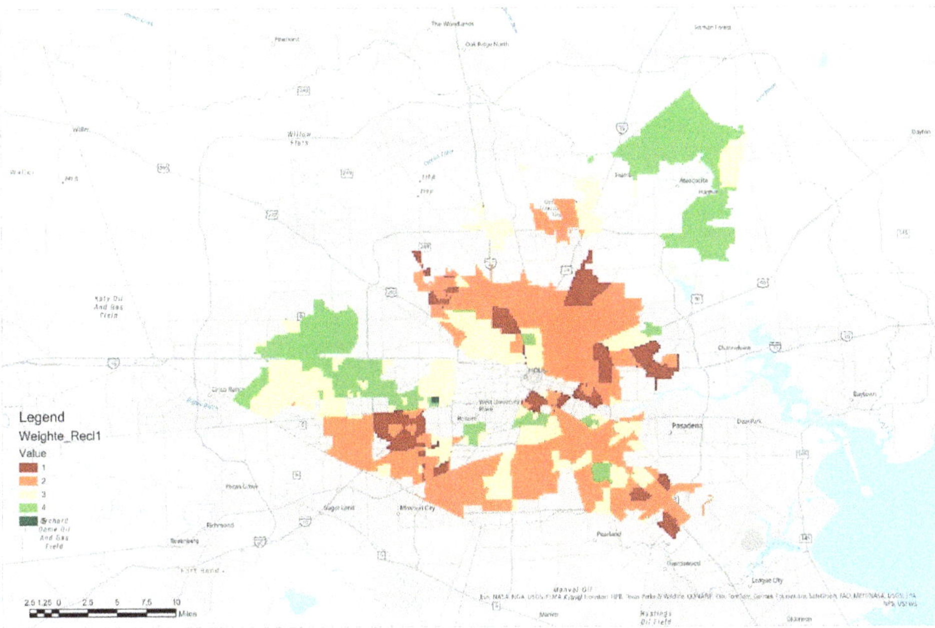

Figure 7. Weighted overlay of total social vulnerability and completion rate of home assistance program projects. Note: Map constructed by the research team using data from Transparency and Reporting Portal (https://recovery.houstontx.gov/dr17/transparency, accessed on 5 March 2024), City of Houston Recovery Programs, n.d. Retrieved 5 March 2024. (https://recovery.houstontx.gov/dr17/transparency, accessed on 5 March 2024). The value represents a combination of social vulnerability and program completion rate. A value of 1 indicates the area with the highest social vulnerability and the lowest completion rate.

3.2. Surveys

3.2.1. Demographics

Table 1 presents the demographic breakdown of the 50 surveyed homeowners who participated in the study. The sample was predominantly older, with 68% of respondents aged 60–79, while only 10% were under 60. Most participants (68%) were female, and 64% identified as Black or African American, followed by 20% identifying as "Other" and 10% as Mixed-Race.

Regarding ethnicity, 28% of respondents identified as Hispanic, while 66% identified as Non-Hispanic. The survey was conducted primarily in English (84%), with 16% of responses collected in Spanish, reflecting the linguistic diversity of the affected communities.

In terms of income levels, most respondents reported monthly incomes between USD 1000 and USD 2999 (50%), with 14% earning below USD 1000 per month, indicating that many participants were from lower-income households. The average monthly income among respondents was USD 1510.45.

Table 1. Selected characteristics of research participants.

Characteristic	Variable	Number of Respondents (N)	%
Age (a)	20–39	2	4%
	40–59	3	6%
	60–79	34	68%
	80+	9	18%
	Not Provided	2	4%
Gender	Female	34	68%
	Male	14	28%
	Not Provided	2	4%
Race	White	0	0%
	Asian	0	0%
	Native Hawaiian and Other Pacific Islander	0	0%
	Black or African American	32	64%
	American Indian or Alaska Native	3	6%
	Mixed-Race	5	10%
	Other	10	20%
Ethnicity	Hispanic	14	28%
	Non-Hispanic	33	66%
	Not Provided	3	6%
Survey Language	English	42	84%
	Spanish	8	16%
Income (b)	USD 0 to USD 499	2	4%
	USD 500 to USD 999	7	14%
	USD 1000 to USD 1499	11	22%
	USD 1500 to USD 1999	4	8%
	USD 2000 to USD 2499	7	14%
	USD 2500 to USD 2999	3	6%
	USD 3000 to USD 3499	2	4%
	USD 3500 to USD 3999	4	8%
	USD 4000 to USD 4499	2	4%
	USD 4500 to USD 4999	1	2%
Disability (c)	Has a Disability	26	52%
	Does Not Have a Disability	19	38%
	Not Provided	5	10%
Elevated Home	Yes	37	74%
	No	13	26%

Note: N = 50. (a) The average age of respondents was 70.1 years. (b) The average monthly income among respondents was USD 1510.45. (c) Disability was auto-designated by participants; they were simply asked whether they had a disability or someone with a disability in their household composition.

A significant proportion of respondents (52%) reported having a disability, while 38% did not, and 10% chose not to disclose this information. This finding suggests that accessibility and mobility concerns may be important considerations in home elevation decisions.

Finally, 74% of respondents had elevated their homes, while 26% had not, highlighting varying adoption rates of home elevation across different communities. These findings provide critical insights into the socioeconomic and demographic factors influencing recovery and home elevation decisions in flood-prone neighborhoods.

3.2.2. Diverse Home Elevation Practices and Their Relationship with Flooding Frequency

Initially, our research team expected that homeowners would only consider pier and beam homes as elevated, given that this method—raising homes on stilts or pilings—is a well-established strategy for flood protection. However, fieldwork revealed that residents recognized additional methods of home elevation, broadening our understanding of how elevation is perceived and implemented in flood-prone communities.

Figure 8 categorizes the four types of home elevation identified by residents. In Figure 8a, the top left image shows a non-elevated slab home constructed directly on a concrete foundation at ground level. In Figure 8b, the top right image presents a standard pier and beam home raised above the floodplain, often with ramps or stairs for accessibility. The bottom images (Figure 8c,d) depict slab homes that residents considered elevated due to modifications made by construction companies. Figure 8c (bottom left) showcases a slab home raised by 4–6 inches, while Figure 8d (bottom right) illustrates a home elevated using fill material beneath the slab foundation to increase elevation above the flood zone. These alternative elevation techniques highlight the variety of approaches used to mitigate flood risks, demonstrating that elevation is not solely limited to traditional pier and beam construction.

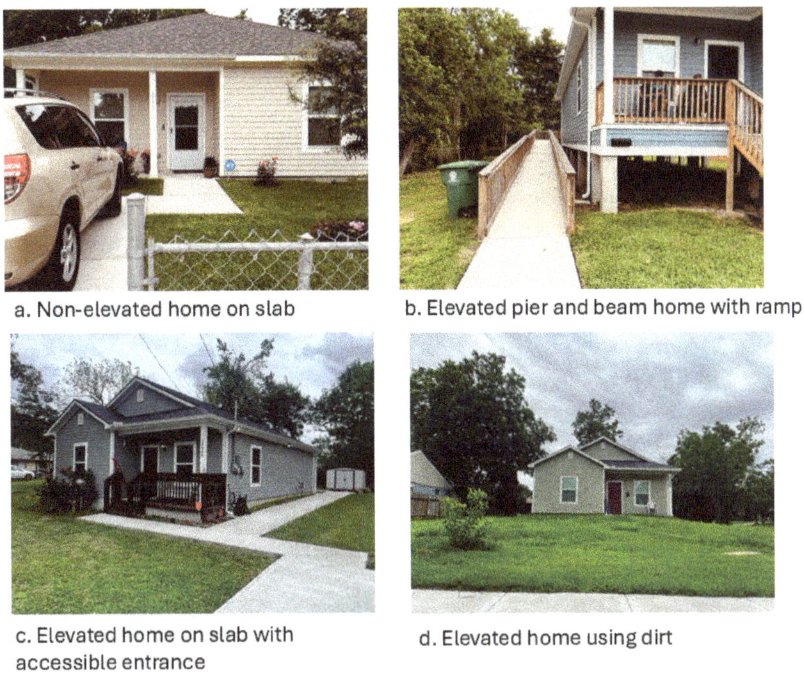

Figure 8. How residents categorized types of non-elevated and elevated homes. Photographs of homes showcasing the different types of elevations that were observed as part of the fieldwork: (**a**) a non-elevated home on a slab; (**b**) a property elevated in the pier and beam style with a ramp; (**c**) a property elevated on a slab with an accessible entrance; and (**d**) a home elevated using dirt. Photographs by the research team, 2024.

3.2.3. Survey Findings on Home Elevation and Flooding Frequency

Survey responses provide additional insights into home elevation decisions and flooding experiences among homeowners. As shown in Table 2, most homeowners applied for elevation assistance between 2019 and 2021, aligning with post-Hurricane Harvey recovery efforts. Of the 37 homes that were elevated, 30% were located in designated flood zones, while 70% were not, suggesting that some homeowners pursued elevation even outside high-risk areas. Similarly, among the 13 non-elevated homes, nearly 93% were located outside of flood zones, which may indicate a perception that elevation is unnecessary unless the home is within a high-risk area.

Table 2. Survey responses regarding home elevation and flooding frequency.

Category	Variable	Number of Respondents	
		N	%
Application Year	2017	2	4%
	2018	8	16%
	2019	17	34%
	2020	5	10%
	2021	8	16%
	2022	4	8%
	2023	3	6%
	Not Provided	3	6%
Elevation Decision	Yes (Elevated Home)	37	74%
	No (Declined Home Elevation)	13	26%
Location of Elevated Homes [a]	In Flood Zone	11	30%
	Not in Flood Zone	26	70%
Location of Non-Elevated Homes [b]	In Flood Zone	1	7.69%
	Not in Flood Zone	12	92.34%
Homeowner Knowledge of Home's Location	Certain in Flood Zone	20	40%
	Certain was Not in Flood Zone	9	18%
	Unsure of Whether in Flood Zone	20	40%
	Not Provided	1	2%
Flooding Frequency Group	Frequent Flooding Group	9	18%
	Harvey Flooding Group	29	58%
	Other Flooding Group	5	10%
	Never-Flooded Group	5	10%
	Not Provided	2	4%

Note: $N = 50$ survey and interview respondents. [a] $N = 37$ homeowners in our sample who elevated their homes. [b] $N = 13$ homeowners in our sample who did not elevate their homes.

Homeowners' awareness of their flood risk played a role in their decision-making. While 40% of respondents were certain that their homes were in flood zones, an equal percentage was unsure of their flood zone status, pointing to gaps in public awareness and risk communication.

3.2.4. Home Elevation Heights and Flooding History

Survey data also revealed differences in elevation height based on prior flooding experiences. As Table 3 illustrates, homeowners who experienced frequent flooding raised their homes by an average height of 44.2 inches, significantly higher than those who experienced flooding only during Hurricane Harvey (31.7 inches) or from other minor events (26.4 inches). This pattern suggests that frequent flood victims are more inclined to pursue higher elevations as a long-term flood mitigation strategy.

Table 3. Average home elevation height by flooding frequency group.

Flooding Frequency Group	Average Home Elevation Height	Standard Deviation (SD)
All Groups	35 inches	9.14 inches
Frequent Flooding Group	44.2 inches	28.6 inches
Harvey Flooding Group	31.7 inches	25.8 inches
Other Flooding Group	26.4 inches	17.3 inches

Note: $N = 37$ homeowners in our sample who elevated their homes.

3.2.5. Demographic Factors Influencing Home Elevation Decisions

The analysis of respondent demographics revealed statistically significant relationships between age, disability status, and home elevation decisions. The following points were indicated by chi-squared test results.

Older homeowners were more likely to elevate their homes. Approximately 70.27% of respondents aged 60–79 elevated their homes, compared to only 5.41% of those aged 40–59 and one respondent under 39. This suggests that older homeowners, possibly influenced by past flooding experiences and long-term homeownership, are more proactive about flood mitigation.

Disability status influenced elevation choices. Among respondents who did not report a disability, 96% elevated their homes, compared to 68% of those who identified as having a disability or living with someone with a disability. The disparity suggests that physical mobility concerns, financial limitations, or accessibility challenges may discourage some households from pursuing elevation. However, given the small sample size, further research with a larger dataset is recommended to validate these findings.

Other demographic factors, such as education level and income, did not show statistically significant relationships with home elevation decisions. Additionally, there was no substantial difference in the average number of years that respondents had lived in their neighborhoods—those who elevated had lived there for an average of 38.41 years, while those who did not had lived there for 39.08 years.

3.3. *Interviews*

3.3.1. The Role of Government Assistance in Home Elevation

Discussing the impact of government assistance programs, an elevation specialist highlighted the financial support available for homeowners' considering elevation: *"There are funds available to assist homeowners in paying for this, and a lot of the time financial assistance will cover 100% of the cost. Sometimes homeowners have a cost share, but a lot of the time it can be 100% cost for them."* This statement underscores the critical role of financial aid in making elevation accessible to homeowners, particularly those directly affected by disasters and lacking the financial resources to fund such modifications independently.

3.3.2. Perceptions of Home Elevation Effectiveness

While financial support can incentivize home elevation, community perspectives on its effectiveness remain divided. Some residents were skeptical about whether elevation would provide long-term flood protection and questioned the structural durability of elevated homes. One homeowner expressed concern, stating, *"I've seen a couple of elevated homes around here that still experienced issues during floods. It makes you wonder how effective this is."*

Additionally, some residents worried about potential stability issues, particularly regarding water pooling beneath elevated homes and long-term foundation degradation. Others voiced concerns about future maintenance costs, particularly the need for repairs to elevated supports over time. Contrasting these doubts, a professional in elevation construction emphasized the peace of mind that elevation provides: *"Once it gets their house off the ground, they're not susceptible to flooding anymore. It brings peace of mind."*

This perspective suggests that while skepticism exists, direct experiences with flooding and assurances from professionals may influence homeowners to reconsider their stance on elevation. A growing awareness of climate change and increasing flood risks also shaped residents' perceptions. One homeowner reflected on the uncertainty of future weather patterns, stating: *"With the climate changing and all these weather patterns shifting, I'm not sure what will happen in the next few years. It makes you think if elevation might be a necessity.*" Similarly, past flood experiences prompted some homeowners to actively consider elevation: "*Every time it rains hard, I just brace myself because this area is known for flooding. That's why I considered elevation now.*"

3.3.3. Community Influence and Esthetic Preferences

Beyond financial and structural considerations, community esthetics played a significant role in homeowners' decisions. Some residents were reluctant to elevate their homes due to concerns about neighborhood character. One participant expressed hesitation, stating, "*I don't want my house to stick out like a sore thumb in this neighborhood.*"

This sentiment reflects a broader desire to maintain visual continuity in residential areas. Another homeowner emphasized the importance of preserving traditional architectural styles: "*Most of us moved here because of the character of the houses.*" Additionally, concerns arose about the cumulative effect of multiple elevated homes on neighborhood esthetics: "*I worry that too many elevated homes might change the look.*"

Recognizing these concerns, an elevation expert emphasized the importance of designing elevated homes to align with existing neighborhood styles: "*We have to make sure that everything we do, especially if it's visible, fits in with the character of the neighborhood.*"

3.3.4. Practical and Physical Considerations Regarding Elevated Homes

Practical concerns, especially those related to accessibility and mobility, were major factors in homeowners' elevation decisions. One resident highlighted the daily challenges of living in an elevated home: "*It's not just about flood protection; it's about how we live day-to-day. My concern is accessing my home easily every day without strain.*"

Emergency preparedness also influenced concerns, with one participant questioning how quickly they could evacuate if necessary: "*What worries me is how quickly we can get out in an emergency with these elevated setups. It's something we have to think about seriously.*" Mobility challenges were particularly relevant for older residents, as one participant noted: "*Getting up and down the new stairs can be daunting, especially for older folks like us who've lived here all our lives.*"

To address accessibility concerns, ramps were frequently cited as a crucial feature. Many residents found them beneficial, as one homeowner shared, "*These ramps have made a difference for my mobility. Without them, I couldn't imagine accessing my home comfortably.*"

Even for those without mobility issues, ramps were seen as a necessity for future homebuyers: "*The ramps are not just a convenience; they are a necessity for many of our neighbors who would otherwise struggle with stairs.*" However, some homeowners found ramps impractical due to their placement, particularly when they extended far from driveways or entrances. Additionally, wooden ramps were viewed as less desirable than concrete alternatives, which provide more stability and longevity.

4. Discussion

4.1. Synthesizing Findings: Disparities in Flood Adaptation and Policy Implications

Our findings align with existing research on flood adaptation, social vulnerability, and housing policy, reaffirming that disaster mitigation efforts must be tailored to the needs of socioeconomically disadvantaged communities. Prior studies have demonstrated that homeowners' decisions to elevate their homes are influenced by financial constraints, risk perception, and access to government assistance [1,6]. Our study highlights the multifaceted nature of home elevation decisions and the broader implications for flood adaptation and disaster recovery. While financial incentives and risk awareness are crucial motivators, social vulnerability, accessibility concerns, and neighborhood esthetics significantly influence homeowners' decisions. These findings underscore the fact that policy interventions must extend beyond financial support to address structural inequities in program implementation.

Disparities in home elevation completion rates reflect broader patterns of inequitable disaster recovery. The spatial analysis reveals that socially vulnerable communities, particularly those with higher proportions of minority and low-income residents, have lower completion rates for home elevation projects. This suggests that systemic barriers—such as bureaucratic complexity, lack of information, and historical disinvestment—continue to impede equitable access to resilience-building measures. While wealthier communities tend to retain existing housing structures with minor modifications, low-income neighborhoods experience more extensive demolitions and rebuilds, further exacerbating displacement and long-term housing insecurity.

Our findings align with previous research on the challenges faced by low-income communities in accessing disaster mitigation funding. Despite the availability of home elevation assistance, program uptake remains uneven, particularly among homeowners with disabilities and those uncertain about their flood risk. A significant proportion of surveyed homeowners were unaware of their flood zone status, pointing to deficiencies in public communication and outreach efforts. Addressing these gaps requires a more proactive, community-centered approach to disaster preparedness and risk mitigation.

4.2. *Practical Considerations and Homeowner Decision-Making*

Beyond financial and policy-related factors, homeowners' perceptions of home elevation effectiveness vary widely. While many see elevation as a long-term flood protection strategy, others remain skeptical due to concerns about structural integrity, maintenance costs, and neighborhood esthetics. These perceptions influence program participation rates and highlight the need for clearer public messaging on the benefits and trade-offs of home elevation.

The findings also reveal important insights into mobility and accessibility concerns. While ramps and accessibility modifications are available, they are not always practical or desirable for homeowners, particularly in communities where esthetic cohesion is highly valued. Older homeowners, who make up a significant portion of those opting for elevation, often cite past flooding experiences as a primary motivator. Conversely, individuals with disabilities report hesitations due to the logistical challenges of accessing elevated homes. This suggests that current elevation programs may need to incorporate more flexible, adaptive solutions tailored to diverse household needs.

4.3. *Policy Recommendations and Future Research Directions*

Many homeowners remain uncertain about their flood risk status and available mitigation options. More effective communication strategies, including targeted community engagement and multilingual resources, are needed to ensure that residents can make informed decisions.

While elevation can provide crucial flood protection, accessibility remains a major barrier for individuals with disabilities. Future programs should integrate more inclusive design features and offer financial support for adaptive modifications beyond traditional ramps.

The lower completion rates in socially vulnerable neighborhoods suggest systemic inequities in program implementation. Streamlining application processes, reducing bureaucratic hurdles, and providing targeted support for historically marginalized communities can enhance accessibility and participation.

While elevation is a widely recommended flood mitigation strategy, its long-term effectiveness in different socio-economic and environmental contexts remains uncertain. Future research should examine post-elevation outcomes, including maintenance challenges, changes in property values, and shifts in community demographics.

Given the clear disparities in program outcomes, home elevation initiatives should be embedded within broader environmental justice frameworks. Policies should prioritize vulnerable communities, ensuring equitable access to flood protection resources and minimizing displacement risks.

4.4. Limitations and Opportunities for Further Study

While this study provides valuable insights, it is limited in scope and sample size. A larger dataset encompassing a wider geographic area would enhance the generalizability of findings. Additionally, future research should incorporate longitudinal analyses to track changes in home elevation adoption and effectiveness over time. Comparative studies across different cities and regions could further illuminate best practices for equitable disaster recovery and climate adaptation.

5. Conclusions

This study has underscored the complex interplay between social vulnerability, housing policy, and flood adaptation in post-disaster recovery. While home elevation programs offer critical flood mitigation opportunities, their effectiveness is constrained by systemic inequities in program implementation, accessibility, and homeowner perceptions. Older homeowners are more likely to elevate their homes, whereas individuals with disabilities face additional barriers, pointing to the need for more adaptive, inclusive solutions.

Beyond its immediate policy implications, this research contributes to broader discussions on climate resilience, disaster recovery, and equitable urban planning. Addressing disparities in home elevation outcomes requires a holistic approach that integrates financial assistance, targeted outreach, and accessibility considerations. By centering the needs of historically marginalized communities, policymakers and practitioners can develop more effective and equitable flood mitigation strategies that enhance long-term resilience for all homeowners.

Author Contributions: Conceptualization, I.G., Z.T. and J.O.; Methodology, I.G., Z.T. and J.O.; Software, I.G. and Z.T.; Validation, I.G., Z.T., J.O. and L.M.-R.; Formal analysis, I.G. and Z.T.; Investigation, I.G., Z.T., J.O. and W.W.; Resources, I.G., Z.T. and J.O.; Data curation, I.G., Z.T., J.O. and W.W.; Writing—original draft, I.G., Z.T., J.O. and L.M.-R.; Writing—review & editing, I.G., Z.T., J.O. and L.M.-R.; Visualization, I.G. and Z.T.; Supervision, I.G.; Project administration, I.G.; Funding acquisition, I.G., Z.T. and J.O. All authors have read and agreed to the published version of the manuscript.

Funding: This research received external funding from the Natural Hazard Center.

Institutional Review Board Statement: Institutional Review Board (IRB) (ID: STUDY2004-0011).

Data Availability Statement: Data are available upon request.

Acknowledgments: We thank our participants.

Conflicts of Interest: The authors declare no conflicts of interest.

Abbreviations

The following abbreviations are used in this manuscript:

FEMA	Federal Emergency Management Agency
GLO	Texas General Land Office
HOA	Homeowner Assistance Program
HoAP	Harvey Homeowner Assistance Program

References

1. Peacock, W.G.; Zandt, S.V.; Zhang, Y.; Highfield, W.E. Inequities in Long-Term Housing Recovery After Disasters. *J. Am. Plan. Assoc.* **2014**, *80*, 356–371. [CrossRef]
2. Lu, Y. Hurricane Flooding and Environmental Inequality: Do Disadvantaged Neighborhoods Have Lower Elevations? *Socius* **2017**, *3*, 2378023117740700. [CrossRef]
3. Zahran, S.; Brody, S.D.; Peacock, W.G.; Vedlitz, A.; Grover, H. Social Vulnerability and the Natural and Built Environment: A Model of Flood Casualties in Texas. *Disasters* **2008**, *32*, 537–560. [CrossRef] [PubMed]
4. Grothmann, T.; Reusswig, F. People at Risk of Flooding: Why Some Residents Take Precautionary Action While Others Do Not. *Nat. Hazards* **2006**, *38*, 101–120. [CrossRef]
5. Poussin, J.K.; Botzen, W.J.W.; Aerts, J.C.J.H. Factors of Influence on Flood Damage Mitigation Behaviour by Households. *Environ. Sci. Policy* **2014**, *40*, 69–77. [CrossRef]
6. Botzen, W.J.W.; Aerts, J.C.J.H.; van den Bergh, J.C.J.M. Individual Preferences for Reducing Flood Risk to near Zero through Elevation. *Mitig. Adapt. Strat. Glob. Chang.* **2013**, *18*, 229–244. [CrossRef]
7. Danh, V.T.; Mushtaq, S. Living with Floods: An Evaluation of the Resettlement Program of the Mekong Delta of Vietnam. In *Environmental Change and Agricultural Sustainability in the Mekong Delta*; Stewart, M.A., Coclanis, P.A., Eds.; Advances in Global Change Research; Springer: Dordrecht, The Netherlands, 2011; pp. 181–204, ISBN 978-94-007-0934-8.
8. Al Assi, A.; Mostafiz, R.B.; Friedland, C.J.; Rohli, R.V.; Rahim, M.A. Homeowner Flood Risk and Risk Reduction from Home Elevation between the Limits of the 100- and 500-Year Floodplains. *Front. Earth Sci.* **2023**, *11*, 1051546.
9. Mobley, W.; Atoba, K.O.; Highfield, W.E. Uncertainty in Flood Mitigation Practices: Assessing the Economic Benefits of Property Acquisition and Elevation in Flood-Prone Communities. *Sustainability* **2020**, *12*, 2098. [CrossRef]
10. Cohen, S.; Brakenridge, G.R.; Kettner, A.; Bates, B.; Nelson, J.M.; McDonald, R.R.; Huang, Y.-F.; Munasinghe, D.; Zhang, J. Estimating Floodwater Depths from Flood Inundation Maps and Topography. *J. Am. Water Resour. Assoc.* **2018**, *54*, 847–858. [CrossRef]
11. Zhang, W.; Villarini, G.; Vecchi, G.A.; Smith, J.A. Urbanization Exacerbated the Rainfall and Flooding Caused by Hurricane Harvey in Houston. *Nature* **2018**, *563*, 384–388. [CrossRef] [PubMed]
12. Texas Division of Emergency Management Hurricane Harvey DR-4332. Available online: https://www.tdem.texas.gov/disasters/hurricane-harvey (accessed on 8 March 2025).
13. National Hurricane Center Costliest U.S. Tropical Cyclones Tables Updated. 2018. Available online: https://data.globalchange.gov/report/costliest-us-tropical-cyclones-tables-updated-2018 (accessed on 1 October 2024).
14. Malecha, M.L.; Woodruff, S.C.; Berke, P.R. Planning to Exacerbate Flooding: Evaluating a Houston, Texas, Network of Plans in Place during Hurricane Harvey Using a Plan Integration for Resilience Scorecard. *Nat. Hazards Rev.* **2021**, *22*, 04021030. [CrossRef]
15. Douglas, B.E. One in Five Texans Lives in a Floodplain, State's First-Ever Analysis Shows. Available online: https://www.texastribune.org/2023/07/28/texas-floodplain-water-development-board-flood-plan/ (accessed on 8 March 2025).
16. Flanagan, K. A Genealogy of Public Housing Production: Practice, Knowledge and the Broadacre Housing Estate. *Hous. Theory Soc.* **2015**, *32*, 407–428. [CrossRef]
17. Ramesh, B.; Jagger, M.A.; Zaitchik, B.; Kolivras, K.N.; Swarup, S.; Deanes, L.; Hallisey, E.; Sharpe, J.D.; Gohlke, J.M. Flooding and Emergency Department Visits: Effect Modification by the CDC/ATSDR Social Vulnerability Index. *Int. J. Disaster Risk Reduct.* **2022**, *76*, 102986. [CrossRef] [PubMed]

land

MDPI

Article

Analysis of the Spatial Pattern of Urban Expansion in African Countries Under Different Shared Socioeconomic Pathway (SSP) Scenarios

Binglin Liu [1,2,†], Shuang Xie [1,2,†], Minru Chen [1,2,*], Nini Yao [3,*] and Weijiang Liu [4,*]

1 School of Geography and Planning, Nanning Normal University, Nanning 530001, China; dg1827018@smail.nju.edu.cn (B.L.); xieshuang@nnsfdx.wecom.work (S.X.)
2 Key Laboratory of Environment Change and Resources Use in Beibu Gulf, Ministry of Education, Nanning Normal University, Nanning 530001, China
3 Department of Architecture and Built Environment, University of Nottingham, Ningbo 315154, China
4 College of Engineering, City University of Hong Kong, Hong Kong 999077, China
* Correspondence: 15627568896@163.com (M.C.); saxny1@nottingham.edu.cn (N.Y.); lliuweiji2-c@my.cityu.edu.hk (W.L.)
† These authors contributed equally to this work.

Abstract: Exploring the urban spatial pattern and expansion characteristics of African countries under shared socioeconomic pathways (SSPs) is crucial to optimizing urban development in Africa and ensuring ecological sustainability. We use land and socioeconomic panel data and the least squares dummy variable regression method to predict the urban land increment in African countries from 2030 to 2060, we use the FLUS model to simulate the urban spatial layout in 2060, and we analyze from the perspective of the relationship between population density and urban expansion. The results show that the urban space of African countries will show a significant expansion trend from 2020 to 2060, with stronger growth under the SSP1 and SSP5 scenarios and relatively weaker growth under the SSP3 scenario; the urban land expansion patterns of different countries under different SSP scenarios are significantly different, and countries with rapid urbanization and economic growth are mostly urban patch agglomeration and extended expansion, while urban patches are relatively evenly distributed; a large number of cities in Africa show specific expansion patterns, with large cities mostly showing loose expansion and small- and medium-sized cities mostly showing compact expansion; and cities in different regions such as North Africa and sub-Saharan Africa have their own expansion characteristics in terms of population density and urban form. Our research provides important data support and inspiration for promoting the rational development of African cities and enhancing regional ecological resilience.

Keywords: African cities; urban expansion; shared socioeconomic pathways (SSPs) scenarios; land use; population density; spatial analysis

1. Introduction

Against the backdrop of the continued advancement of global urbanization, urban expansion has become one of the hot topics that has attracted much attention in the world today [1]. As the core carrier of human social and economic activities, the expansion of cities not only profoundly reflects the dynamic changes in many aspects such as population growth, economic development, and social change [2], but also has a close connection with the sustainability of the ecological environment [3]. In-depth exploration of the spatial pattern of urban expansion is undoubtedly of great theoretical and practical significance for optimizing urban planning layout, rationally allocating land resources, effectively protecting the ecological environment, and achieving sustainable development of cities and regions [4].

Many internationally renowned scholars have carried out a lot of in-depth and fruitful work in the field of urban expansion research and achieved fruitful results. Globally, scholars have used complex system dynamics models to conduct a comprehensive and in-depth analysis of urban expansion in North America, Europe, and other regions. For example, in a study of North American cities, scholars constructed a system dynamics model to analyze in detail the dynamic relationship between industrial development, transportation infrastructure construction and urban expansion, and found that the large-scale construction of factories and the extension of transportation lines in the process of industrialization attracted a large number of people to migrate, thereby promoting the continuous outward expansion of cities in space [5]. In Europe, another study used system dynamics models to explore how the construction of transportation hubs changes the flow of urban population and the distribution of economic activities, thereby affecting the direction and scale of urban expansion [6]. The research reveals the close internal connection between urban expansion and the process of industrialization, transportation infrastructure construction, and population migration. In Asia, research on rapidly urbanizing countries such as China and India has focused on several key aspects. In China, many scholars have conducted in-depth studies on the patterns and characteristics of land use conversion. For example, Xu et al. integrated system dynamics and cellular automata models to predict land use and land cover changes, demonstrating the in-depth exploration of land use conversion in China [7]. In addition, Chaudhuri and Clarke reviewed the SLEUTH land use change model, which has been widely used in urban expansion simulations in China [8–10]. Jiang and O'Neill also provided a framework for global urbanization projections, which includes an in-depth analysis of urbanization trends in China [11]. Similarly, in India, a large amount of research has focused on land use conversion [12]. For example, Kantakumar et al. explored the factors driving urban growth in Pune through logistic regression and relative importance analysis, providing important insights into land use conversion in the urbanization process of India [13]. In addition, Gibson et al.'s research not only focused on the expansion and density changes in urban land in India but also on the impact of urban expansion on food security, further enriching the research content of land use conversion in India [14].

Some research used multivariate regression analysis methods based on long-term land use data and found that with the rapid development of the economy, urban construction land has continued to expand, the area of cultivated land has gradually decreased, and land use has shown a rapid transformation from agricultural land to construction land. This transformation shows obvious differences in different regions and is affected by factors such as economic development level, industrial structure adjustment, and policy systems [15]. At the same time, another study used geographically weighted regression methods to analyze the interactive relationship between the evolution of urban construction land and economic growth. The results show that in the eastern coastal areas with faster economic growth, urban construction land has expanded more rapidly, and the upgrading of industrial structure has played an important role in promoting the expansion of urban space [16]. Similar studies have also emerged in India. Through surveys and the data analysis of multiple cities in India, some scholars have found that with the advancement of industrialization, a large number of rural people have poured into cities, resulting in a sharp increase in urban population density, a significant increase in the demand for urban construction land, and the continuous outward expansion of urban space [17]. In addition, policy systems also play a key role in the process of urban expansion in India. For example, the adjustment of some urban planning policies directly affects the direction and intensity of urban land development [18]. In addition, there are studies focusing on the ecological and environmental issues in urban expansion in other Asian countries. For example, in Southeast Asian countries, as cities expand, forest resources are destroyed and wetland areas are reduced, which has led to a series of ecological and environmental problems, which in turn affect the sustainable development of cities [19]. In Japan, scholars have found through research that a balance needs to be sought between the efficient use of land resources and the protection of the ecological environment during urban expansion. By formulating strict land use planning and environmental protection policies, the negative impact of urban expansion has been alleviated to a certain extent [20].

In the development history of urban land use scenario simulation research, early research methods mainly included trend extrapolation and Markov chain models [21,22]. Trend extrapolation predicts future trends based on historical data [23]. In contrast, the Markov chain model predicts future land use patterns through state transition probabilities [24]. These models have provided the theoretical foundation for urban land use scenario simulation and have been widely applied in practice. They laid the foundation for research in this field for a certain period of time, but they also had obvious shortcomings. As a relatively intuitive prediction method, the core principle of trend extrapolation is to make simple extended predictions based on the linear or nonlinear trends presented by historical data. There are many cases of using this method in the research of some European and American cities. For example, in a study of New York City in the United States, scholars used the linear trend extrapolation method to predict the scale changes in construction land in the future based on the growth data of urban construction land area in the past few decades [25]. In research on London, UK, a study used nonlinear trend extrapolation to predict land use changes in urban fringe areas in an attempt to grasp the trend of urban expansion [6]. However, these studies all face a common problem, that is, the trend extrapolation method only relies on the surface trend of historical data and hardly considers the dynamic changes in socio-economic factors in depth. Social and economic development is a complex dynamic system affected by the interaction of many factors such as population growth, economic structure adjustment, and policy changes. As time goes by, the development of cities may be impacted by factors such as the rise of new industries, major infrastructure construction projects, and adjustments to population migration policies. This leads to the accuracy of predictions based solely on historical trends having great limitations, and it is difficult to accurately reflect the real changes in urban land use.

The Markov chain model also occupies an important position in the study of urban land use. It mainly focuses on analyzing the transition probability between land use types. In the land use research of some small- and medium-sized cities, the Markov chain model has been widely used [26]. In some small- and medium-sized cities in Germany, such as Heidelberg, similar studies have also used this model to explore the transformation laws of different land use types during urban expansion [27]. Although the Markov chain model can reveal the transformation trend of land use types to a certain extent, its defects are also obvious. When constructing the transition probability matrix, it often assumes that the transformation of land use types depends only on the current state, while ignoring the influence of other complex socio-economic driving factors. In fact, the change in land use is a process affected by the combined effects of multiple factors, such as the guidance of urban planning policies, the improvement of transportation infrastructure to enhance land value, and the pull of industrial development on land demand. These factors are difficult to be fully and deeply reflected in the Markov chain model, resulting in its relatively weak ability to describe complex socio-economic driving factors and the inability to accurately simulate the dynamic change process of urban land use. With the continuous deepening of research and the improvement in our understanding of the complexity of urban land use, researchers gradually realized that more advanced and comprehensive methods are needed to compensate for the shortcomings of early models in order to more accurately simulate urban land use scenarios. This has also prompted the rise and development of new methods such as integrated assessment models (IAMs) and cellular automata (CA) models [4,28]. These new methods have stronger capabilities in considering the diversity and dynamics of socio-economic factors, and can better simulate the complex interrelationships between urban land use and socio-economic, environmental, and other factors, bringing new breakthroughs and development opportunities to the study of urban land use scenario simulation [25].

Although significant progress has been made in urban expansion research, our existing research still has notable limitations. In terms of research regions, a disproportionate focus has been placed on economically developed or rapidly urbanizing regions such as Europe, America, and Asia, while the African continent has received relatively scant attention. This imbalance can be partially attributed to the limited data availability in many African countries, as well as the relative lack of research interest and resources directed towards African urbanization. Furthermore, the unique socio-economic and geographical contexts of African countries pose additional challenges for researchers seeking to understand and model urban expansion in this region. Recognizing these limitations, future research should strive to address the data gaps in Africa, foster international collaboration, and develop models that are better suited to the complex realities of African urbanization. As an important region for global urbanization development, Africa's unique socioeconomic background, complex geographical environment, and rapidly changing population structure present urban expansion characteristics and laws that are completely different from other regions. In terms of research content, existing studies are still in the initial stage of simulating the urban spatial pattern of long-term series in Africa, and lack the systematic and in-depth analysis of the future expansion characteristics of African cities. This not only limits our comprehensive understanding of the diversity of global urban expansion, but also means there is a lack of scientific and effective theoretical guidance and practical reference in urban planning and ecological and environmental protection in Africa. In view of the above situation, we closely focus on shared socioeconomic scenarios (SSPs), select land and socioeconomic panel data of African countries (regions) from 2000 to 2020, and carefully construct a regression model to predict the urban land area at the national scale from 2030 to 2060. Then, the FLUS model is used to simulate the urban spatial pattern with a spatial resolution of 1 km under five SSP scenarios in 2030, 2040, 2050, and 2060, and the spatial pattern characteristics of key potential cities in African countries are deeply analyzed from the unique perspective of the relationship between population density and urban expansion. The theoretical contributions of this study include (1) providing a comprehensive framework for analyzing long-term urban expansion in Africa, which has been largely overlooked in previous research; (2) developing an integrated modeling approach that combines socioeconomic scenarios with spatially explicit land use simulations, offering new insights into the drivers and patterns of urban growth; and (3) highlighting the role of population density in shaping urban expansion, which can inform future urban planning and policy making. These innovations collectively enhance our understanding of urban dynamics in Africa and provide a robust scientific basis for sustainable urban development and ecological resilience, aligning with the goals of the 'Africa 2063 Agenda'.

2. Materials and Methods

2.1. Study Area

As the second largest continent in the world, Africa has a rich and diverse geographical environment, culture, and economic form. This study covers all African countries (regions), which have a vast land use area and geographically span multiple climate zones such as tropical, subtropical, and temperate zones, including vast deserts, dense tropical rainforests, vast grasslands, and numerous rivers, lakes, and other natural landscapes. The level of economic development in Africa varies greatly among different countries. There are both emerging economies with rapid economic growth and underdeveloped countries facing many development challenges. In terms of urbanization, Africa as a whole has been in a stage of rapid urbanization in recent years, with a large number of people migrating from rural areas to cities and the size of cities continuing to expand. For example, in populous countries such as Nigeria and Egypt, their major cities such as Lagos and Cairo have seen a sharp increase in population in the past few decades, and the demand for urban land has continued to rise. We selected 266 major cities in Africa as the research objects of future urban expansion characteristics. In order to facilitate readers' identification, we used ArcGIS to highlight all urban land in the study area in 2020 (Figure 1).

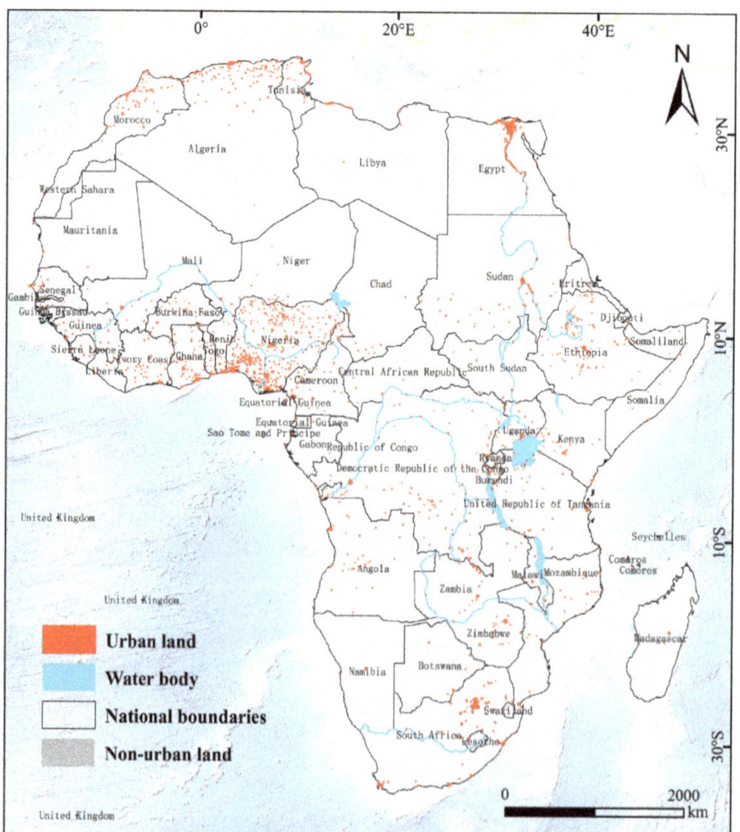

Figure 1. Location of the study area.

2.2. Data Sources

The data sources of this study are extensive. The statistics on the urban land area, population, and gross domestic product (GDP) of various countries over the years are derived from the Food and Agriculture Organization of the United Nations (FAO). Future population data and Purchasing Power Parity (PPP) data for different countries are obtained from the SSP database (https://tntcat.iiasa.ac.at/SspDb (accessed on 15 January 2025)). Land use data are sourced from the European Space Agency (ESA) (http://www.esa-landcover-cci.org/ (accessed on 15 January 2025)).

To conduct more precise simulation analyses using the FLUS model, the study selected 11 types of data from four dimensions—land use, social, economic, and environmental—as driving factors for the model. The detailed information relating to these data is shown in Table 1. The reasons for selecting each factor are as follows:

Table 1. Main data of scenario simulation.

Factor	Data	Years	Resolution	Data Source
Land Use (A)	Land use data (A1)	2020	300 m	https://cds.climate.copernicus.eu/cdsapp#!/dataset/satellite-land-cover?tab=form (accessed on 15 January 2025)
Social Factors (B)	Distance to Airport (A2)	2015	1 km	HUANGZ, WU
	Distance to road (B1)	2020	1 km	NASA, Socioeconomic Data, and Applications Center
	Distance to the administrative center (B2)	2020	1 km	United Nations, Department of Economic and Social Affairs
	Distance to river (B3)	2020	1 km	ESA-CCL (http://www.esa-landcover-cci.org/) (accessed on 16 January 2025)
Economic factors (C)	Population Density (C1)	2020	1 km	https://www.nature.com/articles/s41597-022-01675-x (accessed on 16 January 2025)
	Global Human Impact Index (C2)	2005	0.5′	NASA, Socioeconomic Data, and Applications Center
	Night light data (C3)	2020	1 km	National Oceanic and Atmospheric Administration (NOAA)
Environmental factors (D)	Elevation (D1)	2010	1 km	USGeological Survey (USGS), Globalmulti-resolution TerrainElevationData2010 (GmTED2010)
	Slope (D2)	2010	1 km	
	Slope aspect (D3)	2010	1 km	

Land Use (A): The 2015 land use data (A1) with a resolution of 1 km from the European Space Agency (ESA-CCL) (http://www.esa-landcover-cci.org/ (accessed on 15 January 2025)) are used to provide the initial land cover conditions, which serve as the basis for simulating future land use changes. These data are crucial for understanding the starting point of land use transformation [29]. The distance to the airport (A2) (1 km in 2015) is selected because it reflects the accessibility of air transportation in a region, which has an impact on urban development and economic activities. Previous studies on urban development have shown that areas close to airports often have more development opportunities and higher land values [25].

Social Factors (B): The distance to roads (B1) (1 km in 2018), provided by the National Aeronautics and Space Administration (NASA) and the Socioeconomic Data and Applications Center, is an important factor. Good road connectivity can promote the movement of people, goods, and information and is crucial for urban expansion. As some studies have proven, areas with better road infrastructure often experience faster urban growth [30]. The distance to administrative centers (B2) (1 km in 2014), from the United Nations Department of Economic and Social Affairs, can reflect the impact of administrative management on urban development. Areas close to administrative centers may receive more policy support and resource allocation [31]. The distance to rivers (B3) (1 km in 2015), from the European Space Agency, is included because rivers can provide water resources and transportation functions, which have a significant impact on urban location and development [32].

Economic Factors (C): Population density (C1) (1 km in 2015) is a key indicator reflecting the concentration of population in a region. It is closely related to urban expansion because areas with higher population density typically have a greater demand for urban land. In many urban studies, population density is an important variable [25]. The Global Human Impact Index (C2) (0.5′ in 2005), provided by the Socioeconomic Data and Applications Center of the National Aeronautics and Space Administration (NASA), can measure the overall impact of human activities on the environment and is related to urban development patterns [33]. Although this is archival data, it still has a certain reference value. Despite significant changes that may have occurred since then, the data can provide historical context and relative reference for understanding long-term trends of human activities in the study area. In the case of scenario simulation requiring baseline data, historical data helps analyze the evolution of urban development and provides a reference for understanding the background of human activities in the study area, a view supported by the research of Gao et al. [34]. The nighttime light data (C3) (1 km in 2015), provided by the National Oceanic and Atmospheric Administration (NOAA), can reflect the intensity of human activities and the level of economic development. Areas with brighter nighttime lights usually indicate more prosperous economic activities and higher levels of urbanization, and have been widely used in urban studies [35].

Environmental Factors (D): Elevation (D1) (1 km in 1996), slope (D2) (1 km in 1996), and aspect (D3) (1 km in 1996) can affect the difficulty of urban construction and the suitability of land use. Steep slopes and high elevations may restrict urban expansion, while gentle slopes and suitable elevations are more conducive to development [36].

2.3. Shared Socioeconomic Scenarios (SSPs)

SSPs represent a new scenario framework introduced by the IPCC that integrates socio-economic scenarios and climate scenarios [28], covering five different scenarios: SSP1 focuses on sustainable development and the achievement of the Millennium Development Goals, while ensuring socio-economic progress, reducing the dependence on resources such as fossil fuels, promoting rapid development in low-income countries, and presenting an open, fair, and economically globalized world [37]; SSP2 depicts that the world will continue the development trajectory of recent decades, achieve certain results in the process of achieving development goals, and gradually reduce the dependence on fossil fuels, which is a medium development scenario [27]; SSP3 depicts the situation of local advancement or uneven development in the world, failing to achieve global development goals and relying heavily on resources such as fossil fuels, which is a scenario with a significant tendency towards deglobalization [38]; SSP4 shows a situation of extremely uneven development within and between countries, with a small number of wealthy groups generating the vast majority of emissions, which is a scenario facing severe adaptation challenges [39]; SSP5 focuses on addressing mitigation challenges, and is a conventional development scenario with large greenhouse gas emissions caused by rapid economic development [40].

2.4. Prediction Model of Urban Land Area in Africa

In this study, we selected the least squares dummy variable (LSDV) regression model to predict changes in urban land area in African countries. The LSDV regression model can quantify the impact of various socio-economic factors (such as per capita GDP, urbanization rate, etc.) on urban land area. By incorporating these factors as independent variables into the model, we can estimate their relative contributions to changes in urban land area, thereby providing a solid basis for understanding the drivers of urban expansion in Africa [41]. Moreover, the model can predict future urban expansion trends based on historical data. By introducing time-series data (such as data from 1993 to 2020), the model can capture the long-term trends and relationships between urban land area and socio-economic indicators, thus providing a scientific basis for future urban planning [42]. It is worth noting that our data cover the time series from 1993 to 2020, and the LSDV regression model performs well in handling such data. It can capture the dynamic relationship between urban land area and socio-economic indicators, thereby providing predictions for future urban expansion [43]. In addition, the LSDV regression model can simultaneously consider the impact of multiple independent variables (such as GDP, urbanization rate, etc.) on the dependent variable (urban land area). This multivariate analysis capability allows the model to more comprehensively reflect the complexity of urban expansion [41]. The LSDV regression model also provides statistical significance test results, which help assess the contribution of different factors to urban expansion. This statistical rigor ensures the reliability and credibility of the research results [42]. Moreover, the model has high flexibility and can adapt to different types of data structures and research objectives. For example, by introducing dummy variables, the unique characteristics of different countries or regions can be considered, thereby improving the accuracy and relevance of the model [43].

In summary, the LSDV regression model is an appropriate and effective tool for analyzing urban expansion in Africa. Based on the socio-economic data of African countries and urban land area data from 1993 to 2020, we constructed a regression model using the least squares dummy variable (LSDV) regression method in the Stata 18.0. The model sets the per capita urban land area (B) as the dependent variable, taking GDP per capita (G) and urbanization rate (P) as independent variables. In order to highlight the differences between African countries, a regional dummy variable (Z, generally with a value of 0 or 1) is introduced to reflect the different attributes of the variables. For an independent variable with n categorical attributes, one of the categories is selected as a reference to generate n-1 dummy variables. By inputting the GDP per capita (G) and urbanization under different shared socioeconomic scenarios (SSPs), The rate (P) data can be used to estimate the urban land area of African countries under different scenarios.

$$B = \beta_0 + \beta_1 \times G_{r,t} + \beta_2 \times P_{r,t} + \sum_{i=1}^{n} a_i \times Z_{r,t} + \varepsilon_{r,t} \quad (1)$$

In the above formula, ε and β_0 represent the error term and intercept term, respectively, r refers to the countries in Africa, and t represents the year.

2.5. FLUS Model

The FLUS model uses an artificial neural network model to extract cells from the land use data of the first phase and related driving factors such as social, economic and environmental factors (such as topography, GDP, population density, and night lights), and converts them into the development probability of various land use types. At the same time, the inertia coefficient and competition mechanism are used to present the interaction and competition between urban and non-urban land in the dynamic simulation process. In addition, the model combines factors such as suitability probability, restrictive development conditions and future total land demand to simulate the spatial distribution of land use under specific scenarios in the future, effectively avoiding the limitations of traditional cellular automata in terms of cell morphology and neighborhood rules.

2.6. African Urban Expansion Indicators

We analyze the process and pattern characteristics of future urban expansion in Africa by selecting the urban expansion index (I) and population density. Among them, I represents the difference between the growth rate of urban land area (γ) and the growth rate of urban population (α), which is used to reflect a period of the degree of expansion of urban land over time; population density is expressed as the ratio of urban population to urban land area.

$$\alpha_{(t_1,t_2)} = [(\frac{A_{t_1}}{A_{t_2}})^{\frac{1}{t_2-t_1}} - 1] \times 100\% \quad (2)$$

$$\gamma_{(t_1,t_2)} = [(\frac{R_{t_1}}{R_{t_2}})^{\frac{1}{t_2-t_1}} - 1] \times 100\% \quad (3)$$

$$I_{(t_1,t_2)} = \alpha_{(t_1,t_2)} - \gamma_{(t_1,t_2)} \quad (4)$$

3. Results

3.1. Change of Urban Land Area in African Countries Under the SSPs Scenario

In order to explore the spatial-temporal differences of urban land growth in African countries, we conducted a detailed analysis of the change of urban land area in different SSPs scenarios during the period 2030–2060.

From the overall data (Figure 2), the urban land area in Africa showed a continuous growth trend during the time period of 2030–2060. In the SSP1 scenario, In 2020, the urban land area is 42,272 km^2, By 2030, the annual growth increased to 53,249 km^2, Was further increased to 64,525 km^2 in 2040, To 69,960 km^2 in 2050, To 2060, the annual growth is to 75,079 km^2; In the SSP2 scenario, The area of the corresponding years is 42,272 km^2, 52,997 km^2, 64,508 km^2, 70,345 km^2 and 76,132 km^2; In the SSP3 scenario, 42,272 km^2, 54,547 km^2, 68,797 km^2, 76,604 km^2 and 84,863 km^2 respectively; The SSP4 scenario is 42,272 km^2, 55,316 km^2, 70,141 km^2, 78,112 km^2 and 86,397 km^2; SSP5, the scenario is 42,272 km^2, 55,031 km^2, 68,343 km^{22}, 74,634 km^2 and 80,485 km^2.

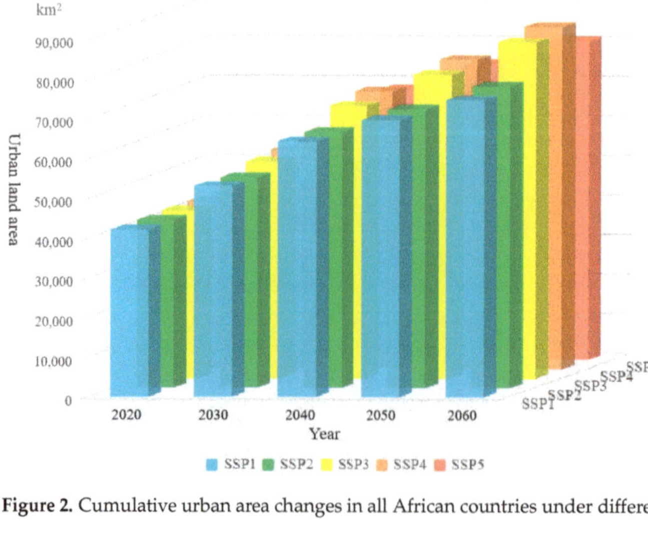

Figure 2. Cumulative urban area changes in all African countries under different SSP scenarios.

We further observe the specific data of each country (Figures 3 and 4). For example, Nigeria, which has a relatively prominent growth in urban land area, increases by 2853 km^2 from 2030 to 2040, 1491 km^2 from 2040 to 2050, and 1459 km^2 from 2050 to 2060 under the SSP1 scenario. Its growth area is relatively high among African countries. This significant urban expansion in Nigeria can be attributed to a variety of factors. First, Nigeria has a large population base. Second, the large and growing population creates a natural impetus for people to seek better living conditions and opportunities in urban areas [44]. Additionally, rapid economic development in recent years has played a key role. The growth of industries such as oil, manufacturing, and services has created a large number of job opportunities in cities. As an important oil-producing country in Africa, Nigeria's oil industry occupies a pivotal position in its economic and urbanization processes. The development of the oil industry has led to the agglomeration of related upstream and downstream industries in cities, attracting a large number of people to migrate to cities in search of employment [44]. For example, in areas of Nigeria rich in oil resources, the development of the oil industry has led to the establishment of a large number of related service and manufacturing enterprises in urban centers, attracting a large influx of labor. A study by Obi-Ani and Isiani on the city of Onitsha shows that the expansion of oil-related industries has significantly changed the city's population structure and economic landscape, with a large number of workers attracted by job opportunities converging from surrounding areas [45]. The oil industry has always been the main driving force for Nigeria's economic development and urbanization. In addition, studies have found that urban economic growth driven by the oil industry is an important factor in attracting internal migration, further promoting urban expansion and development [46]. Climate change has also had an impact on population migration in Nigeria [47]. With climate change leading to more extreme weather events and declining agricultural productivity in some rural areas, farmers are increasingly forced to abandon agricultural activities and move to cities in search of alternative livelihoods [48]. Studies have shown that there is a clear trend of rural–urban migration in areas of Nigeria affected by drought and flooding [49]. In addition, the influx of foreign investment should not be overlooked. The cheap labor force in Nigerian cities has attracted foreign investors to set up factories. For example, in the textile and manufacturing sectors, foreign companies have established production facilities in Lagos and other major cities, further promoting urban expansion [49].

Another example is Ethiopia. Under the SSP3 scenario, it increases by 470 km^2 from 2030 to 2040, 260 km^2 from 2040 to 2050, and 287 km^2 from 2050 to 2060. Its growth trend is also quite significant. This is due to the country's continued economic growth and accelerated urbanization process. In recent years, Ethiopia has experienced steady economic growth driven by sectors such as agriculture, manufacturing, and infrastructure development [50]. The government's investment in infrastructure projects, including roads, railways, and energy facilities, has not only improved connectivity but also stimulated the development of urban areas along these corridors [51]. For example, the construction of the Addis Ababa–Djibouti Railway has led to the growth of urban settlements in surrounding areas [52]. The expansion of industrial parks and manufacturing zones, such as the Hawassa Industrial Park, has attracted an increasingly large number of workers from rural areas [53].

The growth of urban land area in some countries under certain scenarios presents unique stage characteristics. Take South Africa as an example. In the SSP1 scenario, it increases by 789 km^2 from 2030 to 2040, but the growth rate is significantly smaller from 2050 to 2060, with an increase of only 316 km^2. Then, it increases by 251 km^2 in 2060 compared with 2050. This can be attributed to several factors related to South Africa's development context. Firstly, after experiencing rapid urbanization in the early stages, urban land expansion has gradually slowed down with the adjustment of economic structures [25].

As economic development progresses, there is a shift from a primary focus on extensive urban expansion to a more balanced approach that considers the quality and efficiency of urban growth [54]. This trend aligns with the general pattern of urban development in many regions, where the demand for additional urban land may decrease as the economy matures, while the emphasis on optimizing existing urban areas increases [44,55]. Secondly, changes in land use policies have also played a key role. South Africa has implemented policies to better manage urban growth and protect agricultural and ecological land [56]. For example, in some African countries, urban planning policies have been established to guide the rational use of land [57]. In South Africa, the government has adopted similar policies to control urban expansion. The government could implement zoning regulations or land conservation measures [58]. In some regions, specific areas may be designated for different uses, such as industrial, commercial, and residential purposes, thereby restricting the arbitrary expansion of urban land. These policy changes often have a significant impact on the speed of urban land expansion. By regulating the supply and use of land, the government can influence the pace and pattern of urbanization, ensuring that it is consistent with broader social, economic, and environmental objectives. This is an important aspect of sustainable urban development, as it helps to balance the need for urban growth with the protection of natural resources and the provision of basic services [59].

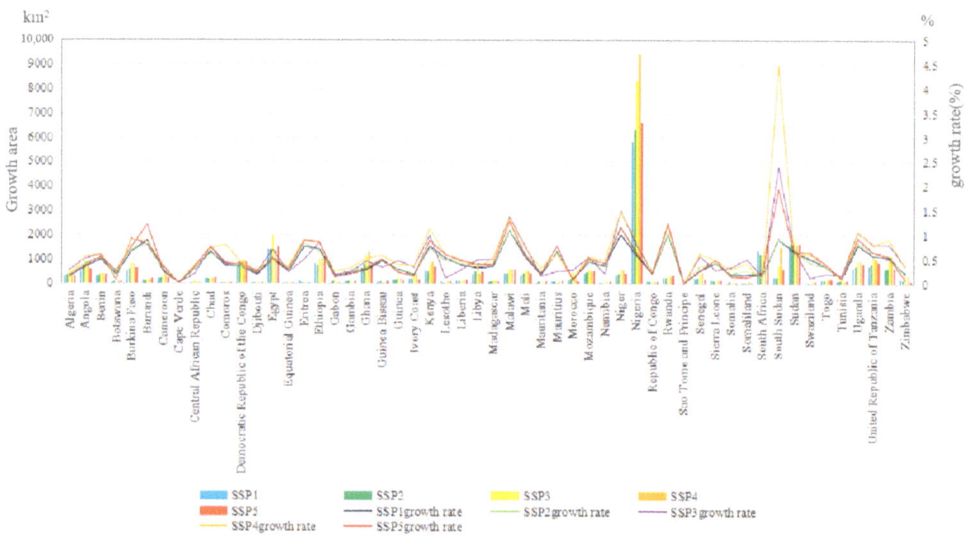

Figure 3. Urban land growth area and growth rate in African countries.

From the perspective of growth rate, Burkina Faso has a higher urban land growth rate under the SSP1 scenario, reaching 0.648521. This can mainly be attributed to the fact that Burkina Faso is in a stage of rapid urbanization [60]. The population of Burkina Faso's major cities is increasing, driven by natural growth and rural–urban migration, leading to strong demand for urban land infrastructure and industrial development [61]. Due to the lack of comprehensive urban planning policies and land management strategies, it is difficult to effectively control the expansion of urban land. In contrast, Morocco's growth rate under the SSP5 scenario is only 0.055344, which is relatively slow. This is related to the country's relatively mature urban planning and land use management system. Relevant studies have shown that the Moroccan government has been implementing measures to manage urban growth, such as promoting the development of satellite cities and improving land use efficiency in existing urban areas [62]. In addition, the Moroccan government has also focused on protecting historical and cultural areas, which has limited the extent of urban expansion. Compared with Burkina Faso, Morocco's economic structure is more diversified, which may also lead to a slowdown in urban land growth rates. Morocco's service and tourism industries play an important role in the economy, and these sectors do not require extensive land expansion like heavy industry or large-scale manufacturing [63].

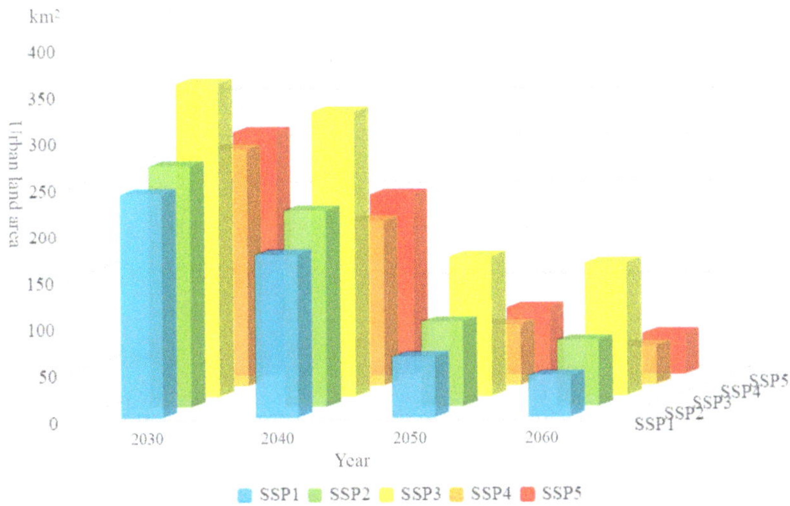

Figure 4. Urban land growth in African countries every 10 years.

In general, the growth of urban land area varies among African countries under different SSP scenarios, which is closely related to various factors such as population growth, economic development level, policy orientation, and resource endowment [64].

3.2. Spatial Distribution of Future Urban Land Use in African Countries Under the SSP Scenario

Under the SSP1 scenario, full consideration is given to achieving the Sustainable Development Goals (SDGs) and the Millennium Development Goals (MDGs) [65]. This scenario is committed to achieving sustainable development and ensuring social and economic progress, while reducing the dependence on resources such as fossil fuels. In this scenario, an emphasis is placed on a sustainable economic growth model, focusing on the efficient use of resources and environmental protection, and actively promoting the development and application of renewable energy to reduce the reliance on traditional fossil fuels. As shown in Figure 5, overall, urban land in Africa shows a relatively orderly expansion trend.

In some countries with a good economic foundation and active participation in global development, urban expansion is more obvious [66]. For example, in South Africa, the urban land area will increase from 5678 km^2 in 2020 to 8765 km^2 in 2060. This is because South Africa can fully utilize its own resource and technological advantages to attract more investment and industrial development in this open and equal development environment, thereby promoting urban expansion. Its industrial upgrading and economic diversification prompt cities to need more land for commercial, industrial, and residential purposes, which further accelerates the process of urban expansion [67]. In addition, South Africa's urban expansion is also influenced by global development trends and domestic economic policies. Against the backdrop of globalization, South Africa actively participates in international economic cooperation, promotes the upgrading of domestic industries and economic diversification through attracting foreign investment and technology transfer [68]. At the same time, the South African government has also formulated a series of economic policies, such as infrastructure construction and tax incentives, to provide strong support for urban expansion [69]. These measures not only promote South Africa's economic growth but also drive the process of urban expansion. With the increase in the urban population and the continuous development of the economy, South Africa's cities need more land to meet the demands of commerce, industry, and housing, which further promotes urban expansion [70].

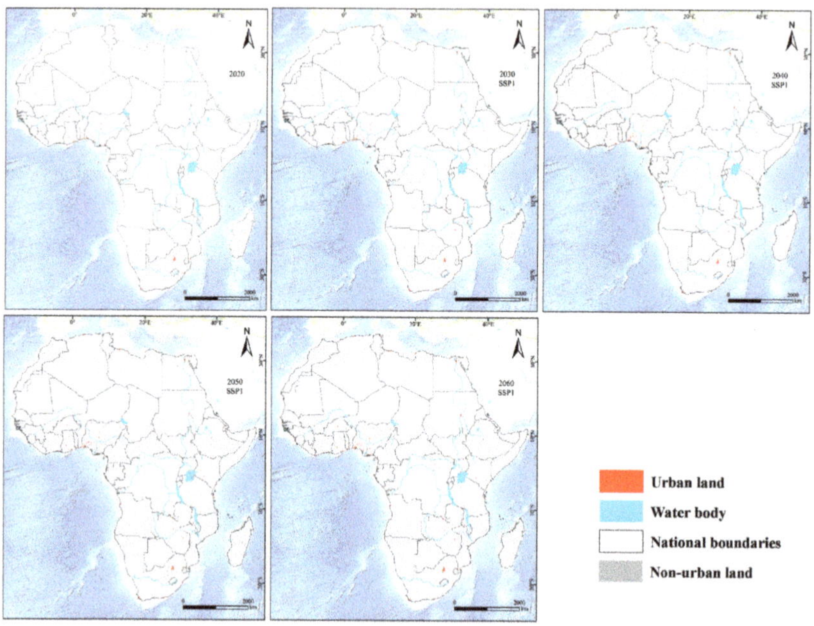

Figure 5. Simulation results of urban expansion in African countries under SSP1 scenario.

For low-income countries, such as Burkina Faso, the urban land area will increase significantly from 987 km^2 in 2020 to 3456 km^2 in 2060. This is because under the SSP1 scenario, it can obtain more international aid and cooperation opportunities, domestic infrastructure construction will be accelerated, the industrialization process will be promoted, and a large number of rural population will move to cities, which will promote the rapid expansion of urban land.

SSP2 describes a scenario in which the world maintains the development pattern of the past few decades, making certain progress in achieving development goals and gradually reducing the dependence on fossil fuels [27]. This is a medium development scenario. In SSP2, economic growth is relatively stable, but it also faces some challenges, such as resource shortages and environmental pollution. As can be seen from Figure 6, the expansion rate of urban land in African countries is relatively stable.

For traditional regional powers like Egypt, the urban land area increases from 7890 km^2 in 2020 to 11,234 km^2 in 2060. Relying on its existing development foundation and relatively stable economic growth model, Egypt gradually promotes urbanization while maintaining its original development trajectory. In some small- and medium-sized countries, such as Benin, urban land area increases from 567 km^2 in 2020 to 1012 km^2 in 2060.

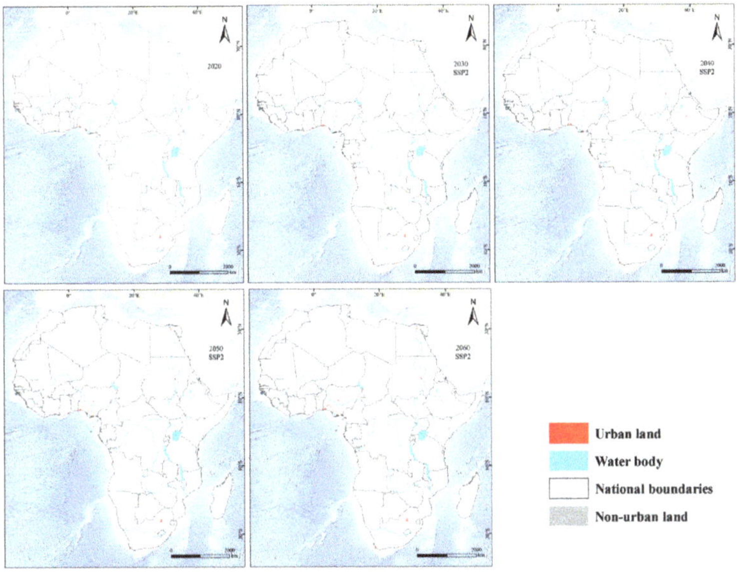

Figure 6. Simulation results of urban expansion in African countries under SSP2 scenario.

Under the SSP2 scenario, there are significant differences in urban land expansion across different regions in Africa. North African countries, such as Egypt, with a stronger economic base and an industrial structure dominated by tourism and manufacturing, experience faster and larger-scale urban land expansion [71]. In contrast, sub-Saharan African countries like Benin, with weaker economic foundations and agriculture-based economies, have slower and smaller-scale urban land expansion [72]. Additionally, the construction and layout of transportation infrastructure have also had a significant impact on urban land expansion. For example, the construction of the Addis Ababa–Djibouti Railway in Ethiopia has significantly propelled urban expansion along the railway [52]. Under the SSP2 scenario, urban land expansion in African countries is significantly influenced by the structure of population growth. Africa's population growth is dominated by young labor forces, who are more inclined to migrate to large cities in search of employment opportunities and better living conditions. This migration preference drives urban land expansion, especially in large cities. For example, cities like Cairo and Alexandria in Egypt have higher demands for urban land due to the influx of young labor forces [73]. In contrast, countries like Benin, with relatively slower population growth and a predominantly rural population, rely more on natural growth and infrastructure development to drive urban expansion [74]. Additionally, according to the World Population Prospects, the trend of population aging in Africa's urbanization process is relatively low, meaning that urban land expansion is more driven by the demands of the young population [75]. Meanwhile, under this scenario, the economic development models of African countries have also had a significant impact on urban land expansion. For example, Egypt's economic growth model, dominated by tourism, agriculture, and manufacturing, creates new demands for urban land. In particular, the development of the tourism industry requires more land for hotels, transportation, and entertainment facilities, thus driving urban land expansion [76]. For countries like Benin, with weaker economic foundations and single industrial structures, urban expansion relies more on the slow development of traditional industries and the gradual improvement of infrastructure [77]. Additionally, resource-based economies such as Nigeria, where the oil industry plays a significant role, also have a notable impact on urban land expansion. The layout and development of various segments of the oil industry in cities, especially the construction of oil processing and transportation facilities, further drive urban land expansion [45]. At the same time, environmental factors may also have an impact on urban land expansion under this scenario. For example, extreme climate events and sea-level rise caused by climate change have a more significant impact on coastal cities, while inland cities are more affected by droughts and floods. These environmental issues prompt population migration to cities, further increasing the demand for urban land [78]. Additionally, according to the IPCC report, African countries' relatively weaker capacity to address climate change further exacerbates the pressure on urban land expansion [79].

SSP3 depicts a scenario of partial or inconsistent global development, with a failure to achieve global development goals and a high dependence on fossil fuels and other resources [38]. In this scenario, various reasons lead to unbalanced and inconsistent development across countries. Some regions may fail to achieve development goals effectively due to political, economic, and social issues. Meanwhile, the high dependence on fossil fuels also exerts significant pressure on the environment. As can be seen from Figure 7, there is a clear differentiation in urban expansion among African countries. In some resource-rich countries that rely on fossil fuel industries, such as Libya, urban land area increases from 4567 km^2 in 2020 to 7891 km^2 in 2060. However, for some resource-poor countries that lack international cooperation opportunities, such as Somalia, urban land area will grow slowly between 2020 and 2060, increasing only from 345 km^2 to 456 km^2.

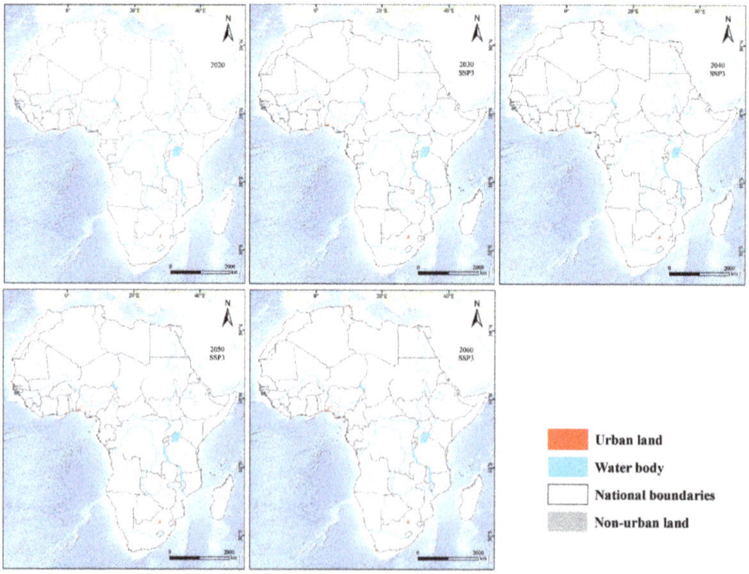

Figure 7. Simulation results of urban expansion in African countries under SSP3 scenario.

Under the SSP3 scenario, there are significant differences in urban land expansion across different regions in Africa. Resource-rich countries in North Africa, such as Libya, experience faster and larger-scale urban land expansion due to their stronger economic foundations and dependence on fossil fuel industries [80]. In contrast, resource-poor countries in sub-Saharan Africa, such as Somalia, have slower and smaller-scale urban land expansion due to weaker economic foundations and a lack of international cooperation opportunities [81]. These regional differences reflect the impact of economic development levels, resource endowments, and international cooperation opportunities on urban land expansion. In this scenario, differences in population growth structures across African countries also significantly affect urban land expansion. For example, resource-rich countries like Libya, assuming internal peace and the avoidance of civil wars, are likely to experience faster economic development due to abundant oil and natural gas resources. This attracts a large influx of young labor into cities, driving urban land expansion [82]. In contrast, resource-poor countries like Somalia, with weaker economic foundations, slower population growth, and a predominantly rural population, rely more on natural growth for urban expansion [83]. These differences in population structure lead to significant variations in the speed and scale of urban land expansion. The economic development models of African countries also play an important role in urban land expansion under this scenario. Resource-rich countries, which rely heavily on fossil fuel industries, have relatively single economic structure, urban expansion mainly around resource development and processing industries [84]. However, this resource-dependent economic model lacks diversification and is vulnerable to fluctuations in international energy market prices. In contrast, resource-poor countries like Somalia, lacking an industrial base, rely more on traditional agriculture and fisheries for urban expansion, resulting in slower expansion rates [85].

SSP4 describes a scenario of highly uneven development both within and between countries, where a small number of wealthy groups generate the majority of emissions, posing significant adaptation challenges [39]. In SSP4, the widening gap between the rich and the poor exacerbates social inequality. As shown in Figure 8, the imbalance of urban expansion in African countries is very prominent.

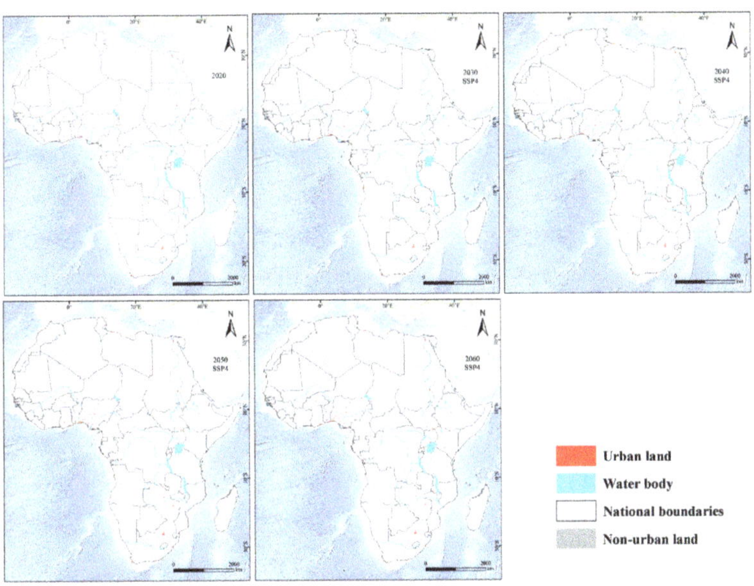

Figure 8. Simulation results of urban expansion in African countries under SSP4 scenario.

In some countries with developed economies and where resources are concentrated in the hands of a few groups, such as Nigeria, the land area in urban core areas increases significantly, from 9876 km^2 in 2020 to 15,678 km^2 in 2060. This is because industrial development and investment led by a small number of wealthy groups are mainly concentrated in urban centers, driving the rapid expansion of urban core areas [86]. These industries are often capital-intensive and high-emission, further exacerbating the imbalance in urban development [87]. For the vast rural areas and some marginal areas, urban land has hardly increased or even shrunk. For example, in some remote areas of Niger, the urban land area will decrease from 234 km^2 in 2020 to 189 km^2 in 2060.

Under the SSP4 scenario, urban land expansion in African countries is influenced by a variety of factors, including economics and population dynamics. The imbalance in economic development affects urban land expansion to some extent. Regions with more developed economies tend to attract more investment and population inflows, thereby driving urban land expansion. For example, cities with abundant natural resources or important ports may experience rapid expansion due to the development of trade and industry [88]. As the economy develops, the adjustment of industrial structures also impacts urban land expansion. The rise of industries such as manufacturing and services may require more land for the construction of factories, office buildings, and commercial facilities, further fueling urban land expansion [89]. However, economic factors do not always promote urban land expansion. In some economically underdeveloped regions, urban development is slow and urban land expansion is relatively limited due to a lack of capital and technology. Moreover, economic instability and financial crises can also lead to stagnation or a reversal in urban land expansion [90]. Population growth is one of the key drivers of urban land expansion in African countries. Under this scenario, Africa's population is projected to continue growing rapidly. As the population increases, cities require more housing, infrastructure, and public services, which inevitably leads to urban land expansion. In particular, the concentration of populations in large cities makes urban land expansion more pronounced [91]. In addition, population movement also impacts urban land expansion. The migration of rural populations to cities is one of the main features of urbanization in Africa. A large influx of rural residents into cities in search of better living conditions and employment opportunities further intensifies the pressure on urban land. Meanwhile, international migration may also have an impact on urban land expansion in some African countries [92].

SSP5 is not a business-as-usual scenario; it focuses on addressing mitigation challenges and is characterized by high energy and resource intensity. Rapid economic development in this scenario leads to significant greenhouse gas emissions [40]. While economic growth is swift under SSP5, it also brings substantial environmental pressures. As shown in Figure 9, urban land use in African countries is generally expanding rapidly.

Taking Kenya as an example, the urban land area will grow rapidly from 3456 km^2 in 2020 to 8901 km^2 in 2060. Under the SSP5 scenario, Kenya's economy develops rapidly, and large-scale industrialization and urbanization are accelerating. Emerging industries continue to emerge, attracting a large number of people from rural areas to cities. In order to meet the needs of population residence, industrial development, and infrastructure construction, urban land has expanded rapidly. At the same time, this rapid development is also accompanied by tremendous pressure on the environment, such as large-scale land development and energy consumption leading to increased greenhouse gas emissions.

Under the SSP5 scenario, urban land expansion in African countries is the result of multiple factors working together. First, SSP5 is an energy- and resource-intensive scenario where rapid economic development leads to significant greenhouse gas emissions. In this scenario, African countries experience swift economic growth, which dramatically increases the demand for energy and resources. On the one hand, this drives the development of industries and the urbanization process; on the other hand, it exerts substantial environmental pressure. For example, large-scale energy development and consumption can lead to the degradation of land resources and the deterioration of the ecological environment, which in turn affects urban land expansion [40]. Rapid economic growth is a key driver of urban land expansion in African countries. As the economy develops, new industries continue to emerge, attracting a large rural population to move to cities. To meet the needs of population housing, industrial development, and infrastructure construction, urban land expands rapidly. The case of Kenya demonstrates that under this scenario, economic growth creates more job opportunities and business prospects, drawing a large population into cities [93]. In addition, economic development also prompts cities to require more land for commercial, industrial, and residential purposes. For example, in South Africa, industrial upgrading and economic diversification drive cities to need more land for commercial, industrial, and residential purposes, further accelerating the process of urban expansion [67]. Under the SSP5 scenario, population growth is also an important factor in urban land expansion in African countries. With economic development and improved living standards, the population growth rate in African countries is relatively high. A large rural population moves to cities, leading to a rapid increase in urban population. To meet the housing needs of the population, cities need to continuously expand land. For example, in some African countries, the growth rate of urban population far exceeds the construction speed of urban infrastructure, resulting in problems such as urban traffic congestion and housing shortages. This further drives the expansion of urban land [91].

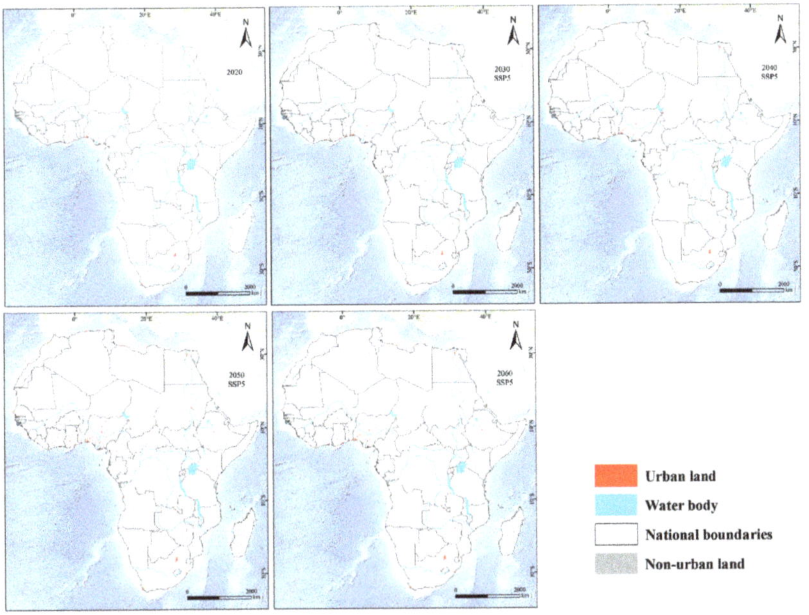

Figure 9. Simulation results of urban expansion in African countries under the SSP5 scenario.

Since the area of urban land is relatively small in the total land area, in order to show the simulation results more clearly, we used focal statistical analysis in ArcGIS to obtain the amount of new urban land in Africa within a radius of 10 km (Figure 10). The figure shows the increase in new urban land in cities within a radius of 10 km in Africa from 2020 to 2060 under the five scenarios of SSP1–SSP5, reflecting the spatial characteristics of urban expansion in Africa under different scenarios, especially the representative areas composed of Nigeria (the core country), Togo, Benin, Cameroon, etc., in West Africa, which show obvious differences under different scenarios.

Under the SSP1 scenario, urban land expansion in some regions of Africa is relatively significant. For example, in Nigeria, some areas have a large amount of new urban land, with colors ranging from dark red to purple, indicating more pronounced urban land expansion within a radius of 10 km. This is particularly evident around major cities like Lagos.

This phenomenon can be attributed to the emphasis on sustainable development and global cooperation in the SSP1 scenario. In this scenario, Nigeria may gain more resources and technological support, thereby promoting the urbanization process, especially in regions with a strong economic base and convenient transportation, where urban expansion is more noticeable [94]. In contrast, urban land expansion in Togo, Benin, and Cameroon is relatively moderate, with colors mostly yellow and orange, indicating a slower pace of urban expansion in these countries compared to Nigeria. This may be due to Nigeria's development spillover effect and their own benefits from sustainable development cooperation. Urban expansion under the SSP1 scenario may also be influenced by other factors. For example, population growth and economic development drive urban expansion. As people pursue better living conditions, cities become centers of attraction, leading to an increase in urban land [95]. In addition, government policies and investments may also impact urban expansion. Under the SSP1 scenario, which emphasizes sustainable development, governments may increase investment in infrastructure and urban planning to promote orderly urban development.

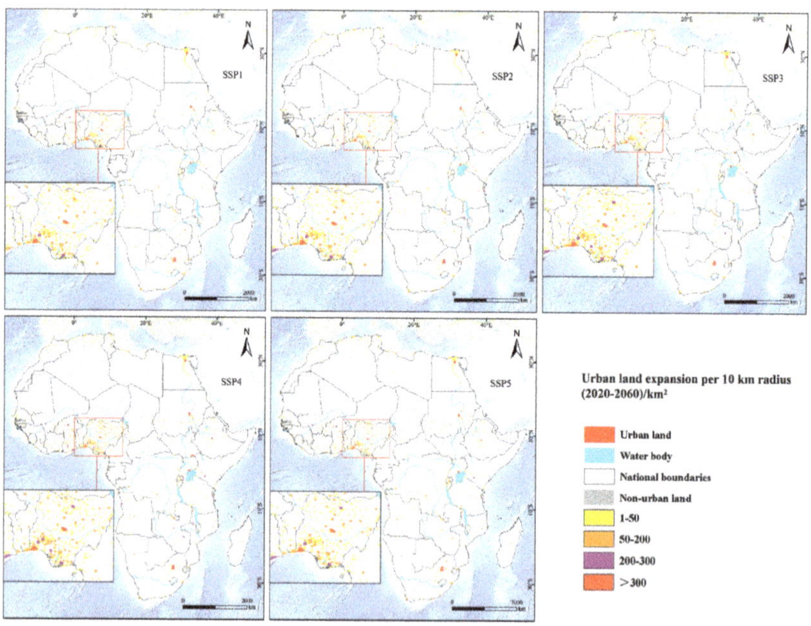

Figure 10. Statistical results of urban land use in African countries and their representative regions under different SSP scenarios.

Under the SSP2 scenario, urban land expansion is relatively stable and evenly distributed. Representative regions such as Nigeria, Togo, Benin, and Cameroon all show a moderate trend in urban land expansion. The colors around some cities in Nigeria are mostly yellow and orange, indicating a moderate amount of new urban land. Togo, Benin, and Cameroon also show similar colors, indicating that urban expansion in these countries is advancing at a relatively stable pace under this medium development scenario, without significant fluctuations in large-scale urbanization or urban expansion. The differences in urban land expansion among countries are relatively small [92]. The stable expansion under the SSP2 scenario may be the result of multiple factors working together. On the one hand, stable economic development provides a continuous driving force for urban construction. Under this scenario, countries may adopt a more balanced development strategy, focusing on the coordinated development of the economy, society, and environment. On the other hand, the role of urban planning and management cannot be ignored. Rational urban planning can guide orderly urban expansion and avoid problems caused by disorderly development [96].

Under the SSP3 scenario, urban land expansion shows significant differentiation. Nigeria, as a resource-rich country, has darker colors around its oil-producing regions, indicating a large amount of new urban land. Urban expansion is more pronounced around the Niger Delta and other oil-producing areas. This is because, under the scenario of deglobalization and high resource dependence, Nigeria continues to rely on the development of the oil industry, which drives urban expansion in surrounding areas [97]. In contrast, Togo and Benin, with relatively scarce resources and limited international cooperation, experience very slow urban land expansion, with lighter colors and little change in some areas. Cameroon, though having some resources, also faces slower urban land expansion under this scenario, forming a stark contrast with Nigeria. The differentiation under the SSP3 scenario reflects the importance of resource allocation and international cooperation. Resource-rich countries are more likely to rely on resource development to drive urban expansion, while resource-poor countries face greater development challenges. Moreover, the trend of deglobalization may lead to a reduction in international cooperation, further limiting development opportunities for these countries [98].

Under the SSP4 scenario, the imbalance in urban land expansion is further exacerbated. In Nigeria, regions around major cities like Lagos and Abuja have darker colors and concentrated ranges, indicating a large amount of new urban land within a 10 km radius. This is due to the highly unequal development within and between countries under the SSP4 scenario, where resources and development opportunities are concentrated in a few wealthy groups and large cities, leading to rapid urban land expansion around major cities in Nigeria [99]. However, most areas in Togo, Benin, and Cameroon have little urban land expansion, with only a few color changes around individual cities, indicating fewer development opportunities and urban expansion concentrated in a few areas in these countries, further widening the gap with Nigeria. The imbalanced development under the SSP4 scenario highlights the issue of socio-economic inequality. The rapid development of a few wealthy groups and large cities contrasts sharply with the stagnation in other regions, which may lead to increased social instability and resource allocation inequality [100]. At the same time, the lack of effective policy intervention and international cooperation may also perpetuate this imbalance.

Under the SSP5 scenario, urban land expansion in Africa is rapid overall. Representative regions such as Nigeria, Togo, Benin, and Cameroon all show varying degrees of rapid urban land expansion. The urban land expansion range in Nigeria further expands, with multiple areas in deep purple and red, indicating a large amount of new urban land within a 10 km radius. Togo, Benin, and Cameroon also show clear trends of urban expansion, with colors mostly dark yellow, orange, or even red, indicating that large-scale industrialization and urbanization under the conventional development scenario characterized by rapid economic development and significant greenhouse gas emissions lead to rapid urban land expansion in these countries over a broader area. However, Nigeria's urban expansion scale and speed still lead in the region [101]. The rapid expansion under the SSP5 scenario may be an inevitable result of economic development and urbanization. As industrialization and urbanization progress, the demand for land continues to increase, leading to rapid urban land expansion. However, this rapid expansion also brings a series of challenges, such as environmental issues, resource shortages, and social inequality.

In general, urban expansion in African countries presents quite different characteristics under different SSP scenarios, which is closely related to factors such as the socio-economic development model, resource utilization mode, and degree of globalization set in each scenario.

3.3. Expansion Characteristics of Cities in African Countries Under Different SSP Scenarios

In our study of urban expansion in African countries, we revealed their unique expansion characteristics based on the analysis under the SSP scenario. This study uses the expansion index I value as a key indicator to divide urban expansion characteristics into four categories: loose growth (I > 0.9), relatively loose growth ($0 \leq I \leq 0.9$), relatively compact growth ($-0.9 \leq I < 0$), and compact growth ($I < -0.9$) [102], and conducts an in-depth analysis of urban expansion in African countries (Figure 11).

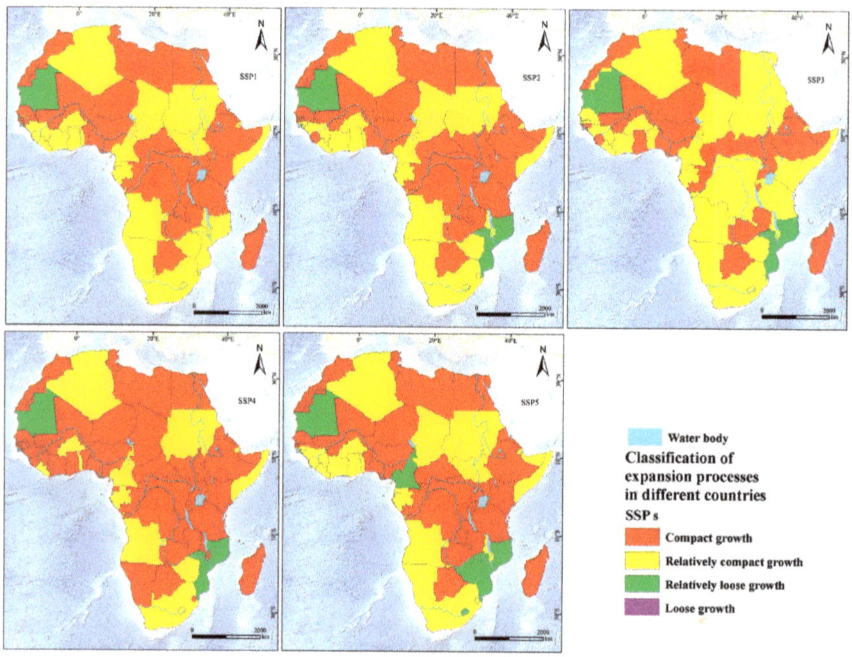

Figure 11. Expansion characteristics of cities in African countries under the SSP scenario.

From the overall data, under the SSP1 scenario, most African countries show a relatively consistent expansion trend. For example, Algeria's I value is −0.207803, which is in a relatively compact growth range. This may be because the country has promoted economic development under this scenario, but it is more cautious in land use planning, focusing on the optimization and integration of existing urban areas, so that urban land growth and population growth are relatively coordinated, and there is no large-scale urban land expansion. Angola's I value is −0.950516, which is in the compact growth category. The reason may be that the economic structure is still relatively dependent on specific industries under the SSP1 scenario, urban development is concentrated in the core area, and population growth is also mainly concentrated in these limited urban areas, resulting in relatively slow growth of urban land and a compact growth trend.

Compared with SSP1, SSP2 shows some new trends. In the SSP2 scenario, most countries still maintain a compact growth trend similar to SSP1, but some countries have also experienced changes to varying degrees. This may be because Ghana has adjusted its economic development path in this scenario, such as increasing investment in some emerging industries, attracting more people to flow into cities, thereby increasing the demand for urban land and accelerating the expansion rate. Botswana's I value is −0.305012, which is still a relatively compact growth. This reflects that while maintaining a certain development speed, the country has better controlled the disorderly expansion of urban land, which may be due to its reasonable urban planning policies and relatively stable population growth model. In general, although the overall pattern in the SSP2 scenario is similar to that in SSP1, the new changes in some countries reflect the development differences under different scenarios.

Under the SSP3 scenario, urban expansion in African countries presents a complex situation. For example, Cameroon's I value is −0.582311, which is in a relatively compact growth range. This may be because in the context of local development or inconsistent development, the development of some areas is restricted, urban expansion is concentrated in a few advantageous areas, and the overall urban land growth is relatively conservative. Looking at Libya, it presents a loose growth trend. This may be due to its economy being highly dependent on oil resources. Under the regional competition and deglobalization trend of SSP3, the fluctuation of the oil industry has a greater impact on urban development. Some areas have rapid urban land expansion due to resource development or trade activities, while other areas have slow development, thus forming the complexity of overall urban expansion.

In the SSP4 scenario, the compact growth characteristics of urban expansion are further highlighted. It can be clearly seen from the figure that the number of red areas has increased. For example, the I value of South Africa is 2.066804, which represents a loose growth trend. This shows that in the scenario of highly uneven development within and between countries, some parts of South Africa may form a loose growth pattern due to the concentrated investment and development of a small number of wealthy groups, rapid expansion of urban land, and relatively lagging population growth. The I value of Burundi is −3.779169, which represents a compact growth state. This may be due to its disadvantage in resource allocation and access to economic development opportunities, which severely restricts urban development and makes it difficult for urban land to expand effectively under the pressure of population growth.

Under the SSP5 scenario, the background of rapid economic development has led to significant urban expansion in some countries. Nigeria's I value is −0.978423, which is in the compact growth range. Although the economy has developed, the rapid population growth may not keep up with the construction and planning of urban infrastructure, resulting in relatively slow growth of urban land and compact growth. Mauritius's I value is 0.224376, showing loose growth. This may be due to its active promotion of urban modernization and industrial upgrading during its rapid economic development, which has attracted a large number of people to flow into cities, but the urban land planning is relatively loose, resulting in a faster growth rate of urban land.

In general, under different SSP scenarios, the urban expansion characteristics of African countries are different. This is not only closely related to the economic development level and population growth trend of each country, but also deeply influenced by factors such as the socio-economic development model, resource utilization mode, and degree of globalization set by the SSP scenario.

4. Discussion

4.1. Comparative Analysis of Urban Land Growth in African Countries Under Different SSP Scenarios

In the study of urban land growth changes in African countries, different SSP scenarios show different urban land growth patterns. The reasons behind this are multifaceted and complex. We believe that this is related to economics, resources, policy planning, and population factors being closely related.

The patterns of urban land expansion in Africa vary significantly under the five SSP scenarios, primarily due to differences in economic development, resource utilization, and policy assumptions across these scenarios. Under the SSP1 scenario, urban land expansion is characterized by sustainable development and global cooperation. For example, South Africa, with increased international investment, especially in infrastructure construction and industrial upgrading, has seen steady growth in urban land. This scenario emphasizes investment in renewable energy and environmental technologies, driving related industrial development and attracting population aggregation, thereby promoting urban land expansion [27]. This scenario assumes a global shift towards sustainable development practices, resulting in balanced urban land expansion and moderate growth rates. In contrast, the SSP2 scenario follows a middle-of-the-road development path, with more stable urban land expansion in countries like Botswana. This scenario reflects historical development patterns, where urban land growth relies on natural economic growth and population movement, lacking the strong external driving factors present in the SSP1 scenario [103]. As a result, urban land expansion is slower and more stable. The SSP3 scenario is characterized by regional competition and deglobalization, leading to stark contrasts in urban land expansion. Resource-rich countries like Libya, with their dependence on the fossil fuel industry, experience significant urban land expansion, driving the development of urban infrastructure and related service industries [38]. In contrast, resource-poor countries with limited international cooperation, such as Somalia, see very slow urban land expansion, or even stagnation. The SSP4 scenario highlights extreme socio-economic inequality, with rapid urban land expansion in some regions of South Africa controlled by a few wealthy groups and large cities. This scenario leads to highly imbalanced urban land expansion, exacerbating urban–rural disparities and regional development imbalances [39]. Finally, the SSP5 scenario focuses on fossil fuel-driven economic development, resulting in significant urban land expansion. For example, Nigeria's rapid industrialization and urbanization process has led to a large influx of rural populations into cities, increasing the demand for urban land [40]. This scenario assumes a continued reliance on fossil fuels, leading to high greenhouse gas emissions and significant environmental impacts. Overall, the differences in urban land expansion under these scenarios mainly stem from different economic growth models, resource utilization methods, and policy directions. For example, SSP1 and SSP2 emphasize sustainable and stable development, leading to moderate urban land expansion; SSP3 and SSP4 highlight resource dependence and inequality, resulting in a rapid expansion in resource-rich areas and stagnation in resource-poor areas. SSP5, driven by fossil fuels, leads to rapid urban land expansion but at the cost of environmental sustainability.

Considering economic, environmental, and policy factors comprehensively, we predict that urban land expansion in Africa will follow multiple paths, depending on the interaction of economic, demographic, policy, and environmental conditions. Under the sustainable development scenario (SSP1), urban land expansion will be more compact and efficient, driven by effective planning and green technology investment. In the middle-of-the-road development scenario (SSP2), urban land expansion will be moderate, reflecting stable economic growth and gradual urbanization. Under the resource-dependent and unequal scenarios (SSP3 and SSP4), urban land expansion will be highly imbalanced, with rapid expansion in resource-rich areas and stagnation in resource-poor areas. Finally, under the fossil fuel-driven scenario (SSP5), urban land expansion will be rapid but potentially unsustainable, with significant environmental impacts.

4.2. Comparison of This Study with the Studies in Other Regions

In the context of global urban development, by comparing the urban expansion in Africa with that in Europe, America, Asia, and other regions under different SSP scenarios, we can clearly find their similarities and differences and the underlying logic behind them. In the SSP1 scenario, Europe and America, such as the United States, with its strong scientific and technological research and development and abundant financial support, vigorously develop clean energy and intelligent transportation systems, and closely combine urban land growth with ecological protection, presenting a compact and efficient expansion model. According to the research of Van Vuuren et al. (2017), European and American cities focus on the optimal layout of urban functions in this scenario, and guide the distribution of population and industry through reasonable planning, effectively avoiding the disorderly expansion of cities [37]. Asian countries such as Japan and South Korea focus on industrial upgrading and urban renewal driven by scientific and technological innovation. Through the transformation and efficient use of old urban areas, and the construction of scientific and technological research and development centers and innovative industrial clusters around cities, they achieve the intensive growth of urban land. At the same time, they attach great importance to the protection and restoration of urban ecological space and maintain a good ecological balance.

In the SSP1 scenario, although Africa is influenced by the concept of sustainable development, due to factors such as weak economic foundation, relatively backward technological level and lagging infrastructure construction, urban expansion presents a relatively loose and disorderly state, making it difficult to achieve compact and efficient development like some cities in Europe, America, and Asia.

In the SSP2 scenario, European and American cities continue their historical development trajectory based on their stable political and economic environment, with a moderate urban expansion rate. They focus on the improvement of core urban functions and moderate suburban development. Through reasonable planning and policy guidance, they ensure balanced coverage of urban infrastructure and public services, and the urban spatial structure is relatively stable [104]. In this scenario, urban expansion in developing countries in Asia is driven by rapid industrialization and urbanization. A large number of people flow into cities from rural areas, which promotes the rapid expansion of urban built-up areas. However, they also face the problem that urban planning and management cannot keep up with the pace of development. There are some phenomena of waste of urban land and ecological environment damage [105]. In the SSP2 scenario, Africa is also limited by economic and technological factors, and its growth rate is relatively slow. However, some cities also experience local rapid growth due to resource development or the development of specific industries, and the overall expansion is less balanced.

Under the SSP3 scenario, Europe and the United States, due to their strong economic resilience and resource allocation capabilities, can alleviate the pressure of urban development to a certain extent through internal adjustments and technological innovation when facing regional competition and deglobalization trends. Urban land growth is relatively stable, but some traditional industrial cities may face problems in adjusting the structure of urban land due to difficulties in industrial transformation [106]. Cities in some resource-dependent countries or regions in Asia may face development difficulties due to resource market fluctuations and trade protectionism, and urban land growth will be suppressed. Some emerging economies may seek new urban development opportunities by strengthening regional cooperation and domestic market development, and urban land growth will present a complex situation. Under the SSP3 scenario in Africa, urban land growth in resource-rich countries is closely linked to resource industries, while urban development in resource-poor countries is more difficult, and urban land growth is slow or even stagnant, further exacerbating the imbalance of urban development in the region.

Under the SSP4 scenario, although there is socioeconomic inequality in Europe and the United States, the extreme polarization of urban land use has been alleviated to a certain extent by virtue of a sound welfare system and redistribution mechanism. Urban expansion is generally orderly. Although the land use of large cities has increased rapidly due to resource concentration, they focus on optimizing and updating the internal space and improving sustainability. For example, in some European cities, reasonable planning has promoted the balanced development of urban functions. In Africa, under this scenario, socioeconomic inequality has a more significant impact on the growth of urban land use. Taking some parts of South Africa as an example, the rapid growth of urban land use is concentrated in a few wealthy areas and large cities, while the vast rural and marginal areas lag behind in development, and the growth of land use is almost stagnant. The urban–rural gap has widened sharply, seriously hindering the overall coordinated development of cities, which is in sharp contrast with Europe and the United States [39].

In the SSP5 scenario, Europe and the United States rely on advanced technology and management experience to focus on energy efficiency improvements and environmental protection while maintaining large-scale fossil fuel-driven economic development. While urban land use has increased significantly, scientific planning and strict supervision have been used to reduce negative environmental impacts and improve infrastructure construction. For example, some American cities have built high-efficiency environmental protection facilities in the process of energy industry development [40]. In contrast, in African countries such as Nigeria, although urban land use has increased significantly in the SSP5 scenario, due to insufficient technology and management, many environmental problems and instances of disorderly expansion have occurred, land resources have been wasted, and the ecological damage has been serious. They are in a disadvantageous position in the global urban development pattern and urgently need to explore sustainable development paths.

From the perspective of global urban expansion, factors such as economic development level, technological innovation capability, policy planning orientation, and resource endowment are universal factors affecting urban expansion. However, due to their unique historical, cultural, geographical, and socioeconomic backgrounds, different regions also show obvious regional specificity in the process of urban expansion. African urban expansion has a unique position in the global urban development pattern. The many difficulties and challenges it faces not only affect its own sustainable development, but also provide special cases and valuable experience for global urban development research.

4.3. Model Limitations and the Direction of Improvement

In this study, the model we used has certain limitations in the study of urban expansion in Africa. As for the least squares dummy variable regression model, its assumptions may not be fully met in the complex environment of African cities. The particularity of African urban data, such as the difficulty in collecting data in some regions and inconsistent caliber of statistical data, may lead to the destruction of the independence and homoscedasticity of the model assumptions. In terms of parameter setting, due to the large differences between African cities and other regions in economic structure, social culture and other factors, even if we have set rigorous parameters, it may not be possible to accurately reflect the actual situation of African cities. For example, in the traditional parameter setting of the relationship between economic growth and urban expansion, a large number of informal economic activities in African cities are difficult to accurately quantify and include in the model, thus affecting the model's accurate judgment of the driving factors of urban expansion [103]. The FLUS model also faces similar problems. In the study of urban expansion in Africa, the parameter setting of its land use type conversion rules may not fully consider the special situation in Africa. There are a large number of special land use forms such as tribal land and traditional agricultural land in African cities, which may not be properly reflected in the parameters of the model. Moreover, in the rapid urbanization process of African cities, the impact of sudden political and social events on urban expansion is difficult to accurately reflect in the model. In view of these problems, in the future, in the direction of model improvement, considering the characteristics of African cities, we believe that we should strengthen field research and obtain more accurate local data to optimize parameter settings. For example, we should conduct in-depth research on the relationship between informal economic activities and urban land expansion in African cities and establish an indicator system specifically for African cities. At the same time, we can learn from the successful experience of other similar urban expansion research cases in developing regions, such as the methods used by some developing Asian countries to deal with data problems and model optimization in the process of rapid urbanization [38].

In terms of the uncertainty factors in the research process, we realize that although the SSP scenario assumptions provide a macro framework, the uncertainty is still high for the complex and changing situation of African cities. The model predictions themselves also have a certain range of errors, especially in the special socio-economic environment of African cities, the errors may be magnified. To reduce these uncertainties, future research can strengthen international cooperation, integrate multi-source data, and improve the accuracy and completeness of data. In the SSP scenario assumptions, more detailed adjustments and localization should be made in combination with the actual situation in Africa. At the same time, more actual cases should be used to repeatedly verify and optimize the model, such as selecting different types of cities in Africa for comparative analysis, so as to improve the scientificity and robustness of the research and provide more valuable conclusions for urban expansion research.

4.4. Relevant Policy Recommendations

Based on the results of this study, we propose the following recommendations for urban planning and land use policies in Africa under different SSP scenarios to promote the sustainable development of African cities and strive to minimize the negative impacts of urbanization.

The SSP1 scenario emphasizes sustainable development and global cooperation. African countries should make full use of international cooperation opportunities to promote the optimization of urban planning. For example, countries with a good economic foundation such as South Africa can learn from the experience of Europe and the United States and increase investment in green infrastructure, such as building public transportation systems such as urban light rail, guiding urban compact development, and reducing the dependence on private cars, thereby optimizing land use structure and alleviating the pressure of disorderly expansion on the ecological environment [37]. For countries with abundant resources but which still need to develop their economies, such as Angola, strict environmental standards should be formulated to ensure ecological restoration and land reclamation during resource development, reasonably define urban growth boundaries, and prevent excessive expansion from eroding ecological land. In addition, Angola can use international cooperation funds and technical support to promote the development and utilization of renewable energy, reduce the dependence on fossil fuels, and achieve a win–win situation for economic development and ecological protection.

The SSP2 scenario is based on historical development patterns. African countries should steadily promote urban planning based on their own economic growth and population migration trends. Taking Kenya as an example, the country has stable economic growth and rapid population growth. It should focus on the construction of affordable housing and supporting infrastructure around cities, guide the orderly agglomeration of population, avoid the emergence of slums similar to those in some Asian cities, optimize urban land use structure, and improve land use efficiency [103]. At the same time, African countries should also strengthen the transformation of existing urban built-up areas, improve the level of infrastructure services, and enhance the urban carrying capacity. In addition, the government can promote urban renewal projects and use modern technologies to improve urban functions, such as intelligent transportation systems and green buildings, to meet the needs of rapid urbanization.

Under the SSP3 scenario, Africa faces regional competition and deglobalization trends. Resource-rich countries such as Nigeria should strengthen the coordinated planning of resource industries and urban development, ensure that resource transportation channels are connected with urban infrastructure construction, reasonably lay out related industrial parks, promote orderly urban expansion through industrial development, and strengthen ecological and environmental protection to avoid ecological damage caused by resource development [38]. For resource-poor countries such as Somalia, the focus should be on developing characteristic agriculture and light industry, supporting industrial development, rationally planning urban growth boundaries, preventing blind expansion, focusing on ecological protection, and ensuring the stability of urban ecosystems. In addition, Somalia can use international aid and technical support to enhance urban planning capabilities and improve infrastructure to meet the challenges brought by deglobalization.

Under the SSP4 scenario, the imbalance of urban land use in Africa has intensified, and the government should formulate policies to promote balanced regional development. In large cities and wealthy areas, such as Johannesburg in South Africa, land use efficiency should be improved, urban renewal and redevelopment should be strengthened, and excessive expansion should be restricted [39]. At the same time, the government should increase investment in rural and marginal areas, improve infrastructure and public services, guide the rational layout of industries, promote balanced population distribution, and reduce the polarization of urban land use. In addition, countries such as South Africa, Egypt, Nigeria, and Morocco can use their technological and economic advantages to promote regional cooperation, help neighboring countries improve urban planning capabilities, and narrow regional development gaps.

Under the SSP5 scenario, African countries such as Angola and Kenya have rapid economic development and strong demand for urban land, and should strengthen environmental supervision and land use planning, and formulate strict industrial access standards to ensure that pollution emissions from high-energy-consuming industries meet standards. At the same time, urban functional areas should be rationally planned, the proportion of ecological land should be guaranteed, and attention should be paid to the protection and restoration of ecological space to achieve a balance between economic development and environmental protection [40]. In addition, African countries can also promote green technology innovation, develop a low-carbon economy, and reduce the dependence on traditional fossil fuels. For example, Kenya can use its economic position in East Africa to promote regional ecological cooperation and jointly address climate change and environmental degradation.

In addition to policy recommendations for different scenarios, African countries should also strengthen regional cooperation and exchanges, and share successful experiences and advanced technologies. For example, an African urban development alliance or cooperation platform can be established to jointly study common problems in urban expansion and promote the sustainable development of cities on the African continent. At the same time, African countries should focus on improving the professional capabilities of urban planning and land management departments, strengthen talent training and introduction, and ensure the effective implementation and execution of policies. Through these comprehensive measures, African countries can better respond to urban development challenges under different SSP scenarios, achieve healthy, orderly, and sustainable urban development, find a global urban development path suitable for themselves, improve the quality and competitiveness of urban development, and create a better life and ecological environment for the African people.

5. Conclusions

Through a series of comprehensive analyses, this study deeply explores the urban expansion patterns of African countries under different SSP scenarios.

First, between 2020 and 2060, African countries as a whole show a significant trend of urban spatial expansion. Urban expansion is particularly strong under scenarios such as SSP1 and SSP5. In SSP1, countries such as South Africa and Angola achieve growth due to factors such as increased international cooperation and investment opportunities related to sustainable development. Under SSP5, countries such as Nigeria and Kenya expand rapidly due to rapid economic development and large-scale industrialization. In contrast, in SSP3, expansion is relatively weak. Countries such as Somalia face difficulties in urban development due to limited resources and lack of international cooperation, and the growth rate is sluggish. In addition, although the urban land area of most African countries has shown an upward trend during this period, a few countries may experience negative growth in urban land area after experiencing previous growth under certain future scenarios due to factors such as population decline and economic downturn.

Second, there are significant differences in the simulation results of the spatial pattern of urban land in 2060 under different scenarios. In countries with rapid urbanization and economic growth, expansion is mainly manifested as the agglomeration and extension of urban patches. For example, in some rapidly developing African countries, new urban areas tend to gather around existing urban cores to form continuous built-up areas. On the contrary, when a country's urbanization and economic development slow down, the expansion pattern shows the characteristics of relatively uniform distribution of urban patches. This shows that the development momentum and resource allocation of different countries play a key role in determining the specific forms of urban expansion.

Finally, a considerable proportion of African cities show specific expansion patterns. A large number of large cities in Africa follow a loose urban spatial expansion pattern, which is usually due to factors such as the existence of a large number of informal settlements, the lack of effective urban planning, and an economic structure that is not conducive to compact development. On the other hand, a considerable number of small- and medium-sized cities show a compact growth pattern, which may be related to their relatively stable population growth and limited land resources, prompting them to use land more efficiently. In addition, different regions in Africa also show unique expansion characteristics. Cities in North Africa may be affected by historical and cultural factors and regional economic structure, and show different expansion trends from cities in sub-Saharan Africa.

In summary, this study reveals in detail the urban expansion patterns of African countries under different SSP scenarios. It not only enriches the understanding of urban development in Africa, but also provides an important reference for African governments to formulate scientific and reasonable urban planning and land use policies. By considering the specific characteristics of different scenarios, African countries can better manage urban growth, promote sustainable development, and enhance regional ecological resilience. At the same time, it also calls for further research to continue to explore and respond to the challenges and opportunities in urban development in Africa.

Author Contributions: Conceptualization, B.L., S.X., N.Y. and W.L.; Software, B.L.; Validation, B.L., S.X. and N.Y.; Formal analysis, B.L., S.X., M.C. and W.L.; Investigation, S.X.; Resources, M.C. and N.Y.; Data curation, B.L.; Writing—original draft, S.X. and M.C.; Writing—review & editing, B.L.; Project administration, M.C., N.Y. and W.L.; Funding acquisition, B.L. All authors have read and agreed to the published version of the manuscript.

Funding: This research was funded by the National Natural Science Foundation of China Youth Science Fund Project "Study on the interactive mechanism between informal settlement expansion and urbanization: A case study of typical African cities" (No. 42301227), the Research funding for the 2024 Green seedling Program of the Human Resources and Social Security Department of Guangxi Zhuang Autonomous Region, China (60203038919630213), Nanning Normal University Demonstration Modern Industrial College (No. 6020303891920), Nanning Normal University Characteristic Undergraduate College Construction and College Teaching Quality and Reform Engineering Project—Undergraduate Education and Teaching Key Project (No. 6020303891924), Nanning Normal University Doctoral Research Startup Project (No. 602021239447).

Data Availability Statement: The original contributions presented in this study are included in the article. Further inquiries can be directed to the corresponding authors.

Conflicts of Interest: The authors declare no conflicts of interest.

References

1. Angel, S.; Parent, J.; Civco, D.L.; Blei, A.M. *Making Room for a Planet of Cities*; Lincoln Institute of Land Policy: Cambridge, MA, USA, 2011.
2. Brenner, N.; Schmid, C. Towards a New Epistemology of the Urban. *City* **2014**, *18*, 151–182. [CrossRef]
3. Grimm, N.B.; Faeth, S.H.; Golubiewski, N.E.; Redman, C.L.; Wu, J.; Bai, X.; Briggs, J.M. Global Change and the Ecology of Cities. *Science* **2008**, *319*, 756–760. [CrossRef] [PubMed]
4. Batty, M. *The New Science of Cities*; MIT Press: Cambridge, MA, USA, 2013.
5. Knaap, G.J.; Hopkins, L.D. The inventory approach to urban growth boundaries. *J. Am. Plan. Assoc.* **2001**, *67*, 314–326. [CrossRef]
6. Hall, P.; Pfeiffer, U. Urbanization in Europe 1950–2000: An overview. *GeoJournal* **2008**, *71*, 1–14.
7. Xu, X.; Du, Z.; Zhang, H. Integrating the System Dynamic and Cellular Automata Models to Predict Land Use and Land Cover Change. *Int. J. Appl. Earth Obs. Geoinf.* **2016**, *52*, 568–579. [CrossRef]
8. Chaudhuri, G.; Clarke, K. The SLEUTH land use change model: A review. *Environ. Resour. Res.* **2013**, *1*, 88–105.
9. Luo, H.; Li, Y.; Gao, X.; Meng, X.; Yang, X.; Yan, J. Carbon Emission Prediction Model of Prefecture-Level Administrative Region: A Land-Use-Based Case Study of Xi'an City, China. *Appl. Energy* **2023**, *348*, 121488. [CrossRef]
10. Li, X.; Liu, Z.; Li, S.; Li, Y.; Wang, W. Urban land carbon emission and carbon emission intensity prediction based on patch-generating land use simulation model and grid with multiple scenarios in Tianjin. *Land* **2023**, *12*, 2160. [CrossRef]
11. Jiang, L.; O'Neill, B.C. Global Urbanization Projections for the Shared Socioeconomic Pathways. *Glob. Environ. Change* **2017**, *42*, 193–199. [CrossRef]
12. Karutz, R.; Klassert CJ, A.; Kabisch, S. On farmland and floodplains—Modeling urban growth impacts based on global population scenarios in Pune, India. *Land* **2023**, *12*, 1051. [CrossRef]
13. Kantakumar, L.N.; Kumar, S.; Schneider, K. What Drives Urban Growth in Pune? A Logistic Regression and Relative Importance Analysis Perspective. *Sustain. Cities Soc.* **2020**, *60*, 102269. [CrossRef]
14. Gibson, J.; Boe-Gibson, G.; Stichbury, G. Urban Land Expansion in India 1992–2012. *Food Policy* **2015**, *56*, 100–113. [CrossRef]
15. Liu, Y.; Liu, X.; Chen, Y.; Long, H. Land use transitions in China: Insights from the integrated analysis of satellite remote sensing data. *J. Clean. Prod.* **2020**, *262*, 121273.
16. Tian, L.; Qiao, X. Spatial-temporal evolution and influencing factors of urban land expansion in China: Based on the geographical weighted regression model. *Habitat Int.* **2018**, *74*, 1–10. [CrossRef]
17. Datta, A. Urbanization and economic development in India. *J. Asian Econ.* **2012**, *23*, 237–245.
18. Sivaramakrishnan, K.C.; Kundu, A. Urbanization and urban governance in India. *Econ. Political Wkly.* **2004**, *39*, 4545–4552.
19. Satterthwaite, D. Sustainable Cities or Cities that Contribute to Sustainable Development? In *The Earthscan Reader in Sustainable Cities*; Routledge: London, UK, 2021; pp. 80–106.
20. Fujii, S.; Managi, S. Land use change and its impacts on ecosystem services in Japan. *Ecol. Econ.* **2016**, *127*, 241–249.
21. Jantz, C.A.; Goetz, S.J.; Shelley, M.K. Using the SLEUTH Urban Growth Model to Simulate the Impacts of Future Policy Scenarios on Urban Land Use in the Baltimore-Washington Metropolitan Area. *Environ. Plan. B Plan. Des.* **2004**, *31*, 251–271. [CrossRef]
22. Liu, Y.; Wang, L.; Long, H. Spatio-temporal analysis of land-use conversion in the eastern coastal China during 1996–2005. *J. Geogr. Sci.* **2008**, *18*, 274–282. [CrossRef]

23. Seto, K.C.; Reenberg, A.; Boone, C.G.; Fragkias, M.; Haase, D.; Langanke, T.; Marcotullio, P.; Munroe, D.K.; Olah, B.; Simon, D. Urban land teleconnections and sustainability. *Proc. Natl. Acad. Sci. USA* **2012**, *109*, 7687–7692. [CrossRef]
24. Gidey, E.; Dikinya, O.; Sebego, R.; Segosebe, E.; Zenebe, A. Cellular automata and Markov Chain (CA_Markov) model-based predictions of future land use and land cover scenarios (2015–2033) in Raya, northern Ethiopia. *Model. Earth Syst. Environ.* **2017**, *3*, 1245–1262. [CrossRef]
25. Seto, K.C.; Güneralp, B.; Hutyra, L.R. Global forecasts of urban expansion to 2030 and direct impacts on biodiversity and carbon pools. *Proc. Natl. Acad. Sci. USA* **2012**, *109*, 16083–16088. [CrossRef] [PubMed]
26. Popp, A.; Calvin, K.; Fujimori, S.; Havlik, P.; Humpenöder, F.; Stehfest, E.; Bodirsky, B.L.; Dietrich, J.P.; Doelmann, J.C.; Gusti, M.; et al. Land use futures in the shared socio-economic pathways. *Glob. Environ. Change* **2017**, *42*, 331–345. [CrossRef]
27. Fricko, O.; Havlik, P.; Rogelj, J.; Klimont, Z.; Gusti, M.; Johnson, N.; Kolp, P.; Strubegger, M.; Valin, H.; Amann, M.; et al. The marker quantification of the shared socioeconomic pathway 2: A middle-of-the-road scenario for the 21st century. *Glob. Environ. Change* **2017**, *42*, 251–267. [CrossRef]
28. O'Neill, B.C.; Kriegler, E.; Riahi, K.; Ebi, K.L.; Hallegatte, S.; Carter, T.R.; Mathur, R.; van Vuuren, D.P. A new scenario framework for climate change research: The concept of shared socioeconomic pathways. *Clim. Change* **2014**, *122*, 387–400. [CrossRef]
29. Bian, R.; Zhao, A.; Zou, L.; Liu, X.; Xu, R.; Li, Z. Simulation and Prediction of Land Use Change and Carbon Emission under Multiple Development Scenarios at the City Level: A Case Study of Xi'an, China. *Land* **2024**, *13*, 1079. [CrossRef]
30. Potapov, P.V.; Turubanova, S.A.; Hansen, M.C.; Adusei, B.; Broich, M.; Altstatt, A.; Mane, L.; Justice, C.O. Quantifying forest cover loss in Democratic Republic of the Congo, 2000–2010, with Landsat ETM+ data. *Remote Sens. Environ.* **2012**, *122*, 106–116. [CrossRef]
31. Verburg, P.H.; Soepboer, W.; Veldkamp, A.; Limpiada, R.; Espaldon VO, L.; Sharifa, M.A. Modeling the spatial dynamics of regional land use: The CLUE-S model. *Environ. Manag.* **2002**, *30*, 391–405. [CrossRef]
32. Brown, D.G.; Johnson, K.M.; Loveland, T.R.; Theobald, D.M. Rural land-use trends in the conterminous United States, 1950–2000. *Ecol. Appl.* **2005**, *15*, 1851–1863. [CrossRef]
33. Liu, X.; Li, X. Assessing the impact of urbanization on ecosystem services using land use change scenarios. *J. Environ. Manag.* **2018**, *206*, 1122–1132.
34. Gao, Y.; Jiao, L.; Wang, W. Analysis of the spatial pattern of urban expansion in countries along the Belt and Road under the shared social and economic situation. *Geogr. Geogr. Inf. Sci.* **2023**, *6*, 36–44. [CrossRef]
35. Elvidge, C.D.; Baugh, K.E.; Kihn, E.A.; Kroehl, H.W.; Davis, E.R. Mapping city lights with nighttime data from the DMSP operational linescan system. *Photogramm. Eng. Remote Sens.* **1997**, *63*, 727–734.
36. Peng, K.; Jiang, W.; Deng, Y. Simulating urban land-use changes by incorporating logistic regression and CLUE-S model: A case study of Wuhan city. In Proceedings of the 2021 28th International Conference on Geoinformatics, Nanchang, China, 3–5 November 2021; IEEE: Piscataway, NJ, USA, 2021; pp. 1–5.
37. Van Vuuren, D.P.; Stehfest, E.; Gernaat, D.E.; Doelman, J.C.; van den Berg, M.; Harmsen, M.; De Boer, H.S.; Bouwman, L.F.; Daioglou, V.; Edelenbosch, O.Y.; et al. Energy, land—Use and greenhouse gas emissions trajectories under a green growth paradigm. *Glob. Environ. Change* **2017**, *42*, 237–250. [CrossRef]
38. Fujimori, S.; Hasegawa, T.; Masui, T.; Takahashi, K.; Herran, D.S.; Dai, H.; Hijioka, Y.; Kainuma, M. SSP3: AIM implementation of shared socioeconomic pathways. *Glob. Environ. Change* **2017**, *42*, 268–283. [CrossRef]
39. Calvin, K.; Bond-Lamberty, B.; Clarke, L.; Edmonds, J.; Eom, J.; Hartin, C.; Kim, S.; Kyle, P.; Link, R.; Moss, R.; et al. The SSP4: A world of deepening inequality. *Glob. Environ. Change* **2017**, *42*, 284–296. [CrossRef]
40. Kriegler, E.; Bauer, N.; Popp, A.; Humpenöder, F.; Leimbach, M.; Strefler, J.; Baumstark, L.; Bodirsky, B.L.; Hilaire, J.; Klein, D.; et al. Fossil-Fueled development (SSP5): An energy and resource intensive scenario for the 21st century. *Glob. Environ. Change* **2017**, *42*, 297–315. [CrossRef]
41. Hu, J.; Liu, D.; Zheng, X. Research on Multi-Scenario Simulation of Urban Expansion for Beijing–Tianjin–Hebei Region Considering Multilevel Urban Flows. *Land* **2024**, *13*, 1830. [CrossRef]
42. Sun, Y.; Liu, D.; Wang, P. Urban Simulation Incorporating Coordination Relationships of Multiple Ecosystem Services. *Sustain. Cities Soc.* **2022**, *76*, 103432. [CrossRef]
43. Li, H.; Liu, Y.; He, Q.; Peng, X.; Yin, C. Simulating Urban Cooperative Expansion in a Single-Core Metropolitan Region Based on Improved CA Model Integrated Information Flow: Case Study of Wuhan Urban Agglomeration in China. *J. Urban Plan. Dev.* **2018**, *144*, 05018002. [CrossRef]
44. Cramer, C.; Sender, J.; Oqubay, A. *African Economic Development: Evidence, Theory, Policy*; Oxford University Press: Oxford, UK, 2020.
45. Obi-Ani, N.A.; Isiani, M.C. Urbanization in Nigeria: The Onitsha experience. *Cities* **2020**, *104*, 102744. [CrossRef]
46. Adewuyi, A.O.; Ayoola, O.B. Internal Migration in Nigeria: Positive Contributions to Human Development. *J. Niger. Stud.* **2018**, *16*, 123–140.
47. Omonigho, T. Okhankhuele, Olaniyan Z. Opafunso. Causes and Consequences of Rural-Urban Migration Nigeria: A Case Study of Ogun Waterside Local Government Area of Ogun State, Nigeria. *Br. J. Arts Soc. Sci.* **2025**, *16*, 185–194.

48. Ologeh, I.O.; Akarakiri, J.B.; Sobanke, V.O. Assessment of Climate Change-Induced Migration of Crop Farmers in Northern Nigeria. In *Handbook of Climate Change Management: Research, Leadership, Transformation*; Springer International Publishing: Cham, Switzerland, 2020; pp. 1–17.
49. Kassouri, Y.; Okunlola, O.A. Analysis of spatio-temporal drivers and convergence characteristics of urban development in Africa. *Land Use Policy* **2021**, *109*, 105868. [CrossRef]
50. Pender, J.; Gebremedhin, B.; Benin, S.; Ehui, S. Strategies for sustainable agricultural development in the Ethiopian highlands. *Am. J. Agric. Econ.* **2001**, *83*, 1231–1240. [CrossRef]
51. Collier, P.; Venables, A.J. Urban infrastructure for development. *Oxf. Rev. Econ. Policy* **2016**, *32*, 391–409. [CrossRef]
52. Zhao, S.; Qi, J.; Li, D.; Wang, X. Land use change and its influencing factors along railways in Africa: A case study of the Ethiopian section of the Addis Ababa–Djibouti Railway. *J. Geogr. Sci.* **2024**, *34*, 1128–1156. [CrossRef]
53. Weltbank. *Ethiopia-Toward the Competitive Frontier: Strategies for Improving Ethiopia's Investment Climate*; World Bank: Washington, DC, USA, 2012.
54. Li, X.; Zhou, W.; Ouyang, Z. Forty years of urban expansion in Beijing: What is the relative importance of physical, socioeconomic, and neighborhood factors? *Appl. Geogr.* **2013**, *38*, 1–10. [CrossRef]
55. Angel, S.; Parent, J.; Civco, D.L.; Blei, A.M.; Potere, D. The dimensions of global urban expansion: Estimates and projections for all countries, 2000–2050. *Prog. Plan.* **2011**, *75*, 53–107. [CrossRef]
56. Pienaar, S.M.; Ndlovu, T. Land Use Policy Changes in South Africa for Managing Urban Growth and Protecting Agricultural and Ecological Land. *Land Use Policy* **2023**, *128*, 106585.
57. Aribigbola, A. Urban Land Use Planning, Policies and Management in Sub-Saharan African Countries: Empirical Evidence from Akure, Nigeria. *Urban Forum* **2007**, *18*, 137–153.
58. Goodness, C.A. Effects of Fiscal and Monetary Policy Uncertainty on Economic Activity in South Africa. *Adv. Decis. Sci.* **2021**, *25*, 1–18.
59. Cobbinah, P.B.; Finn, B.M. Planning and Climate Change in African Cities: Informal Urbanization and 'Just' Urban Transformations. *J. Plan. Lit.* **2023**, *38*, 361–379. [CrossRef]
60. Adhikari, S.; Koirala, P.; Ghosh, A.; Henry, M. Planning for Sustainable Cities in Africa: Experiences, Challenges and Prospects of Monitoring Geospatial Indicators. *Remote Sens.* **2022**, *14*, 2821. [CrossRef]
61. Ouedraogo, V.; Hackman, K.O.; Thiel, M.; Dukiya, J. Intensity Analysis for Urban Land Use/Land Cover Dynamics Characterization of Ouagadougou and Bobo-Dioulasso in Burkina Faso. *Land* **2023**, *12*, 1063. [CrossRef]
62. Loulad, S.; Nguyen, T.T.; Simou, M.R.; Rhinane, H.; Buerkert, A. Monitoring rural-urban transformation in the coastal region of Rabat-Sale-Kenitra, Morocco. *PLoS ONE* **2023**, *18*, e0290829. [CrossRef]
63. Berriane, M.; de Haas, H.; Natter, K. Social Transformations and Migrations in Morocco. 2021. Available online: https://hdl.handle.net/1887/3239103 (accessed on 20 February 2025).
64. Assede, E.S.; Orou, H.; Biaou, S.S.; Geldenhuys, C.J.; Ahononga, F.C.; Chirwa, P.W. Understanding drivers of land use and land cover change in Africa: A review. *Curr. Landsc. Ecol. Rep.* **2023**, *8*, 62–72. [CrossRef]
65. Muinzer, T.L.; Mackie, C. The Sustainable Development Goals. In *Energy Law and the Sustainable Development Goals*; Routledge: London, UK, 2024; pp. 15–25.
66. Wei, G.; He, B.-J.; Sun, P.; Liu, Y.; Li, R.; Ouyang, X.; Luo, K.; Li, S. Evolutionary trends of urban expansion and its sustainable development: Evidence from 80 representative cities in the belt and road initiative region. *Cities* **2023**, *138*, 104353. [CrossRef]
67. Rogerson, C.M. Small-scale manufacturing in South Africa's economic heartland: Problems and future prospects. In *Urban Forum*; Leiden University Catalogue: Dordrecht, The Netherlands, 1998; Volume 9, pp. 279–300.
68. Lewis, J.D. *Policies to Promote Growth and Employment in South Africa*; World Bank, Southern Africa Department: Tshwane, South Africa, 2001; Volume 16.
69. Makhathini, M.S.; Mlambo, V.H.; Mpanza, S. Infrastructure provision as a catalyst for local economic development in South Africa. *Strateg. Rev. S. Afr.* **2020**, *42*. [CrossRef]
70. Liu, Z.; Huang, J.; Wang, R.; Zhou, C.; Lu, L. Urban expansion monitoring in South Asia based on SPOT/VGT and DMSP/OLS data. In Proceedings of the 2017 2nd International Conference on Image, Vision and Computing (ICIVC), Chengdu, China, 2–4 June 2017; IEEE: Piscataway, NJ, USA; pp. 1142–1146.
71. Raihan, A.; Ibrahim, S.; Muhtasim, D.A. Dynamic impacts of economic growth, energy use, tourism, and agricultural productivity on carbon dioxide emissions in Egypt. *World Dev. Sustain.* **2023**, *2*, 100059. [CrossRef]
72. Amadou, O. The Effects of Economic Globalization on Developing States: The Case of The Republic of Benin (2016–2021). *Bölge Çalışmaları Derg.* **2023**, *2*, 70–82.
73. Zinkina, J.; Korotayev, A. Urbanization dynamics in Egypt: Factors, trends, perspectives. *Arab Stud. Q.* **2013**, *35*, 20–38. [CrossRef]
74. Nkeki, F.N. Spatio-temporal analysis of land use transition and urban growth characterization in Benin metropolitan region, Nigeria. *Remote Sens. Appl. Soc. Environ.* **2016**, *4*, 119–137. [CrossRef]

75. Dorling, D. World population prospects at the UN: Our numbers are not our problem? In *The Struggle for Social Sustainability*; Policy Press: Bristol, UK, 2021; pp. 129–154.
76. Allen, C.; Nejdawi, R.; El-Baba, J.; Hamati, K.; Metternicht, G.; Wiedmann, T. Indicator-based assessments of progress towards the sustainable development goals (SDGs): A case study of the Arab region. *Sustain. Sci.* **2017**, *12*, 975–989. [CrossRef]
77. Tiando, D.S.; Hu, S.; Fan, X.; Ali, M.R. Tropical coastal land-use and land cover changes impact on ecosystem service value during rapid urbanization of Benin, West Africa. *Int. J. Environ. Res. Public Health* **2021**, *18*, 7416. [CrossRef]
78. Change, O.C. Intergovernmental panel on climate change. *World Meteorol. Organ.* **2007**, *52*, 1.
79. Kikstra, J.S.; Nicholls, Z.R.J.; Smith, C.J.; Lewis, J.; Lamboll, R.D.; Byers, E.; Sandstad, M.; Meinshausen, M.; Gidden, M.J.; Rogelj, J.; et al. The IPCC Sixth Assessment Report WGIII climate assessment of mitigation pathways: From emissions to global temperatures. *Geosci. Model Dev.* **2022**, *15*, 9075–9109. [CrossRef]
80. Badi, I.; Stević, Ž.; Bouraima, M.B. Overcoming obstacles to renewable energy development in Libya: An MCDM approach towards effective strategy formulation. *Decis. Mak. Adv.* **2023**, *1*, 17–24. [CrossRef]
81. Strangio, D. Analysis of the Economic System of Somalia. In *The Reasons for Underdevelopment. Contributions to Economics*; Springer: Cham, Switzerland, 2012; pp. 39–74. [CrossRef]
82. Irhiam, H.; Schaeffer, M.; Watanabe, K. (Eds.) *The Long Road to Inclusive Institutions in Libya: A Sourcebook of Challenges and Needs*; World Bank Publications: Washington, DC, USA, 2023.
83. Statista. Population Growth in Somalia 2013–2023. 22 November 2024. Available online: https://www.statista.com/statistics/452246/population-growth-in-somalia/ (accessed on 22 November 2024).
84. Aamir, M.; ur Rehman, J. Dynamic nexus among fossil fuels utilization, economic growth, and urbanization: A tri-regional selected countries analysis. *Environ. Sci. Pollut. Res.* **2024**, *31*, 43381–43395. [CrossRef]
85. Le Sage, A.; Majid, N. The livelihoods gap: Responding to the economic dynamics of vulnerability in Somalia. *Disasters* **2002**, *26*, 10–27. [CrossRef]
86. Ologunde, O.H.; Kelani, M.O.; Biru, M.K.; Olayemi, A.B.; Nunes, M.R. Land Use and Land Cover Changes: A Case Study in Nigeria. *Land* **2025**, *14*, 389. [CrossRef]
87. Bloch, R.; Fox, S.; Monroy, J.; Ojo, A. *Urbanisation and Urban Expansion in Nigeria*; Urbanisation Research Nigeria (URN) Research Report; ICF International: London, UK, 2015.
88. Xu, J.; Wang, X. Reversing Uncontrolled and Unprofitable Urban Expansion in Africa through Special Economic Zones: An Evaluation of Ethiopian and Zambian Cases. *Sustainability* **2020**, *12*, 9246. [CrossRef]
89. Kaluwa, B. Industrial development in sub-Saharan Africa: The impact of structural adjustment. In *Structural Adjustment, Reconstruction and Development in Africa*; Routledge: London, UK, 2019; pp. 71–81.
90. Chamberlin, J.; Jayne, T.S.; Headey, D. Scarcity amid plenty? Reassessing the potential for cropland expansion in Africa. *Food Policy* **2014**. [CrossRef]
91. Jiang, S.; Zhang, Z.; Ren, H. Spatiotemporal characteristics of urban land expansion and population growth in Africa from 2001 to 2019: Evidence from population density data. *ISPRS Int. J. Geo Inf.* **2021**, *10*, 584. [CrossRef]
92. Kamaldeen, Y.; Eric, B.D.; Alexandros, G. Sustainability of urban expansion in Africa: A systematic literature review using the Drivers–Pressures–State–Impact–Responses (DPSIR) framework. *Sustain. Sci.* **2023**, *18*, 1459–1479.
93. Iyakaremye, V.; Zeng, G.; Yang, X. In 21st century mid-century increased high temperature extremes and associated population exposure in Africa. *Sci. Total Environ.* **2021**, *790*, 148162. [CrossRef]
94. Owoeye, J.O. Assessing Urban Growth Dynamics and Depletion of Agricultural Land Use in Akure Region, Nigeria. *Land Sci.* **2019**, *1*, 43. [CrossRef]
95. Adzandeh, E.A.; Alaigba, D.; Nkemasong, C.N. Application of Geospatial Techniques and Logistic Regression Model for Urban Growth Analysis in Limbe, Cameroon. *Niger. J. Environ. Sci. Technol. (NIJEST)* **2020**, *4*, 138–151. [CrossRef]
96. Xu, G.; Dong, T.; Cobbinah, P.B.; Jiao, L.; Sumari, N.S.; Chai, B.; Liu, Y. Urban expansion and form changes across African cities with a global outlook:Spatiotemporal analysis of urban land densities. *J. Clean. Prod.* **2019**, *224*, 802–810. [CrossRef]
97. Enuoh, R. Business ethics and the Nigerian oil industry:a study of the niger delta. *Int. J. Res. Innov. Soc. Sci. (IJRISS)* **2019**, *9*, 285–293.
98. Abo El Wafa Moustafa, H. Spatial Scenario Modelling for the Sustainable Planning of Urbanising Regions in Sub-Saharan African Cities. Doctoral Dissertation, Technische Universität Munich, Munich, Germany, 2019.
99. Hashidu, B.R.; Abbas, A.M.; Kamaludeen, A.M. Urban growth pattern and agricultural land use dynamics in Gombe City, Nigeria. *J. Adv. Res. Agric. Sci. Technol.* **2019**, *2*, 43–49.
100. Tadjie, R. Expansion urbaine et acuité des problèmes fonciers au Cameroun. *RiA Recht Afr. Law Afr. Droit En Afr.* **2018**, *20*, 171–191. [CrossRef]
101. Chen, G.; Li, X.; Liu, X.; Chen, Y.; Liang, X.; Leng, J.; Xu, X.; Liao, W.; Qiu, Y.; Wu, Q. A global urban land expansion product at 1-km resolution for 2015 to 2100 based on the SSP scenarios. *PANGAEA* **2019**. [CrossRef]
102. Angel, S.; Blei, A.M. *Atlas of Urban Expansion*; Lincoln Institute of Land Policy: Cambridge, MA, USA, 2012.

103. Liu, X.; Li, X.; Liu, L.; He, J.; Ai, B. A bottom-up approach to discover transition rules of cellular automata using ant intelligence. *Int. J. Geogr. Inf. Sci.* **2017**, *22*, 1247–1269. [CrossRef]
104. Merkle, M.; Alexander, P.; Brown, C.; Seo, B.; Harrison, P.A.; Harmáčková, Z.V.; Pedde, S.; Rounsevell, M. Downscaling population and urban land use for socio-economic scenarios in the UK. *Reg. Environ. Change* **2022**, *22*, 106. [CrossRef]
105. Jing, C.; Tao, H.; Jiang, T.; Wang, Y.; Zhai, J.; Cao, L.; Su, B. Population, urbanization and economic scenarios over the Belt and Road region under the Shared Socioeconomic Pathways. *J. Geogr. Sci.* **2020**, *30*, 68–84. [CrossRef]
106. Reimann, L.; Merkens, J.L.; Vafeidis, A.T. Regionalized Shared Socioeconomic Pathways: Narratives and spatial population projections for the Mediterranean coastal zone. *Reg. Environ. Change* **2018**, *18*, 235–245. [CrossRef]

land

MDPI

Article

Evaluation of Urban Infrastructure Resilience Based on Risk–Resilience Coupling: A Case Study of Zhengzhou City

Wenli Dong [1], Yunhan Zhou [1,*], Dongliang Guo [2], Zhehui Chen [1] and Jiwu Wang [1]

1 College of Civil Engineering and Architecture, Zhejiang University, Hangzhou 310058, China; wenlidong@zju.edu.cn (W.D.); beiming@zju.edu.cn (Z.C.); wangjiwu@zju.edu.cn (J.W.)
2 Zhengzhou Urban Planning Design and Survey Research Institute Co., Zhengzhou 450052, China; guo21000210@163.com
* Correspondence: yunhanzhou@zju.edu.cn

Abstract: The frequent occurrence of disasters has brought significant challenges to increasingly complex urban systems. Resilient city planning and construction has emerged as a new paradigm for dealing with the growing risks. Infrastructure systems like transportation, lifelines, flood control, and drainage are essential to the operation of a city during disasters. It is necessary to measure how risks affect these systems' resilience at different spatial scales. This paper develops an infrastructure risk and resilience evaluation index system in city and urban areas based on resilience characteristics. Then, a comprehensive infrastructure resilience evaluation is established based on the risk–resilience coupling mechanism. The overall characteristics of comprehensive infrastructure resilience are then identified. The resilience transmission level and the causes of resilience effects are analyzed based on the principle of resilience scale. Additionally, infrastructure resilience enhancement strategies under different risk scenarios are proposed. In the empirical study of Zhengzhou City, comprehensive infrastructure resilience shows significant clustering in the city area. It is high in the central city and low in the periphery. Specifically, it is relatively high in the southern and northwestern parts of the airport economy zone (AEZ) and low in the center. The leading driving factors in urban areas are risk factors like flood and drought, hazardous materials, infectious diseases, and epidemics, while resilience factors include transportation networks, sponge city construction, municipal pipe networks, and fire protection. This study proposes a "risk-resilience" coupling framework to evaluate and analyze multi-hazard risks and the multi-system resilience of urban infrastructure across multi-level spatial scales. It provides an empirical resilience evaluation framework and enhancement strategies, complementing existing individual dimensional risk or resilience studies. The findings could offer visualized spatial results to support the decision-making in Zhengzhou's resilient city planning outline and infrastructure special planning and provide references for resilience assessment and planning in similar cities.

Keywords: infrastructure resilience; evaluation system; risk–resilience coupling; resilience transmission; resilience enhancement

1. Introduction

Cities are constantly facing disruption due to several natural and man-made calamities during their growth and operation, including natural disasters like heat waves, torrential rains, earthquakes, and mudslides; public health issues like epidemics; and unintentional disasters like fires and transportation accidents. A variety of disasters happen frequently as a result of the city's ongoing social and economic development and concentration, as well as the increasing conflict between human activity and the environment. Internal and external disturbances are becoming more frequent, intense, and unpredictable in cities [1]. The population, scale, functional agglomeration, and inherent structural complexity of cities all increase during the urban development process, which makes them more convenient and prosperous while also increasing the impact of disasters on cities [2]. The idea of resilient cities has drawn tremendous attention from governments and organizations in order to address the growing risks and disruptions both inside and outside of cities. Resilient city planning and construction has emerged as a new paradigm for dealing with environmental issues and disasters. The term "resilience" originally meant "to return to the original state". It has since been used in a variety of fields. In order to handle the uncertainty of disruptions, the resilient city forgoes the single concept of resistance under the conventional concept of disaster prevention. Instead, it stresses the comprehensive enhancement of the city's ability to resist, absorb, recover, adapt, and transform based on a reasonable establishment of defense standards [3,4]. The conceptual framework [5,6] and evaluation methods [7,8] for urban resilience have been extensively studied worldwide. Following the COVID-19 pandemic, increased focus has been placed on both overall urban resilience and subsystem resilience, particularly in addressing public health challenges [9,10] and multi-hazard scenarios [11], to improve urban resilience and sustainability.

At the G20 Summit in November 2020, China presented a series of sustainable development projects aimed at creating a resilient, sustainable, and inclusive future [12]. Enhancing territorial spatial resilience was one of the guiding requirements in the 2020 Guidelines for the Preparation of Provincial Territorial Spatial Planning. The development of resilient cities was suggested as a visionary target for China's 14th Five-Year Plan for the Development of the National Economy and Society in 2021, making resilient cities a national strategy. Since then, Chinese scholars have increasingly focused on urban resilience [13–15]. The efficient functioning of urban infrastructure systems is essential to the cities. The normal operation of the urban system's activities will be directly impacted by the speed of recovery and the integrity of the infrastructure functions when the city experiences unknown shocks. Most studies on infrastructure resilience have focused on individual subsystems, such as transportation [16], or examined it within the context of urban resilience [13]. Recently, research studies have also investigated the cascading effects of infrastructure systems under disaster conditions [17]. The "7.20" disaster of torrential rainstorms that struck Zhengzhou City in 2021 caused significant economic losses and fatalities, seriously devastated the city's infrastructure system, and presented significant challenges for the creation of resilient municipal infrastructure. Resilience planning is a critical approach for more resistant, adaptive, and sustainable cities in the face of multiple internal and external risks. This process must be grounded in a thorough evaluation of existing risk resistance and resilience capacities.

However, in terms of evaluation scales, the current resilience evaluations mostly focus on urban agglomerations, provincial regions, or entire cities. Examples include the resilience evaluation of the Yangtze River Delta city cluster [13], the vulnerability assessment of climate change hotspots in the Eastern Mediterranean and Middle East region [18], and the Shanghai resilient city planning index system, which primarily targets the city as a scale-free network [12]. Because there are few resilience evaluations based on urban districts, counties, streets, and even communities as the basic unit, it is challenging to provide spatial analysis support for urban resilience planning when it comes to the spatial characterization of urban resilience under the influence of urban disasters and risks. In order to support urban infrastructure resilience planning, thorough research is required for indicator selection, level assessment, and coupling analysis.

In constructing infrastructure resilience evaluation systems, most frameworks are designed based on resilient city characteristics, infrastructure system components, and urban disaster phases, with indicators tailored to the study area's scale and specific features. For example, studies have assessed infrastructure resilience in Northeast China using economic, social, ecological, and infrastructural dimensions [19,20], while others have focused on energy, transportation, drainage, telecommunications, and medical centers within urban agglomerations [21] and provincial areas [22]. When integrated into urban resilience evaluation systems, infrastructure typically includes water, energy, and transportation sectors [23,24]. Additionally, some scholars have incorporated the network characteristics of infrastructure systems into their analyses [25]. Based on the aforementioned, the pertinent research on infrastructure resilience has been broadly categorized into the following focus areas in recent years: research on flood and rain resilience; research on municipal infrastructure resilience, including the risk of secondary disasters like cascading failures of municipal infrastructure brought on by earthquakes, hurricanes, and other disasters; and research on transportation resilience, which is also more closely related to spatial planning.

Regarding infrastructure resilience evaluation methods, traditional municipal engineering approaches can be categorized into the basic calculation method, the expansion of the basic model, and the interrelated infrastructure resilience calculation method [26,27]. For specific infrastructure resilience evaluations, scholars have employed complex network methods to measure transportation network resilience [16,28]. In resilience planning, normalized resilience assessments [29] mostly use the comprehensive index method, weighted summation, TOPSIS [30], and system dynamics [31]. During the evaluation stage of resilient city planning, it is challenging to conduct systematic spatial research to normalize dimensionless infrastructure resilience with other systems, such as urban public service facilities or land use resilience, and to suggest countermeasures from the perspective of spatial planning.

In urban disaster resilience research, some scholars focus on single hazards, while others adopt multi-scale and multidisciplinary approaches to study multi-hazard composite resilience. Examples of single-hazard studies include assessments of geologic hazard resilience in Beijing [32], flood resilience in Xi'an [33], and flood resilience of critical infrastructure in Torbay [34]. However, research on coupled multi-hazard risk and resilience assessment remains limited. For instance, Beijing's multi-hazard urban risk assessment is based on the superposition of hazard factors and the vulnerability matrix of disaster-affected entities [2], similar to Austin's study combining multiple natural hazard exposure with social vulnerability [35], where the focus is primarily on vulnerability rather than resilience. Although Hefei City's resilience evaluation of municipal infrastructure employed cross-scenario analysis of risk and system vulnerability, the methodology was largely qualitative rather than quantitative [29].

In summary, existing infrastructure resilience research predominantly focuses on single-hazard assessments, limited scales, and specific perspectives, lacking multi-hazard, multi-scale, and multi-system coupled evaluations. Additionally, there is insufficient analysis at spatial scales such as districts, counties, and streets. In contrast to previous studies, this paper focuses on the coupled relationship between multi-hazard risks and the internal multi-system resilience of infrastructure at different spatial scales. This approach reflects the complex risk environment and overall resilience level of urban infrastructure in a more comprehensive way, providing a scientific basis and theoretical foundation for enhancing its resilience and sustainability.

Therefore, as part of the resilient city assessment of Zhengzhou, this study incorporates multi-hazard risks and conducts a coupled assessment of multiple hazards and resilience to establish an infrastructure risk and resilience evaluation index system at two distinct scales and conducts a comprehensive infrastructure resilience evaluation using the "risk-resilience" coupling framework. In order to support future urban resilience planning outlines, infrastructure resilience special plans, and action plans, strategies to increase the resilience of infrastructure systems under various risk disturbance scenarios are proposed based on the empirical analysis of Zhengzhou City. Due to their ease of planning transmission, urban districts, counties, and streets are considered the fundamental units of evaluation in terms of scale. The comprehensive indicator evaluation approach, which makes it conducive to spatial systematic research with other urban systems, is used as the evaluation method. In terms of its structure, the evaluation system is separated into three infrastructure resilience research target areas: transportation, flood control and drainage, and lifelines. This approach provides a comprehensive understanding of the complex risk environment and overall resilience level of infrastructure.

2. Materials and Methods

In order to achieve the comprehensive resilience evaluation results of infrastructure, this study first establishes the infrastructure risk and resilience evaluation index system. Next, it builds the "risk-resilience" coupling grading matrix (Figure 1). To more accurately reflect the state of the infrastructure and external pressure, different indicators must be used based on the differences between the external environment and the internal system of infrastructure at different scales. At the same time, a differentiated evaluation system for the city area and the urban area must be built, taking into account the spatial distribution of the data and the challenges of obtaining it. At the spatial scale, districts and counties are adopted as the minimum analysis units for the metropolitan area, and streets are the minimum units for urban areas. This approach facilitates the implementation of subsequent planning while, to some extent, reducing data acquisition challenges.

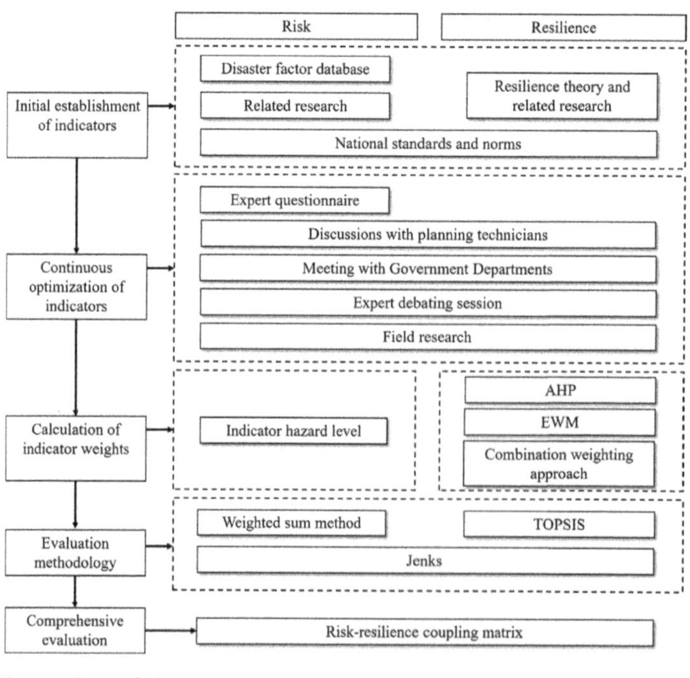

Figure 1. Research design.

2.1. Selection of Indicators

The criteria of operability, typicality, and comprehensiveness are used in the selection of indicators for the infrastructure resilience assessment system. Resilience theory, relevant research, and national standards serve as the foundation for the initial screening and summary processes. Following that, the indicators are regularly improved by field research, expert validation meetings, and conversations with government agencies and planning technicians.

In order to choose the indicators for the risk evaluation system, the disaster factor database, the "China's Comprehensive Natural Hazard Risk Classification System" classification standards, and other literature are first considered. Then, taking into account the disaster's frequency, intensity, scope, and correlation with urban and rural planning, among other factors, a variety of disaster-causing factors are compiled. Second, the disaster-causing elements with a danger rating of 4 or 5 are used as the evaluation indices of infrastructure risk utilizing the expert-assigned questionnaire approach. Lastly, the risk evaluation system for the transportation, flood control and drainage, and lifelines subsystems is built based on the degree of association between the indicators and the subsystems.

2.2. Calculation of Indicator Weights

The weights of the resilience evaluation indicator system are established by combining the Analytic Hierarchy Process (AHP) with the Entropy Weight Method (EWM). Although the weights of the indicators are consistent with their actual importance due to the subjective judgment and experience of the decision maker, the subjective weighting approach is susceptible to the subjective randomness of the decision maker. Although the objective assignment approach is based on original data and makes full use of objective actual information, it is susceptible to interference from sample data and may have indicator weights that are out of proportion to the indicators' true significance. The combined assignment method integrates subjective and objective weights through mathematical optimization, balancing expert experience with objective data. This approach addresses the limitations of single-method weighting systems, offering enhanced robustness and reliability in comprehensive evaluations [36,37]. However, the effect of the combined assignment method depends on the rationality of the weight combination method and the suitability of the application scene. This paper adopts the principle of minimum information entropy after Lagrange multiplier optimization to balance subjective and objective information by constraining optimization and minimizing information loss. Applied to infrastructure resilience evaluation, where it effectively integrates expert experience with objective data to establish a scientifically sound basis for weight allocation. The formula is shown in Equation (1).

$$\omega(j) = \frac{[\omega_1(j)\omega_2(j)]^{0.5}}{\sum_{j=1}^{n}[\omega_1(j)\omega_2(j)]^{0.5}} \tag{1}$$

where n is the number of indicators, and the subjective weight $\omega_1(j)$ from AHP and objective weight $\omega_2(j)$ from EWM are combined to optimize combined weight $\omega(j)$.

In the risk evaluation index system, the weights for risk indicators in the transportation, flood control and drainage, and lifeline subsystems are assigned based on their hazard levels, given the intuitive nature of disaster-causing factors. Specifically, each indicator's weight is proportional to its hazard level, maintaining the same ratio as the hazard level ratio. The average of the weights of each risk indicator in the three systems is used to calculate the total weight of the infrastructure.

2.3. Evaluation Methodology

Technique for Order Preference by Similarity to Ideal Solution (TOPSIS) is a widely used multi-attribute decision analysis method known for its intuitive concept of positive and negative ideal solution distances, flexible weight assignment, and efficient computational process, with applications across various fields [13,30]. TOPSIS is utilized to evaluate and rank the resilience states in this study. The method requires using the matrix D composed of n indicators of m solutions, constructing the positive ideal solution z^+ and negative ideal solution z^-, which are the optimal and the worst solutions. Then, the Euclidean distances D_i^+ and D_i^- between each solution and the positive and negative ideal solutions are calculated in order to obtain the relative affinity close to the positive ideal point and far from the negative ideal point C_i, which is used as the basis for evaluating the advantages and disadvantages. The main calculation formula is as follows:

$$D_i^+ = \sqrt{\sum_{j=1}^{n} \omega(j)\left(z_j^+ - z_{ij}\right)^2} \tag{2}$$

$$D_i^- = \sqrt{\sum_{j=1}^{n} \omega(j)\left(z_j^- - z_{ij}\right)^2} \tag{3}$$

$$C_i = \frac{D_i^-}{D_i^+ + D_i^-} \tag{4}$$

2.4. Risk–Resilience Coupling Matrix

To visualize the coupling between infrastructure risk and resilience, this study develops a risk–resilience coupling matrix, adapting the widely used risk matrix framework from risk management. The "risk-resilience" coupling grading matrix in Figure 2 is built for the comprehensive resilience evaluation based on the four-grade categorization evaluation results of infrastructure risk and resilience. The "resilience" score ranges from 1 to 4 points; the higher the score, the better the resilience. The "risk" score ranges from 1 to 4 points; the higher the score, the lower the risk. Consistent with the risk matrix consistency axioms [38], results are classified into levels I to IV, with higher levels indicating greater comprehensive resilience.

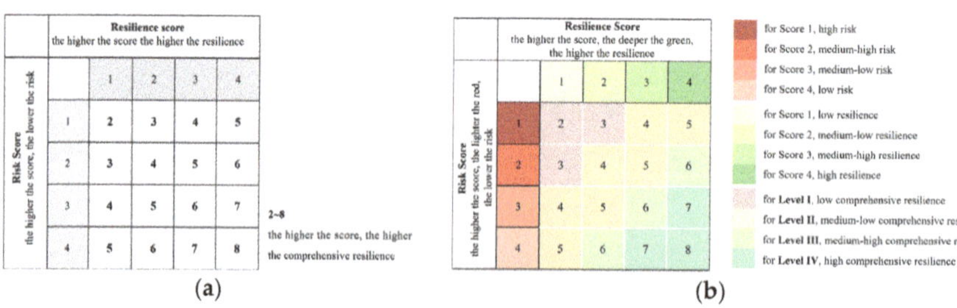

Figure 2. (**a**) Result matrix of risk–resilience coupling. (**b**) Corresponding color legends on the map [2].

2.5. Spatial Autocorrelation

Spatial autocorrelation analysis is a spatial statistical method for examining the distribution characteristics of geospatial data. In this study, spatial correlation analysis is conducted using ArcGIS 10.8. Global spatial autocorrelation was assessed through the global Moran's I index and the General G index, with significance tests indicating the overall spatial correlation. For local spatial autocorrelation, we employed Getis-Ord G* and Anselin Local Moran's I. The former identifies statistically significant hotspot areas (higher-than-expected values) and coldspot areas (lower-than-expected values), while the latter evaluates the spatial correlation between a region and its neighboring areas [39].

2.6. Geo Detector

Geo Detector is a statistical method for detecting spatial heterogeneity and determining the driving factors [40,41]. This study utilizes Geo Detector to analyze the driving causes of geographic variability of infrastructure resilience in the Zhengzhou city region, with a focus on factor detection and interaction detection outcomes. Factor detection can determine each factor's explanatory power for spatial differentiation of resilience. Interaction detection can determine the interaction between different factors. The main calculation formula is as follows:

$$q = 1 - \frac{\sum_{h=1}^{L} N_h \sigma_h^2}{N\sigma^2} \tag{5}$$

where q is the explanatory power of the driving factor—taking the value range [0, 1], a larger value indicates that the factor has higher explanatory power for resilience; L is the number of classifications; N_h and N are the number of sample units of the stratum h and the whole region; and σ_h^2 and σ^2 are the variance of the stratum h and the whole region [41].

2.7. Data Sources

The objective data come from various special planning and statistical documents, Zhengzhou City Statistical Yearbook (2011–2017), and the open platform of Amap. The great majority of the indicator data are gathered through data processing and calculation. Objective data were collected primarily from official sources for the same or similar years to maximize accuracy. Subjective data come from the expert questionnaire of the resilience indicator system and the hazard evaluation of risk indicators. The structured questionnaire of resilience indicator system employed a nine-level pairwise comparison scale. It was completed by ten experts, including six senior planning technicians and four university professors in the urban planning field. The questionnaire of risk–hazard indicators used a rank evaluation method, inviting eight experts familiar with the context of Zhengzhou to provide five-level rankings for each risk indicator, comprising six senior planning technicians and two university professors. In both of the stages, expert responses from valid questionnaires were aggregated using weighted averaging to enhance data reliability and minimize sampling variability.

3. Results

3.1. Evaluation Indicator System

The first is the initial establishment phase of the indicators. In the establishment of the risk indicator system, 52 disaster factors are derived from consulting the disaster factor database, the classification criteria of "Construction of China's Comprehensive Classification System for Natural Disaster Risk" and other literature, as well as by carefully taking into account the disaster's frequency of occurrence, intensity, scope of influence, and relevance to urban and rural planning (Figure 3). In the establishment of the resilience indicator system, 64 resilience factors were derived from the resilience theory and related research, national standards and norms like the guide for safety resilient city evaluation [42], and the five resilience characteristics—robustness, rapidity, redundancy, resourcefulness, and adaptability—that were separated into the three main systems of transportation, flood control and drainage, and lifeline for the initial indicator selection (Figure 4).

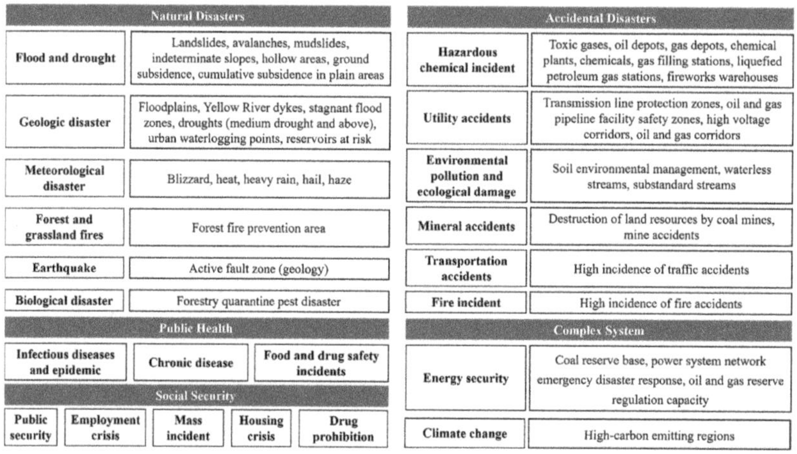

Figure 3. Initial indicator system for infrastructure risk evaluation in Zhengzhou City.

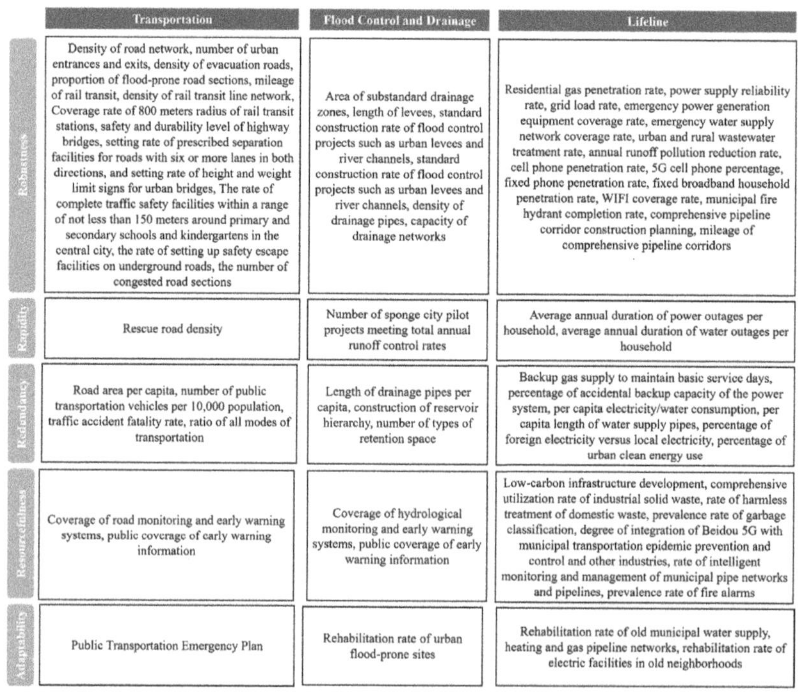

Figure 4. Initial indicator system for infrastructure resilience evaluation in Zhengzhou City.

In order to make the evaluation index system more in line with the actual situation of Zhengzhou City and the requirements of drafting the resilience planning outline, it has undergone an optimization process that has involved adding, removing, and replacing indicators through discussions with planning technicians, conversations with government departments, expert validation meetings, and field research. The resilience evaluation system of Zhengzhou was then established, comprising 23 indicators across the city and urban areas (Tables 1 and 2). Based on the AHP method, 10 questionnaires were distributed, with eight valid responses obtained on both a city and urban scale. The weighted average judgment matrix successfully passed the consistency test (Tables 3 and 4). Concurrently, based on the risk–hazard questionnaire (eight distributed, seven valid responses), two risk evaluation systems were finalized: one for the city area with 15 indicators and another for the urban area with 14 indicators, all selected based on their high risk levels (levels 4 and 5). The weighting of both risk and resilience indicator systems for the city and urban areas was determined using the methodology outlined in Section 2.2 (Tables 1, 2, 5 and 6).

Table 1. Evaluation index system of infrastructure resilience in the Zhengzhou city area.

Target Layer—Infrastructure Resilience		**Unit**	**Resilience Features**	**Weight**		
Criterion Layer—Transportation Resilience						
Sub-Criterion Layer	**Index Layer**			**AHP Weight**	**EWM Weight**	**Combined Weight**
Transportation network	5 km radius coverage of highway entrances and exits	%	Robustness	0.0638	0.0385	0.0546
	Number of external access points	Pcs	Robustness	0.0780	0.0179	0.0412
Transportation hub	Transportation connectivity	%	Redundancy	0.0403	0.0180	0.0297
	Hub accessibility	1/h	Redundancy	0.0727	0.0164	0.0381
Traffic safety	Traffic accident fatality rate per 10,000	%	Robustness	0.0562	0.0157	0.0327
Criterion layer—flood control and drainage resilience						
Flood control and drainage system construction	Drainage network capacity	km/mm	Robustness	0.0339	0.0587	0.0492
	Proportion of stagnant space area	%	Robustness	0.0280	0.0589	0.0448
	Proportion of floodway area	%	Robustness	0.0221	0.0435	0.0342
	Proportion of rivers with flood protection standards	%	Robustness	0.0383	0.0256	0.0345
Criterion layer—lifeline resilience						
Water supply	Water resources per capita	L	Robustness	0.0461	0.1034	0.0761
	Percentage of outgoing water sources	%	Redundancy	0.0247	0.0420	0.0355
	Density of water supply network	km/km^2	Robustness	0.0491	0.0234	0.0374
	Waterwork service capacity	L/person	Redundancy	0.0727	0.0939	0.0911
Electricity supply	Substation service capacity	Pcs/person	Redundancy	0.0578	0.0386	0.0520
	Density of high-voltage grid	km/km^2	Robustness	0.0770	0.0306	0.0535
	Density of medium-voltage grid	km/km^2	Robustness	0.0578	0.0234	0.0405
Heat supply	Density of heating network	km/km^2	Redundancy	0.0125	0.0718	0.0265
Gas supply	Residential gas penetration rate	%	Robustness	0.0102	0.0564	0.0256
	Gas utility service capacity	Pcs/person	Robustness	0.0123	0.0439	0.0170
	Density of high-pressure and sub-high-pressure gas pipeline network	km/km^2	Robustness	0.0102	0.0234	0.0566
	Density of medium-pressure gas network	km/km^2	Robustness	0.0397	0.0664	0.0331
Firefighting	Coverage of 150 m radius of municipal fire hydrants	%	Robustness	0.0265	0.0557	0.0423
	Coverage of fire and rescue 5 min	%	Robustness	0.0701	0.0341	0.0539

Table 2. Evaluation index system of infrastructure resilience in the Zhengzhou urban area.

Target Layer—Infrastructure Resilience		**Unit**	**Resilience Features**	**Weight**		
Criterion Layer—Transportation Resilience						
Sub-Criterion Layer	**Index Layer**			**AHP Weight**	**EWM Weight**	**Combined Weight**
Transportation network	Road area per capita (A1)	m^2	Redundancy	0.0105	0.0578	0.0307
	Density of the backbone network (A2)	km/km^2	Robustness	0.0596	0.0094	0.0295
	Road network density (A3)	km/km^2	Robustness	0.0397	0.0096	0.0244
	Density of rail transit network (A4)	km/km^2	Robustness	0.0182	0.0511	0.038
	800 m radius coverage of rail transit stations (A5)	%	Robustness	0.0105	0.0307	0.0224
Transportation hub	Hub accessibility (A6)	1/h	Redundancy	0.0835	0.0020	0.0161
Emergency transportation	Density of overall emergency road (A7)	km/km^2	Resourcefulness	0.0807	0.0121	0.039
	Proportion of disaster-prone emergency roads (A8)	%	Resourcefulness	0.0322	0.0008	0.0063
Criterion layer—flood control and drainage resilience						
Sponge city construction	Proportion of drainage zoning area in sponge cities meeting standards (B1)	%	Robustness	0.0392	0.0961	0.0765
Flood control and drainage system construction	Proportion of stagnant space area (B2)	%	Robustness	0.0549	0.0349	0.0546
	Proportion of floodway area (B3)	%	Robustness	0.0401	0.0418	0.0511
	Proportion of rivers with flood protection standards (B4)	%	Robustness	0.0568	0.0918	0.09
Criterion layer—lifeline resilience						
Water supply	Density of water supply network (C1)	km/km^2	Robustness	0.1330	0.0111	0.0479
Electricity supply	Substation service capacity (C2)	Pcs/person	Redundancy	0.0425	0.0662	0.0662
	Density of high-voltage grid (C3)	km/km^2	Robustness	0.0567	0.0293	0.0508
	Density of medium-voltage grid (C4)	km/km^2	Robustness	0.0425	0.0111	0.0271
Heat supply	Density of heating network (C5)	km/km^2	Robustness	0.0374	0.0563	0.0573
Gas supply	Gas utility service capacity (C6)	Pcs/person	Redundancy	0.0100	0.1086	0.0412
	Density of high-pressure and sub-high-pressure gas pipeline network (C7)	km/km^2	Robustness	0.0134	0.0862	0.0424
	Density of medium-pressure gas network (C8)	km/km^2	Robustness	0.0100	0.0111	0.0132
Comprehensive pipeline corridor	Density of comprehensive corridors built (C9)	km/km^2	Robustness	0.0361	0.1126	0.0795
Firefighting	Coverage of 150 m radius of municipal fire hydrants (C10)	%	Robustness	0.0317	0.0425	0.0457
	Coverage of fire and rescue 5 min (C11)	%	Robustness	0.0607	0.0267	0.0502

Table 3. Consistency ratios (CRs) of AHP for infrastructure resilience index system in the Zhengzhou city area ('/' means no need for consistency test).

Target Layer	CR	Criterion Layer	CR	Sub-Criterion Layer	CR
Infrastructure resilience	0.036	Transportation resilience	0.051	Transportation network	0.009
				Traffic safety	/
				Emergency transportation	/
		Flood control and drainage resilience	/	Flood control and drainage system construction	0.013
		Lifeline resilience	0.016	Water supply	0.014
				Electricity supply	0.011
				Heat supply	/
				Gas supply	0.021
				Firefighting	/

Table 4. Consistency ratios (CR) of AHP for infrastructure resilience index system in the Zhengzhou urban area ('/' means no need for consistency test).

Target Layer	CR	Criterion Layer	CR	Sub-Criterion Layer	CR
Infrastructure resilience	0.008	Transportation resilience	0.045	Transportation network	0.018
				Transportation hub	/
				Emergency transportation	/
		Flood control and drainage resilience	/	Sponge city construction	/
				Flood control and drainage system construction	0
		Lifeline resilience	0.023	Water supply	/
				Electricity supply	0.015
				Heat supply	/
				Gas supply	0.012
				Comprehensive pipeline corridor	/
				Firefighting	/

Table 5. Evaluation index system of infrastructure risk in the Zhengzhou city area.

Target Layer—Infrastructure Resilience		Unit	Risk Level	Weight
Criterion Layer—Transportation Resilience				
Sub-Criterion Layer	Index Layer			
Flood and drought	Urban flood-prone water points	Pcs	Level 4	0.1141
	Flooding	km^2	Level 5	0.1426
	Dangerous reservoirs	Pcs	Level 5	0.1426
Geologic disaster	Landslides	Pcs	Level 4	0.0439
	Crumble	Pcs	Level 4	0.0439
	Subside	Pcs	Level 4	0.0439
Meteorological disaster	Blizzard	mm	Level 4	0.0439
	Rainstorm	mm	Level 5	0.1426
Criterion layer—accidental disaster				
Mineral accident	Mineral accident	km^2	Level 4	0.0439
Transportation accident	Transportation accident	Pcs	Level 4	0.0205
Facility accident	Electricity facility accident	Pcs	Level 5	0.0549
	Gas facility accident	Pcs	Level 5	0.0549
Hazardous chemical incident	Hazardous chemical incident	Pcs	Level 4	0.0439
Fire accident	Fire accident	m	Level 4	0.0439

Table 5. *Cont.*

Target Layer—Infrastructure Resilience		Unit	Risk Level	Weight
Criterion Layer—Transportation Resilience				
Sub-Criterion Layer	Index Layer			
Criterion layer—public health				
Infectious diseases and epidemic	Infectious diseases and epidemic	Pcs	Level 4	0.0205

Table 6. Evaluation index system of infrastructure risk in the Zhengzhou urban area.

Target Layer—Infrastructure Resilience		Unit	Risk Level	Weight
Criterion Layer—Transportation Resilience				
Sub-Criterion Layer	Index Layer			
Flood and drought	Urban flood-prone water points (D1)	Pcs	Level 4	0.1172
	Flooding (D2)	km^2	Level 5	0.1465
	Dangerous reservoirs (D3)	Pcs	Level 5	0.1465
Geologic disaster	Landslides (D4)	Pcs	Level 4	0.047
	Crumble (D5)	Pcs	Level 4	0.047
	Subside (D6)	Pcs	Level 4	0.047
Meteorological disaster	Blizzard (D7)	mm	Level 4	0.047
	Rainstorm (D8)	mm	Level 5	0.1465
Criterion layer—accidental disaster				
Transportation accident	Transportation accident (D9)	Pcs	Level 4	0.0219
Facility accident	Electricity facility accident (D10)	Pcs	Level 5	0.0588
	Gas facility accident (D11)	Pcs	Level 5	0.0588
Hazardous chemical incident	Hazardous chemical incident (D12)	Pcs	Level 4	0.047
Fire accident	Fire accident (D13)	m	Level 4	0.047
Criterion layer—public health				
Infectious diseases and epidemic	Infectious diseases and epidemic (D14)	Pcs	Level 4	0.0219

3.2. Evaluation Results for Zhengzhou City Area

Based on the TOPSIS and weighting method, the resilience and risk evaluation of the infrastructure in the Zhengzhou city area is carried out, respectively. The "risk-resilience" coupling matrix is used to conduct a comprehensive evaluation of the infrastructure resilience in the Zhengzhou city area (Figure 5).

In the Zhengzhou city area, the overall infrastructure risk shows a spatial pattern of high in the east and west and low in the center, with Gongyi City, Xingyang City, Dengfeng City, Xinmi City, Xinzheng City, Jinshui District, and Zhengdong New District having a higher risk and AEZ (Airport Economy Zone), NHTDZ (National High and New Technology Industries Development Zone), Shangjie District, and ETDZ (Economic and Technological Development Zone) having a lower risk. The overall infrastructure resilience shows a pattern of high urban areas and low surrounding areas, with Shangjie District, Zhongyuan District, and Jinshui District being better; NHTDZ, Erqi District, Guancheng District, ETDZ, and Zhengdong New District being average; and Huiji District, Zhongmou County, Gongyi City, Dengfeng City, Xinmi City, Gongyi City, Xingyang City, Xinzheng City, and AEZ being worse. The overall comprehensive infrastructure resilience shows a spatial pattern of high resilience in the central urban area and low resilience in the periphery, with Xingyang City, Dengfeng City, and Gongyi City being poor, and Zhongyuan District, Shangjie District, NHTDZ, and ETDZ being good.

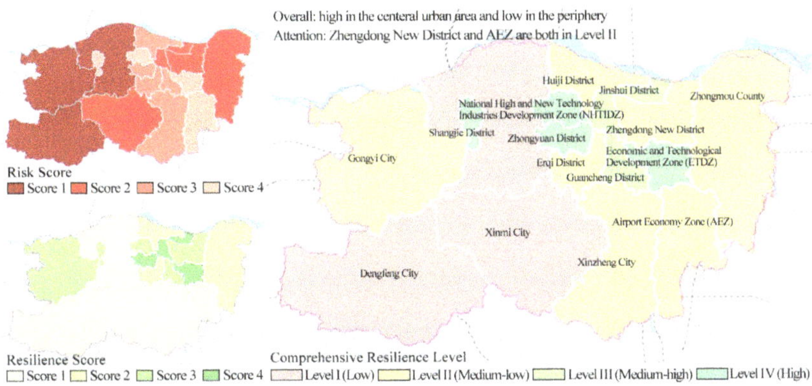

Figure 5. Evaluation results of infrastructure resilience, risk, and comprehensive resilience in the Zhengzhou city area.

An interesting observation is that local professionals frequently expect Zhengdong New District to demonstrate higher resilience. However, the combined resilience results are generally moderate due to significant risks in the area, such as high population density, proximity to the Yellow River, and the presence of 500 KV high-voltage stations. In contrast, although the AEZ is less developed, and its lower risk sources and population density make the comprehensive resilience better than expected.

The risk evaluation results in this study show certain differences with the previous dual-evaluation results of Zhengzhou, which are based on a resource and environmental carrying capacity evaluation and a suitability evaluation for territorial spatial development. The dual evaluation emphasized risk control of natural disasters and conservation of natural resources, whereas the assessment in this paper expands the scope to include accidental disasters and public health risks, resulting in the higher-risk zones in the central city where hazardous facilities are concentrated.

The integrated resilience of the three systems—transportation, flood control and drainage, and lifeline—was computed in order to better investigate the spatial distribution of resilience within the resilience system for each infrastructure specialization (Figure 6). In the city area, the comprehensive resilience of the transportation system as a whole presents a spatial pattern of higher urban area and lower periphery area and higher south and lower north; the comprehensive resilience of the flood control and drainage system as a whole presents a spatial pattern of high urban, low periphery, high central, and low east and west; and the comprehensive resilience of the lifeline system as a whole presents a spatial pattern of higher urban and lower periphery.

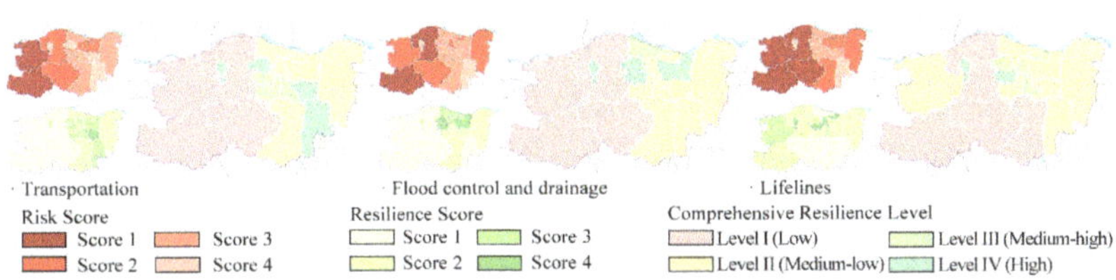

Figure 6. Evaluation results of resilience, risk, and comprehensive resilience of infrastructure subsystems in the Zhengzhou city area (transportation, flood control and drainage, and lifelines).

3.3. *Evaluation Results for Zhengzhou Urban Area*

3.3.1. Spatial Patterns of Resilience

The resilience and risk evaluation of infrastructure in the Zhengzhou urban area is also carried out. The urban area infrastructure risk, resilience, and risk–resilience coupling evaluation results are shown in Figure 7. Further, the comprehensive resilience evaluation results of the three systems of urban transportation, flood control and drainage, and lifeline are shown in Figure 8.

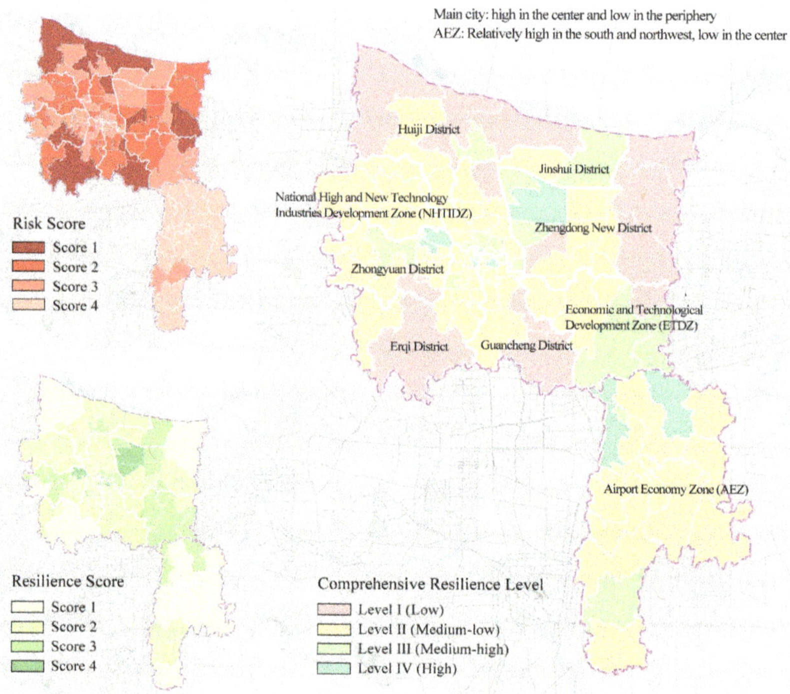

Figure 7. Evaluation results of infrastructure resilience, risk, and comprehensive resilience in the Zhengzhou urban area.

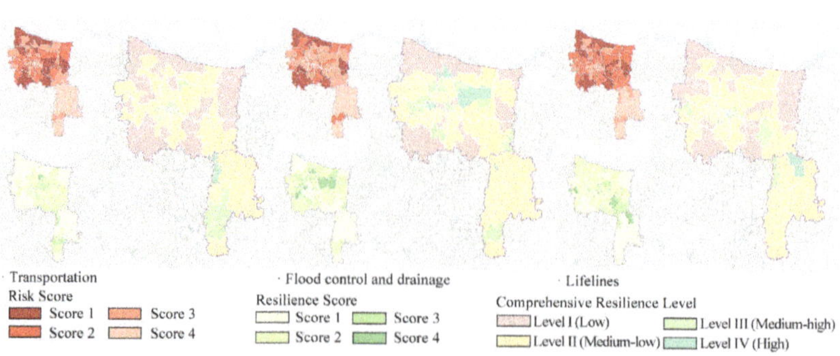

Figure 8. Evaluation results of resilience, risk, and comprehensive resilience of infrastructure subsystems in the Zhengzhou urban area (transportation, flood control and drainage, and lifelines).

In the Zhengzhou urban area, the overall infrastructure risk presents a spatial pattern that is high in the urban area and low in the aerial port area. The transport systems in

Zhengdong New District, the north and east of ETDZ, the south of Guancheng District, the middle of Erqi District, the west of NHTDZ, the south of Zhongyuan District, the north of Huiji District, and the middle of Jinshui District have higher risks; the transport systems in the AEZ and the center of the downtown area have lower risks. The overall infrastructure resilience presents a spatial pattern that is high in the central part of the city, low in the periphery, relatively high in the south and northwest of the aerial port area, and low in the center of the city. The western part of Zhengdong New District, the southern part of Jinshui District, the eastern part of Huiji District, the eastern part of NHTDZ, the central part of Zhongyuan District, the northern part of Guancheng District, ETDZ, and the northern part of AEZ are better; the northeastern part of Zhengdong New District, the central part of Erqi District, the western part of Zhongyuan District, the western part of Huiji District, and the southern and central parts of AEZ are worse.

Infrastructure risk—the overall resilience coupling result shows that the central part of the city is high, the periphery is low, the southern and northwestern parts of the AEZ are relatively high, and the central part is low. The south and north of the AEZ, the northeastern part and the northwestern corner of the Zhongyuan District, the Erqi District, the northern part of the Guancheng District, the southern part of the Jinshui District, the northwestern part of Zhengdong New District, and the northwestern part of the ETDZ are more resilient. The central part of Erqi District, the southern part of Guancheng District, the northern part of Huiji District, the central part of Jinshui District, the periphery of the main city, and the central part of the AEZ are less resilient.

In the urban area, the comprehensive resilience of the transportation system and lifeline system as a whole presents the spatial pattern of the central city as being low and the AEZ as being high; the flood control and drainage system comprehensive resilience as a whole presents the spatial pattern of the central city as being high and the periphery as being low.

3.3.2. Spatial Correlation Analysis

To further evaluate the clustering characteristics of the comprehensive resilience evaluation results in the Zhengzhou city district, we use ArcGIS 10.8 to perform spatial correlation analysis, including global and local spatial autocorrelations.

The global spatial autocorrelation shows a positive Moran's I index of 0.2655, which passes the 1% significant test (p-Value < 0.01, Z-Score > 2.58). The Getis-Ord General G provides a z-score of 4.17, indicating that Zhengzhou urban area comprehensive resilience has a significant positive correlation agglomeration in the global space and a high/high clustering pattern.

The local spatial autocorrelation of Getis-Ord G* provides a distribution map of hot and cold spot clusters (Figure 9a). The hotspot clusters are located in the central city core and the northern portion of the AEZ, while the coldspot clusters are located in the central city's north and south. The findings show that there is a considerable spatial disparity and that the comprehensive resilience of the infrastructure in the central city's core area is strong, while that of the periphery is poor.

Additionally, we acquired Anselin Local Moran I's high/low agglomeration distribution map (Figure 9b). The "High-High" and "Low-High" agglomerations are staggered in the core area of the city. The "Low-Low" agglomerations are distributed in a small number of areas in the middle of Huiji District and the western part of Erqi and Guancheng Districts. The "High-Low" agglomerations are distributed in the middle of Huiji District and the western part of Erqi District. The "High-Low" agglomerations are scattered throughout the eastern section of Huiji District. The findings indicate that while the integrated resilience of infrastructure in the central city's core area is high, there are notable spatial differences with the surrounding areas, whereas there are no significant spatial differences with the integrated resilience of infrastructure in the aerodrome area or the central city's periphery.

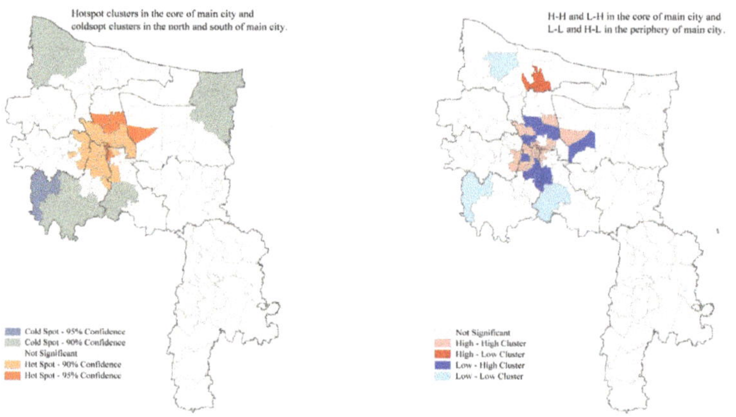

(**a**) (**b**)

Figure 9. Spatial autocorrelation analysis of comprehensive infrastructure resilience in the Zhengzhou urban area ((**a**) Getis-Ord G*, (**b**) Anselin Local Moran's I).

3.3.3. Driving Factor Analysis

Using factor detection with Geo Detector, the infrastructure risk, resilience, and risk–resilience coupling results are analyzed in the Zhengzhou urban area (Table 7). For infrastructure resilience, overall emergency road density (A7), main network density (A2), road network density (A3), 800 m radius coverage of rail transit stations (A5), and hub accessibility (A6) affect transportation system resilience. Water supply network, power grid, medium-pressure gas network density (C1, C3, C4, C8), and fire coverage (C11) influence lifeline resilience. Flood control and drainage factors have little impact on the spatial divergence of infrastructure resilience. For infrastructure risk, heavy rainstorms (D8), traffic accidents (D9), electric utility accidents (D10), diseased reservoirs (D3), hazardous chemical accidents (D12), fires (D13), and infectious disease outbreaks (D14) have a large impact on the spatial divergence of infrastructure risk. Single factors have less significance and explanatory power for comprehensive infrastructure resilience, and the spatial divergence of integrated resilience is more affected by 5 min coverage of fire and rescue (C11) in resilience, diseased reservoirs (D3), and hazardous chemical accidents (D12) in risk.

Interaction detection of Geo Detector is used to undertake an interaction analysis of the driving variables of urban infrastructure comprehensive resilience (Figure 10). The findings show that the factors exhibit varying degrees of two-factor enhancement and nonlinear enhancement, and the explanatory power of the factors after the interaction is significantly increased, implying that the spatial heterogeneity of the comprehensive resilience of urban area infrastructures is the result of multiple factors acting together. Among these, the mean value of the explanatory power of multiple components after interaction with other factors is greater than 0.3, indicating a significant driving factor for the geographic variation of infrastructure resilience in the Zhengzhou urban area. These risk indicators include urban flood-prone water points (D1), dangerous reservoirs (D3), hazardous chemical accidents (D12), and infectious disease outbreaks (D14), while resilience indicators include transportation networks (A1, A2, A3, A4, and A5), the proportion of the areas of sponge city up to the standard drainage zoning (B1), the density of pipeline networks of each municipal facility (C1, C4, C7, and C8), and 5 min coverage of the fire rescue services (C11). These findings share similarities with previous research on resilience factors, such as electricity consumption and gas supply [13], while also incorporating additional considerations, including risk factors, sponge city infrastructure, and fire safety facilities.

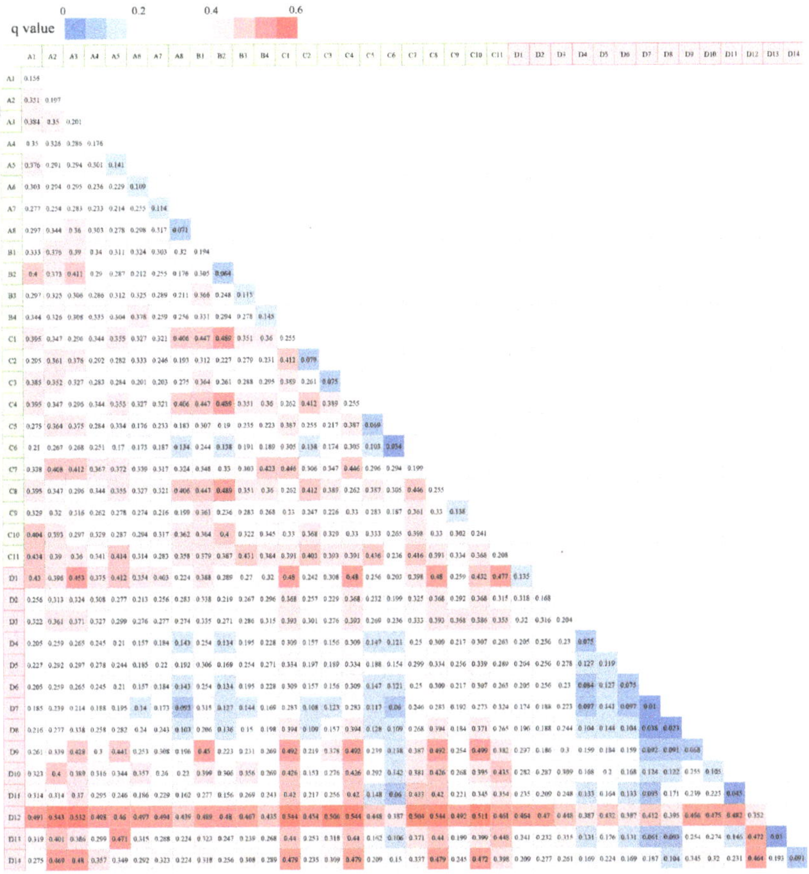

Figure 10. Heat map of the interaction of factors affecting comprehensive infrastructure resilience in the Zhengzhou urban area.

Table 7. Detection of spatial heterogeneity of infrastructure resilience in the Zhengzhou urban area (* means q-value < 0.05).

Resilience Index	Resilience q-Value	Comprehensive Resilience q-Value	Risk Index	Risk q-Value	Comprehensive Resilience q-Value
A1	0.029	0.156	D1	0.14	0.135
A2	0.382 *	0.197	D2	0.107	0.168
A3	0.33 *	0.201	D3	0.313 *	0.204 *
A4	0.278	0.176	D4	0.08	0.075 *
A5	0.247 *	0.141	D5	0.105	0.119
A6	0.215 *	0.109	D6	0.08	0.075 *
A7	0.448 *	0.114	D7	0.063	0.01
A8	0.036	0.071	D8	0.512 *	0.023
B1	0.133	0.194	D9	0.428 *	0.068
B2	0.138	0.064	D10	0.293 *	0.105

Table 7. *Cont.*

Resilience Index	Resilience q-Value	Comprehensive Resilience q-Value	Risk Index	Risk q-Value	Comprehensive Resilience q-Value
B3	0.178	0.115	D11	0.106	0.045
B4	0.192	0.145	D12	0.424 *	0.352 *
C1	0.352 *	0.255	D13	0.192 *	0.03
C2	0.02	0.079	D14	0.313 *	0.091
C3	0.242 *	0.075			
C4	0.352 *	0.255			
C5	0.078	0.069			
C6	0.008	0.034			
C7	0.084	0.199			
C8	0.352 *	0.255			
C9	0.255	0.138			
C10	0.337 *	0.241			
C11	0.249 *	0.208 *			

3.4. Individual Case Study

In order to examine the degree of resilience transmission and the reasons behind resilience effects in the city and urban area units based on the resilience scaling principle, the following case study examines Huiji District, which is situated in the center of the urban area.

Huiji District, a crucial development sector in the Central Plains Economic Zone's core growth pole, is situated on the south bank of the Yellow River and north of Zhengzhou city center. Huiji District has low risk (Score 3), weak resilience (Score 2), and low comprehensive resilience (Level II), and its resilience and comprehensive resilience are ranked low in the central urban area. The analysis is as follows:

Heavy rains (Score 1) and traffic accidents (Score 1) are the primary risk factors in Huiji District (Figure 11). Other risk factors include floodplains, hazardous materials, fires, and infectious disease outbreaks (Score 2). The main metropolitan region is in great danger of severe rainfall due to Zhengzhou's excessive localized precipitation and the "urban rain island" phenomenon. Huiji is more vulnerable to flooding than the districts and counties in the south because of the Yellow River to the north and a lengthy embankment line. In comparison to the county area, Huiji District has a higher risk of hazardous materials, fires, and infectious disease outbreaks due to its oil depots, flammable and explosive production and storage units, urban villages, warehouses, and other fire accidents with poor fire prevention awareness. Additionally, this district also has vegetable markets, health stations, and other areas infectious disease may spread.

Figure 11. Examples of risks faced in Huiji District.

Huiji District has high flood control and drainage resilience (Score 3) but poor lifeline and transportation resilience (Scores 1 and 2). Nearly fifty percent of Scores 1 and 2 are single-indicator grades. In the transportation system, notable weaknesses include the high 10,000 traffic accident death rate (Score 1), insufficient highway exit coverage within a 5 km radius (Score 2), and a limited number of external corridors (Score 2). The percentage of river channels in the drainage and flood control system that satisfy flood control requirements is low (Score 2). Within the lifeline system, significant shortcomings include low per capita water resources (Score 1), high dependence on external water sources (Score 1), sparse high-voltage power grid and heating pipeline networks (Score 2), poor service capacity of water plants (Score 2), and inadequate coverage of municipal fire hydrant within a 150 m radius (Score 2).

As a result, Huiji District needs to strengthen the construction of flood control and drainage systems; improve the early warning capacity for storms, floods, fires, and other disasters in the area of risk prevention and mitigation; and improve the management of infectious agents and hazardous materials in the area. Transportation and lifelines must be prioritized in order to increase resilience. Strengthening the development of safe transportation and continuing to encourage the building of road networks are essential in the transportation sector. River flood control construction needs to be strengthened in the drainage and flood control areas. Promoting water resources, enhancing water safety, and advancing the development of municipal pipeline networks and firefighting facilities are all essential in the lifelines field.

3.5. *Infrastructure Resilience Enhancement Strategies*

Based on the findings of the complete resilience evaluation, a comprehensive enhancement strategy for infrastructure transportation, flood control and drainage, and lifeline systems is suggested (Figure 12). Following that, the risk indicators and weights from the infrastructure risk evaluation index system derived from the survey in this study are used to identify flooding, geologic disasters, fire accidents, infectious diseases, and epidemics as the primary stress risk scenarios. The potential state of transportation, flood control and drainage, and lifeline systems under each disaster risk is stated separately, with matching resilience enhancement methods provided (Figures 13–16).

Figure 12. Resilience enhancement strategies for each system.

Risk	Sub-system	Possible state of affairs	Enhancement strategy
Flooding	Transportation	Disruption of internal and external transportation in the city due to waterlogged roads, water ingress into underground transportation spaces, damaged roads, etc.	Improve drainage and water retention facilities, raise flood prevention standards, and focus on upgrading the waterproofing and drainage capacity of road tunnels, culverts, elevated ramps, etc.
			Optimize the layout of the transportation network, establish a comprehensive three-dimensional disaster relief transportation network, and safeguard internal and external transportation links
		Waterlogged roads and water ingress into underground spaces leading to entrapment. Road collapses threatening personal safety.	Strengthening disaster risk early warning systems to improve public coverage of disaster warnings
			Enhancement of emergency security, command and control and rescue capabilities
			Increase flood prevention and drainage knowledge publicity. Enhance public awareness of disaster prevention and self-rescue and mutual-rescue capabilities
	Flood control and drainage	The amount of rainfall or flooding exceeds the storage capacity of the stagnation space and the drainage capacity of the discharge channel.	Enhance the drainage network capacity of the stormwater pipe network, create a functional composite storage community, and build a high-standard flood relief channel
		Dam burst	Timely elimination of reservoir distress problems. Enhancement of disaster early warning and risk management capacity
		rainwater and sewage mixing	Accelerating stormwater and sewerage rehabilitation
	Lifelines	Pollution of urban water sources	Improvement of water treatment capacity of water plants and safety of water supply
		Major lifeline facilities damaged and out of service	Upgrading the waterproofing and drainage capacity of major lifeline facilities
			Construction of backup facilities for major lifeline facilities
			Improvement of the level of emergency repairs

Figure 13. Resilience enhancement strategies in flooding risk scenarios.

Risk	Sub-system	Possible state of affairs	Enhancement strategy
Geologic disaster	Transportation	Damage to urban roads and disruption of internal and external transportation within the city	Enhance the ability of transportation facilities to withstand geological hazards, and strengthen the geological safety survey work and safety design standards for new transportation facilities.
			Optimize the layout of the transportation network, establish a comprehensive three-dimensional disaster relief transportation network, and safeguard internal and external transportation links
		Waterlogged roads and water ingress into underground spaces leading to entrapment. Road collapses threatening personal safety.	Strengthening disaster risk early warning systems to improve public coverage of disaster warnings.
			Enhancement of emergency security, command and control and rescue capabilities
			Increase geologic disaster defense knowledge publicity. Enhance public awareness of disaster prevention and self-rescue and mutual-rescue capabilities
	Flood control and drainage	Drying up of urban water sources	Establishment of coordination mechanisms for water transfer projects at the regional level. Emphasis on open source and flow control of water resources
	Lifelines	Pollution of urban water sources	Improvement of water treatment capacity of water plants and safety of water supply
		Major lifeline facilities damaged and out of service	Enhancing the ability of major lifeline facilities to withstand geologic hazards
			Construction of backup facilities for major lifeline facilities
			Improvement of the level of emergency repairs

Figure 14. Resilience enhancement strategies in geologic disaster risk scenarios.

Risk	Sub-system	Possible state of affairs	Enhancement strategy
Fire accident	Transportation	Fire on or within a transportation facility that affects roadway traffic	Upgrade the fire protection facilities at transportation facilities, especially major transportation facilities. Enhance fire safety inspections
			Advancement of road improvements to facilitate fire truck access
		Blockage of fire and rescue access	Open up fire and rescue channels in old neighborhoods. Joint traffic police department to regulate parking
	Lifelines	Inadequate fire and rescue facilities and personnel	Promote the construction of fire protection system. Improve fire station categories. Reduce the size of jurisdictions. Improve rescue efficiency
			Strengthening the rescue team and increasing the ratio of firefighters
			Increase fire prevention publicity. Increase public awareness of firefighting and self-rescue and mutual aid capabilities
		Major lifeline facilities damaged and out of service	Upgrading the fire rating of major lifeline facilities
			Construction of backup facilities for major lifeline facilities
			Improvement of the level of emergency repairs

Figure 15. Resilience enhancement strategies in fire accident risk scenarios.

Risk	Sub-system	Possible state of affairs	Enhancement strategy
Infectious diseases and epidemic	Transportation	Strict traffic control leads to reduced transportation capacity	Developing traffic countermeasures under normal and emergency conditions to ensure traffic operation under controlled conditions
			Diversified modes of transportation supply, such as travel reservations
		Public transportation increases the risk of transmission	Develop in-vehicle safety standards. Control passenger density. Increase transit capacity. Enhance public awareness
	Lifelines	Pollution of urban water sources	Improvement of water treatment capacity of water plants and safety of water supply with increased focus on disinfection

Figure 16. Resilience enhancement strategies in infectious diseases and epidemic risk scenarios.

4. Discussion

This study develops an evaluation index system for risk and resilience of infrastructure in Zhengzhou's city and urban areas, uses a combination of AHP and EWM to determine index weights, and evaluates the comprehensive resilience of infrastructure using the TOPSIS method and the coupling matrix of risk–resilience. This study's findings can help to depict the geographical characteristics of Zhengzhou's resilience planning outline and special infrastructure planning. It is beneficial for planning to evaluate at the district, county, and street levels in order to provide ideas and experiences for evaluating safety resilience and resilience planning in similar cities in China, as well as to promote the construction of resilient cities in Zhengzhou City and elsewhere.

Based on Conzen's urban form theory, which emphasizes planning units, this paper employs districts and streets as assessment units to develop unitized resilience enhancement and transmission. The difference in resilience findings between city and urban areas is a spatial manifestation of the resilience scaling principle, which states that infrastructure systems must be resilient at different scales. The influencing factors and their interactions reflect the importance of transportation, flood control and drainage, and lifeline subsystem synergies for resilience enhancement, as well as the principle of dependent coexistence of resilience systems. Finally, resource allocation is adjusted to improve resilience using the resilience augmentation approach.

The limitations of this study are as follows: While the risk–resilience coupling matrix offers an intuitive analytical framework, its limitations may introduce interpretive biases due to the complex interdependence between risk and resilience. The simplified structure of the matrix may also obscure the nuanced dynamics of their relationship. A standardized and scientifically robust urban resilience assessment system is still absent. The current approach relies on static composite indices, which fail to capture the dynamic nature of resilience indicators. Although the AHP-EWM combination weighting method balances subjective expertise with objective data, its accuracy depends on the precision of these methods and may introduce additional uncertainties. Furthermore, the effect of the combined assignment method depends on its rationality and the suitability of the application scenario. Data collection and processing are also constrained, as the timeliness and accuracy of indicator data cannot be fully guaranteed. Data derived from planning documents may only partially reflect real conditions and big data platforms may contain inherent biases. Furthermore, the reliance on relative standards for evaluation is due to limited spatial indicators in national standards and weak inter-city applicability. Additionally, the risk and resilience indicators tailored to Zhengzhou's specific characteristics may not be directly applicable to cities with differing natural environments, economic structures, or risk profiles.

In future research, dynamic system modeling and network analysis will be adopted to explore the complex interactions and intrinsic linkages between risk and resilience, extending the coupling matrix framework. Dynamic evaluation methods that reflect the temporal changes of resilience indicators, as well as differentiated measures for special resilience assessments based on the characteristics of various infrastructure systems, should be introduced, and more efficient data processing and analysis methods should be used. Diverse weighing algorithms will be adopted to improve the scientific and practical relevance of optimization techniques, with additional analysis on the stability of weighting results. Data using the natural resources census should be estimated to be collected, which will broaden data collection methods and procedures. Data from numerous sources should be collected to cross-validate and increase data quality. The establishment of local resilience standards will be promoted based on differences in urban space and metropolitan areas. Further support for the transmission and implementation of resilient city planning will be proposed. Considering the application of this framework on different types of cities, classified resilience evaluation frameworks need to be developed using cross-case study and validation methods, providing more adaptable indicators and approaches to improve its generalizability.

5. Conclusions

The risk and resilience evaluation index system of Zhengzhou's city and urban areas is constructed in this paper using the risk–resilience coupling comprehensive evaluation framework and resilience-related basic theories. The indexes are continuously optimized through discussions with planning technicians, governmental departmental symposiums, expert demonstration meetings, and field research. The weights of the indexes are determined using the combination of the AHP and the EWM. The TOPSIS method and the risk–resilience coupling matrix are used to conduct comprehensive resilience evaluation of infrastructure, and driving factor analysis is carried out using Geo Detector.

The multi-scale resilience of infrastructure exhibits spatial imbalances. In the Zhengzhou city area, the comprehensive infrastructure resilience is shown to have a spatial pattern of high resilience in the city center and low resilience in the city's periphery overall, with higher resilience in Zhongyuan, Shangjie, NHTDZ, and ETDZs and lower resilience in Xingyang, Dengfeng, and Gongyi.

In the Zhengzhou urban area, the overall comprehensive infrastructure resilience result shows that the central part of the city is high, the periphery is low, the southern and northwestern parts of the AEZ are relatively high, and the central part is low. The south and north of the AEZ, the northeastern part and the northwestern corner of the Zhongyuan District, the Erqi District, the northern part of the Guancheng District, the southern part of the Jinshui District, the northwestern part of Zhengdong New District, and the northwestern part of the ETDZ are more resilient. The central part of Erqi District, the southern part of Guancheng District, the northern part of Huiji District, the central part of Jinshui District, the periphery of the main city, and the central part of the AEZ are less resilient.

The results of the spatial autocorrelation analysis show that the comprehensive infrastructure resilience in urban areas is highly concentrated in spatial distribution. According to the Geo Detector results, risk factors like floods and droughts, dangerous chemicals, and infectious disease outbreaks, as well as resilience factors like transportation networks, sponge city development, pipeline networks of various municipal facilities, and fire protection, are the primary factors of the formation of spatial differentiation of resilience. Therefore, enhancing urban infrastructure resilience requires spatial planning tools, including optimized land use, transportation networks, and municipal facility layouts. Concurrently, risks should be reduced through improved risk management and sponge city initiatives, fostering resource efficiency and sustainable urban development.

Huiji District is examined in depth as an example based on the results of the comprehensive evaluation in order to determine the reasons behind the districts' resilience evaluation results in terms of the evaluation indexes, and the corresponding solutions for enhancing resilience and the following suggestions are proposed:

(1) Enhancing infrastructure development in peripheral areas to mitigate regional disparities. The resilience evaluation results of the city and urban areas as a whole show the characteristics of a high center and low periphery, and it is necessary to strengthen the construction of infrastructure in areas with weak resilience, especially in terms of the transportation network, sponge city construction, and municipal facility pipeline network, and to promote the overall coordination of the city and county areas;
(2) Refining disaster risk management and emergency response systems while implementing more stringent environmental conservation measures. Priority should be given to managing critical risk factors, including hydrological extremes, hazardous materials, and infectious disease outbreaks. This requires establishing comprehensive monitoring networks, early warning systems, and contingency plans. Long-term environmental strategies should be emphasized to minimize risk probability and impact, thereby supporting sustainable urban growth;
(3) Facilitating information exchange and cross-sectoral coordination. Using digital technologies such as big data analytics and artificial intelligence can enhance infrastructure resilience monitoring and information sharing, breaking down existing data silos. Strengthening interdepartmental collaboration among transportation, telecommunications, public health, and cooperative enterprises is crucial for improving systemic resilience.

The results reveal the spatial distribution characteristics of infrastructure resilience in Zhengzhou, proposing multi-hazard resilience enhancement strategies. The findings provide a spatial framework and targeted areas for subsequent resilience planning, including risk mitigation zones and resilience improvement districts, while establishing differentiated planning guidelines for local implementation. Furthermore, the research highlights the influence of environmental risks and blue–green infrastructure on system resilience, offering empirical support for stricter environmental regulations and sustainable urban development policies.

Author Contributions: Conceptualization, W.D.; Methodology, W.D. and J.W.; Software, Y.Z. and Z.C.; Validation, Y.Z.; Formal analysis, W.D.; Investigation, W.D., Y.Z., D.G., Z.C. and J.W.; Resources, D.G.; Data curation, Y.Z. and Z.C.; Writing – original draft, W.D. and Y.Z.; Writing – review & editing, W.D. and Y.Z.; Visualization, Y.Z. and Z.C.; Supervision, W.D., D.G. and J.W.; Project administration, W.D., D.G. and J.W.; Funding acquisition, W.D. All authors have read and agreed to the published version of the manuscript.

Funding: This research was funded by the National Social Science Fund of China grant number 21BSH138 and the Natural Science Foundation of Zhejiang Province grant number Y24E080048. The APC was funded by the Natural Science Foundation of Zhejiang Province grant number Y24E080048.

Data Availability Statement: The raw data supporting the conclusions of this article will be made available by the authors on request.

Conflicts of Interest: Author Dongliang Guo was employed by the company Zhengzhou Urban Planning Design and Survey Research Institute Co. The remaining authors declare that the research was conducted in the absence of any commercial or financial relationships that could be construed as a potential conflict of interest.

References

1. Ahern, J. From fail-safe to safe-to-fail: Sustainability and resilience in the new urban world. *Landsc. Urban Plan.* **2011**, *100*, 341–343. [CrossRef]
2. Yang, B.; He, Y.; Zhao, D. Comprehensive risk assessment of multi-hazard cities: A case study of Beijing. In Proceedings of the 2018 China Urban Planning Annual Conference (Vol. 01 Urban Safety and Disaster Planning)—Shared and Quality, Hangzhou, China, 24–26 November 2018; Volume 12, pp. 139–150.
3. Shao, Y.W.; Xu, J. Understanding Urban Resilience: A Conceptual Analysis Based on Integrated International Literature Review. *Urban Plan. Int.* **2015**, *30*, 48–54.
4. Chen, L.; Zhu, X.G.; Sun, J. The Basic Concept, Mechanism and Planning Ideas of Resilient Cities. *Mod. Urban Res.* **2017**, *9*, 18–24.
5. Ribeiro, P.J.G.; Pena Jardim Gonçalves, L.A. Urban resilience: A conceptual framework. *Sustain. Cities Soc.* **2019**, *50*, 101625. [CrossRef]
6. Chelleri, L.; Baravikova, A. Understandings of urban resilience meanings and principles across Europe. *Cities* **2021**, *108*, 102985. [CrossRef]
7. Feofilovs, M.; Romagnoli, F. Dynamic assessment of urban resilience to natural hazards. *Int. J. Disaster Risk Reduct.* **2021**, *62*, 102328. [CrossRef]
8. Datola, G. Implementing urban resilience in urban planning: A comprehensive framework for urban resilience evaluation. *Sustain. Cities Soc.* **2023**, *98*, 104821. [CrossRef]
9. Lin, C.; Chen, R.; Wang, B.; Li, X. Evaluating resilience and enhancing strategies for old urban communities amidst epidemic challenges. *Habitat Int.* **2024**, *153*, 103187. [CrossRef]
10. Hananel, R.; Fishman, R.; Malovicki-Yaffe, N. Urban diversity and epidemic resilience: The case of the COVID-19. *Cities* **2022**, *122*, 103526. [CrossRef]
11. De Fino, M.; Tavolare, R.; Bernardini, G.; Quagliarini, E.; Fatiguso, F. Boosting urban community resilience to multi-hazard scenarios in open spaces: A virtual reality—Serious game training prototype for heat wave protection and earthquake response. *Sustain. Cities Soc.* **2023**, *99*, 104847. [CrossRef]
12. Yan, W.T.; Ren, J.; Zhang, S.W.; He, L. Resilient Urban Planning in Shanghai: Key Issues, General Framework, and Planning Strategies. *Urban Plan. Forum* **2022**, *3*, 19–28.
13. Chenhong, X.; Guofang, Z. The spatiotemporal evolution pattern of urban resilience in the Yangtze River Delta urban agglomeration based on TOPSIS-PSO-ELM. *Sustain. Cities Soc.* **2022**, *87*, 104223. [CrossRef]
14. Ge, Y.; Jia, W.; Zhao, H.; Xiang, P. A framework for urban resilience measurement and enhancement strategies: A case study in Qingdao, China. *J. Environ. Manag.* **2024**, *367*, 122047. [CrossRef] [PubMed]
15. Zhang, Y.; Liu, Q.; Li, X.; Zhang, X.; Qiu, Z. Spatial-temporal evolution characteristics and critical factors identification of urban resilience under public health emergencies. *Sustain. Cities Soc.* **2024**, *102*, 105221. [CrossRef]
16. Najafi, M.R.; Zhang, Y.; Bi, W.; Schooling, J.; Macaskill, K. Assessing flood resilience of urban rail transit systems: Complex network modelling and stress testing in a case study of London. *Transp. Res. Part D Transp. Environ.* **2024**, *134*, 104263. [CrossRef]
17. Martyn, N. A flood risk assessment framework for interdependent infrastructure systems in coastal environments. *Sustain. Cities Soc.* **2021**, *64*, 102516. [CrossRef]

18. Kyprianou, I.; Serghides, D.; Carlucci, S. Urban vulnerability in the EMME region and sustainable development goals: A new conceptual framework. *Sustain. Cities Soc.* **2022**, *80*, 103763. [CrossRef]
19. Zhang, M.D.; Feng, X.Q. Comprehensive evaluation of urban resilience in China. *Urban Probl.* **2018**, *10*, 27–36. [CrossRef]
20. Zhang, M.D.; Li, W.L. Spatial Difference and Convergence of Urban Resilience Level in Northeast China. *J. Ind. Technol. Econ.* **2020**, *39*, 3–12.
21. She, S.; Duan, F. Evaluation of infrastructure resilience in the Yangtze River Delta urban agglomeration based on the TOPSIS entropy weight method. *China Real Estate* **2020**, *27*, 31–35. [CrossRef]
22. Arvin, M.; Beiki, P.; Hejazi, S.J.; Sharifi, A.; Atashafrooz, N. Assessment of infrastructure resilience in multi-hazard regions: A case study of Khuzestan Province. *Int. J. Disaster Risk Reduct.* **2023**, *88*, 103601. [CrossRef]
23. Cutter, S.L.; Ash, K.D.; Emrich, C.T. The geographies of community disaster resilience. *Glob. Environ. Change* **2014**, *29*, 65–77. [CrossRef]
24. Sharifi, A.; Yamagata, Y. Resilient Urban Planning: Major Principles and Criteria. *Energy Procedia* **2014**, *61*, 1491–1495. [CrossRef]
25. Zhu, M.; Zeng, X.; Ying, P.; Bao, L. Network-based resilience assessment of an urban rail transit infrastructure with a multi-dimensional performance metric. *Phys. A Stat. Mech. Its Appl.* **2024**, *656*, 130211. [CrossRef]
26. Hosseini, S.; Barker, K.; Ramirez-Marquez, J.E. A review of definitions and measures of system resilience. *Reliab. Eng. Syst. Saf.* **2016**, *145*, 47–61. [CrossRef]
27. Bi, W.; Macaskill, K.; Schooling, J. Old wine in new bottles? Understanding infrastructure resilience: Foundations, assessment, and limitations. *Transp. Res. Part D Transp. Environ.* **2023**, *120*, 103793. [CrossRef]
28. Yan, W.T.; Li, Z.H.; Fu, L. Assessment of Urban Resilience Based on the Perspective of "Form-Network Function": A Case Study of the Central City of Shiyan. *Shanghai Urban Plan. Rev.* **2023**, *1*, 1–8.
29. Wen, G.; Ji, F. Flood resilience assessment of region based on TOPSIS-BOA-RF integrated model. *Ecol. Indic.* **2024**, *169*, 112901. [CrossRef]
30. Colletta, V.R.; Pagano, A.; Zimmermann, N.; Davies, M.; Butler, A.; Fratino, U.; Giordano, R.; Pluchinotta, I. Socio-hydrological modelling using participatory System Dynamics modelling for enhancing urban flood resilience through Blue-Green Infrastructure. *J. Hydrol.* **2024**, *636*, 131248. [CrossRef]
31. Yu, F.C.; Yan, Y.C.; Li, L.Y.; Huang, J.L.; Hu, C.; Xu, Y.J.; Tong, X.W. Assessing level of resilience to geological hazards in Beijing. *J. Beijing Norm. Univ. Nat. Sci.* **2023**, *59*, 306–312.
32. Yin, Z.G.; Niu, F.Y.; Tang, H. Resilience evaluation of urban flood disasters based on game theory and regret theory: A case study of Xi' an City. *Water Resour. Hydropower Eng.* **2024**, *55*, 74–87. [CrossRef]
33. Vamvakeridou-Lyroudia, L.S.; Chen, A.S.; Khoury, M.; Gibson, M.; Kostaridis, A.; Stewart, D.; Wood, M.; Djordjevic, S.; Savic, D. Assessing and visualising hazard impacts to enhance the resilience of Critical Infrastructures to urban flooding. *Sci. Total Environ.* **2020**, *707*, 136078. [CrossRef] [PubMed]
34. Bixler, R.P.; Yang, E.; Richter, S.M.; Coudert, M. Boundary crossing for urban community resilience: A social vulnerability and multi-hazard approach in Austin, Texas, USA. *Int. J. Disaster Risk Reduct.* **2021**, *66*, 102613. [CrossRef]
35. Wu, H.T.; Zhai, G.F. Resilient City Planning Theory and Method and Its Practice in China: A Case Study of the Improvement Planning of Hefei Infrastructure's Resilience. *Shanghai Urban Plan. Rev.* **2016**, *1*, 19–25.
36. Tian, R.; Zhang, Y.; Peng, L.; Wang, Y.; Wang, W.; Gu, Y. Measurement of flood resilience of metro station based on combination weighting-cloud model. *Int. J. Disaster Risk Reduct.* **2024**, *114*, 104950. [CrossRef]
37. Xiong, F.; Li, P.H.; Yuan, Y.M.; Wu, F. Urban flood disaster risk assessment based on GIS and combined weighting method. *Yangtze River* **2023**, *54*, 60–66. [CrossRef]
38. Anthony Cox, L., Jr. What's Wrong with Risk Matrices? *Risk Anal.* **2008**, *28*, 497–512. [CrossRef]
39. Li, Y.L.; Li, M.L. New urbanization and ecological resilience in the Yangtze River Economic Belt: Coupling coordination, spatial-temporal pattern, and dynamic impact. *Resour. Environ. Yangtze Basin* **2024**, *33*, 2329–2341.
40. Yin, M.S.; Xu, J.G. Characterization and Grouping Analysis of Economic Resilience of County-Level Cities in Jiangsu Province from the Perspective of Complex Adaptive Systems. *Trop. Geogr.* **2024**, *44*, 2001–2014. [CrossRef]
41. Wang, J.F.; Xu, C.D. Geodetector: Principle and prospective. *Acta Geogr. Sin.* **2017**, *72*, 116–134.
42. *GB/T 40947-2021*; Guide for Safety Resilient City Evaluation. State Administration for Market Regulation; National Standardization Administration: Beijing, China, 2021.

land

MDPI

Article

Barriers Affecting Promotion of Active Transportation: A Study on Pedestrian and Bicycle Network Connectivity in Melbourne's West

Isaac Oyeyemi Olayode [1,*], Hing-Wah Chau [2] and Elmira Jamei [2]

[1] School of Science, Computing and Engineering Technologies, Swinburne University of Technology, Hawthorn, VIC 3122, Australia

[2] Institute for Sustainable Industries & Liveable Cities, Victoria University, Footscray, VIC 3011, Australia; hing-wah.chau@vu.edu.au (H.-W.C.)

* Correspondence: olayode89@gmail.com

Abstract: In the last few decades, the promotion of active transport has been a viable solution recommended by transportation researchers, urban planners, and policymakers to reduce traffic congestion and improve public health in cities. To encourage active transport, it is important for cities to provide safe and accessible infrastructure for pedestrians and cyclists, as well as incentives for individuals to choose active modes of transportation over private vehicles. In this research, we focused on the suburb of Point Cook, located within the City of Wyndham in Melbourne's west, owing to its rising human population and private vehicle ownership. The primary aim of this research is to examine the barriers in the interconnectivity of active transport networks for pedestrians and cyclists and to determine the segments of the transportation network that are not accessible to Point Cook residents. Our methodology is enshrined in the use of Social Pinpoint, which is an online interactive survey platform, and ground surveys (face-to-face interviews). In our assessment of the suburb of Point Cook, we utilised the concept of 20-min neighbourhoods to evaluate the accessibility of many important places within an 800-metre walking distance from residents' homes. Based on our online interactive survey findings, approximately one-third of the individuals engaged in regular walking, with a frequency ranging from once a day to once every two days. One-third of the participants engaged in walking trips once or twice a week, whereas the remaining two-thirds conducted walking trips less frequently than once a week. Almost 89% of the participants expressed varying levels of interest in increasing their walking frequency. The findings showed that improving pedestrian and cycling networks that are easily accessible, well-integrated, inclusive, and safe is a prerequisite for achieving active transport and create neighbourhoods in which everything is accessible within a 20-min walking distance.

Keywords: active transport; cycling; walking; 20-min neighbourhood; policy; transportation planning; pedestrian

1. Introduction

Globally, people are encountering growing health issues, such as higher levels of physical inactivity, chronic diseases, road traffic injuries, poor mental health, and obesity [1–4]. There is a consensus that relying heavily on personal vehicles for transportation has significant and far-reaching adverse effects on health, the environment, society, and the economy. Exposures related to traffic, such as air pollution, noise, and increases in

https://doi.org/10.3390/land14010047

local temperature, are linked to negative health outcomes, including overall mortality, cardiovascular illnesses, cardiorespiratory morbidity, cancer, type 2 diabetes, dementia, anxiety, depression, and childhood behavioural difficulties [5]. This adds to the significance of both fatal and nonfatal injuries caused by vehicular accidents [4].

The Australian Bureau of Statistics reported that private vehicles are the primary mode of transportation for most commuting trips in Australia. In addition, 45% of Australian adults and less than 25% of children adhere to the prescribed physical activity standards, as stated by the Australian Institute of Health in 2021 [6]. According to the Australian Bicycle Council in 2017, walking and cycling collectively received only approximately 2% of transport budgets in Australia [7]. Considering that more than 85% of Australians reside in metropolitan regions and over 70% in major cities, there is significant potential for active transport to offer convenient and inexpensive physical activity as a regular part of daily life, leading to an improvement in the overall health of the human population and environment.

Active transport is commonly described as a practical form of physical exercise that entails walking and cycling for functional reasons [8]. Cirianni et al. highlighted the need to assess how active transport goals and planned actions are incorporated into Sustainable Urban Mobility Plans (SUMPs) within the European context [9]. Public policy places significant importance on urban mobility as a key priority, ensuring that individuals have access to essential facilities and services offered by the city [10–12].

Australia's active transport planning and policy have encountered various obstacles, such as the dominance of car-oriented infrastructure, limited accessibility to public transit, and the significance of tackling health concerns linked to physical inactivity [13]. Nevertheless, recent studies have highlighted possible remedies for these difficulties, such as employing comprehensive policy strategies and utilising big data analytics to guide decision making [14]. An important issue in Australia is the limited utilisation of active transport, as fewer than 25% of the population walk and only slightly over 2% use bicycles [15]. These findings have significant implications for the development of sustainable thriving communities. To tackle this issue, policymakers must consider several elements, such as the physical infrastructure, demographic characteristics, affordability and accessibility of public transportation [14].

In rural Australian areas, these difficulties are much more noticeable because there are vast distances to travel, high supply chain expenses, and restricted availability of services and economic opportunities, all of which present considerable obstacles to mobility and overall well-being [16]. Digital and sharing economies provide promising technical advancements that could address these difficulties. However, their suitability for regional Australia is a relatively unexplored topic in research [16]. To address these difficulties, it is necessary to adopt a comprehensive and unified strategy for the design and implementation of active transportation considering all relevant factors.

One strategy for the smooth shift from private vehicles to active forms of transportation is to improve the connectivity of pedestrian and bicycle networks between the key destinations of communities. Owing to the extensive effects of active walking on population and environmental health, governments are focusing on implementing policies that promote walking and cycling through more connected networks [17–20].

1.1. 20-Minute Neighbourhoods, a Pre-Requisite for Active Transport

In recent decades, non-motorised transport, often known as active transport, has gained the interest of transport researchers owing to its advantages, such as improving health conditions among people with varying levels of capabilities and reducing traffic congestion [21–25]. Some European countries, such as Denmark, the Netherlands, and Germany, have been exemplary of adopting active transport. These countries have achieved exceptionally high levels of active transport uptake despite their high-income status and high car ownership rates [26].

A number of studies have extensively examined methods to further promote active modes of transport in cities, with a particular focus on improving walking and cycling infrastructure. These infrastructures mainly include dedicated lanes, trails, and bicycle parking facilities. The implications of such infrastructure on active transport have also been explored in past studies [27–29]. Similar research examined the effects of promoting active modes of transport (walking or cycling) and integrating them with public transport on sustainability and the level of replacement by vehicles [30–34]. Some research has also examined the role that bicycle sharing, renting programs, and urban design can play in improving active transport [35]. The subject of cycling safety and its impact on promoting active transport has been extensively studied in transport research for many years [36]. Other studies have highlighted the important role that urban design and, in particular, accessible, connected, and walkable neighbourhoods can play in promoting active transport; one of these concepts has evolved around the 20-min neighbourhood [37–39].

A 20-min neighbourhood is a crucial framework for promoting active transportation—walking and cycling—in cities. This concept focuses on designing urban spaces where essential services, such as shops, schools, parks, and public transit, are accessible within a 20 min walk or bike ride. This compact layout minimises the need for long car trips, making walking and cycling the most practical and efficient option for travelling short distances. Having a connected pedestrian and cycling network with direct routes and safe pathways is one of the key pre-requisites for creating neighbourhoods where people can walk and cycle.

1.2. Rationale and Study Aims

In Melbourne, the concept of "20-min neighbourhoods" for living locally was first indicated in "*Plan Melbourne: Metropolitan Planning Strategy* in 2014", and then implemented in "*Plan Melbourne 2017–2050*" to create safe, accessible, and well-connected neighbourhoods for pedestrians and cyclists and to optimise active transport [40]. The aim of the "20-min neighbourhood" is to provide safe and convenient access to essential goods and services within 20 min of where people live, accessible by foot, bicycle, or public transport [41]. The concept is based on the 800-metre walkable neighbourhood, in which 800 metres were adopted as a guide for measuring the spatial accessibility of a walkable neighbourhood. Figure 1 shows the key principles of a "20-min Neighbourhood" that guide urban designers in designing infrastructure needed for pedestrians and bikes. It also highlights that walkability and safe, connected networks enable the creation of 20-min neighbourhoods, combining human-scale design with accessibility-oriented planning [42].

Figure 1. Principles of the 20-min neighbourhood [43].

Moreno et al. emphasised the necessity of incorporating six essential land uses within a walkable neighbourhood (e.g., residential, occupational, commercial, healthcare, educational, and recreational) to reduce reliance on cars and promote active transport [44]. O'Gorman and Dillon-Robinson highlighted the important role of accessibility to public transport and appropriate infrastructure for walking and cycling in creating 20-min neighbourhoods [45]. Other researchers also indicate that proximity to green spaces within pedestrian and cycling networks can further promote active transport in 20-min neighbourhoods [39]. Grodach et al. indicated that a catchment size of up to 800 metres is the average distance that can be walked in 10 min, so in designing destinations within land uses, this has to be carefully considered [46].

Melbourne is the capital and most populous city of the Australian state of Victoria and the second-most populous city in Australia, after Sydney. Melbourne's western region is one of the fastest growing in Australia. The area covered by the Melbourne West Growth Corridor Plan is projected to accommodate a population of 377,000 or more and has the capacity to accommodate at least 164,000 jobs [43]. The future development of Melbourne's west will be substantially shaped by transport, access to public transport, and the quality of pedestrian and cycling networks. Point Cook is one of the suburbs situated in Melbourne's west Region.

Point Cook residents exhibit lower levels of physical activity and experience greater rates of obesity than the average population of Victoria. Referring to the Victorian Population Health Survey in 2017, 57.9% of respondents were overweight, compared with 50.8% of all Victorian respondents, and 54.4% of respondents had insufficient physical activity, compared with 44.1% of all Victorian respondents [47]. Convenient and accessible active transport infrastructure encourages residents to walk and cycle.

The Wyndham City Council aims to increase the proportion of walking and cycling trips by three times by 2040; therefore, it is crucial to construct a comprehensive and easily accessible network for pedestrians and cyclists while also improving the existing active transport infrastructure [43]. Point Cook has a vast network dedicated to walking and cycling. Nevertheless, transportation networks exhibit multiple deficiencies, with infrastructure falling short of contemporary benchmarks, necessitating the extension and improvement of existing networks. One possible contributor to the low level of walking and cycling within Point Cook is the condition of the pedestrian and bicycle networks, particularly the networks that connect residents to key destinations (within their 800-metre walk/20-min neighbourhood and public transport train stations. Therefore, this study has the following aims:

- To identify the existing barriers affecting pedestrian and cycling network connectivity that residents may utilise to access their key destinations within their 800-metre catchment zone (20-min neighbourhood concept);
- To develop urban design strategies to promote active transport in the suburb of Point Cook;
- To develop a place-based design guideline master plan using the suburb of Point Cook, located in Melbourne's west, as a case study.

The remainder of this paper is organised into four sections. The first section deals with the introduction, problem statement, theoretical background of the study, and concept of the 20-min neighbourhood. The second section is concerned with the methodology used in this study. The third section presents the findings and discussion of this research, focusing on the significant strategies used in collecting feedback from the residents of Point Cook. The section also comprises policy recommendations based on the results obtained and how these policy recommendations can be implemented and applied by transportation policymakers and urban planners to promote active transport. The fourth section of this study draws upon the entire research, tying together various theories, methodologies, and results, and their contribution to the fields of transportation, urban planning, and active transport.

2. Methodology

2.1. Study Location

Point Cook is a suburb in Melbourne, Victoria, Australia, 22 km (14 mi) south-west of Melbourne's Central Business District, located within the City of Wyndham local government area. Point Cook recorded a population of 66,781 at the 2021 census [2]. Figure 2 shows the location of Point Cook within the Wyndham municipality.

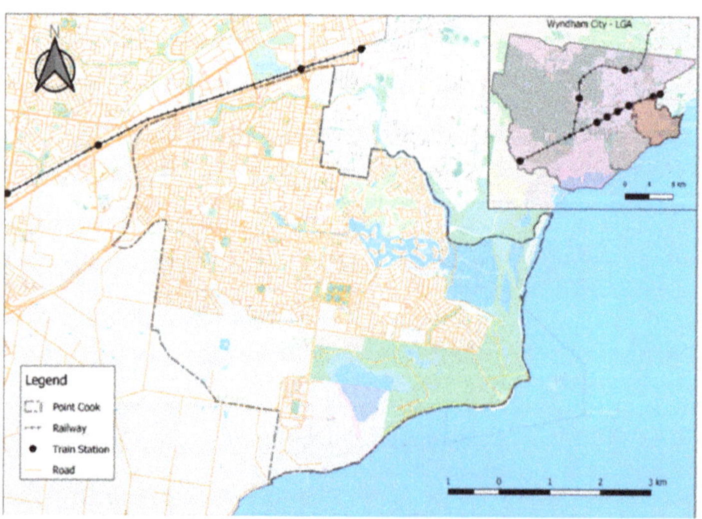

Figure 2. Boundary of Point Cook within Wyndham municipality. Reprinted with permission from ref [48]. Published by MDPI.

2.2. Data Collection, Sampling and Analysis

Social Pinpoint was used to analyse active transport patterns and collect feedback on active transport in the suburb of Point Cook. Social Pinpoint is a digital tool aimed at enhancing community participation, specifically for gathering input, opinions, and suggestions from residents concerning diverse projects, including urban planning and active transportation. Social Pinpoint enables planners and stakeholders to collect public insights, analyse answers, and make data-driven decisions through various interactive tools such as maps, questionnaires, and discussion forums. This is particularly efficacious for extensive urban initiatives, where resident input is essential for comprehending community requirements and inclinations.

The primary objective of Social Pinpoint is to enable anyone to establish a community involvement website that prioritises accessibility. Showcasing a user-friendly drag-and-drop interface, Social Pinpoint facilitates the integration of visually appealing designs that encourage engagement with the people of Point Cook. Social Pinpoint was selected for this research for the following reasons:

- It facilitates precise and pertinent responses during surveys with Point Cook locals;
- It enhances trust and transparency;
- It assesses the community's perception of active transport, stimulates interest, and detects insufficient comprehension of active transport among Point Cook residents.

Table 1 shows the step-by-step process of how Social Pinpoint was used in this study.

Table 1. Procedures for gathering feedback using social pinpoint.

Steps	Application
Establish Objectives	Determination of the exact feedback questions required for residents of Point Cook. The research aims to comprehend Point Cook residents' preferences, challenges, and prospective improvements for walking, cycling, and other non-motorised transportation modes in Point Cook.
Initiate a Project on Social Pinpoint	Registration and establishment of a project page on Social Pinpoint. Incorporating specifics regarding the project's objective, schedule, and how the feedback will influence planning and decision-making processes. Informing Point Cook locals about the intention to improve active transport infrastructure.
Create Interactive Content	Using the tools available in Social Pinpoint interface to generate the following: • Interactive Maps: Chart the Point Cook region and encourage residents to identify sites where they encounter obstacles or perceive potential for active transportation. • Surveys and Polls: Incorporate questions designed to obtain insights on Point Cook residents' walking and cycling behaviours, safety apprehensions, and other pertinent matters.

Table 1. *Cont.*

Steps	Application
Promote the Engagement Platform	Notifying Point Cook residents of the study through several sources, including social media, local community boards, and municipal newsletters, to enhance visibility and engagement. Urge residents to use the Social Pinpoint tool for providing their feedback.
Gather and Evaluate Feedback	Social Pinpoint's dashboard facilitates the visualisation of trends and patterns as residents submit their responses. It presents frequently highlighted regions on the map, prevalent subjects in discussions, and the statistical distribution of survey responses.
Analyse Results and Execute Modifications	Summarising the insights and principal findings in a feedback report for stakeholders engaged in the active transport initiative. This report emphasises Point Cook residents' primary issues, desired active transportation improvements, and existing obstacles they encounter when using different modes of active transportation.

The survey questionnaire for Point Cook residents was designed based on the following key aspects:

- The modes of transportation used by residents in Point Cook;
- The key destinations Point Cook residents travel to;
- Whether Point Cook residents walk or cycle to their destinations;
- The locations in the Point Cook suburb that need improvement in terms of walking and cycling infrastructure.

The feedback information collected through the online interactive survey were classified into three sections:

1. Demographic information about the respondents;
2. Respondents' walking pattern;
3. Respondents' cycling attitude.

Supplementary open-ended questions were incorporated at the end of each section to encourage participants to provide further remarks or recommendations that might not have been addressed in the official survey questionnaire. Observations included the following:

- areas of interest;
- frequently visited locations;
- frequently cycled locations;
- areas of concern; and
- supplementary remarks.

The survey questions were cross-checked with the Steering Group (including key stakeholders in the sector) and received ethical approval from the Victoria University Human Research Ethics Committee under reference number HRE22-108. Information from Point Cook residents was collected using a blend of online and offline modes. The sampling method used in this study was to select a wide variety of respondents of the study area by utilising a website called The Loop (https://theloop.wyndham.vic.gov.au/about, accessed on 2 September to 2 October 2022). The Loop is a digital networking platform for the suburb of Point Cook that facilitates online community engagement. This platform functions as a digital medium for the community to obtain information, engage in discussions, and exert influence on significant initiatives that are relevant to the community.

Prior to the research, postcards, including vital information about the study, such as the website link and the QR code for the online survey, were dispatched by mail to the residents of Point Cook to invite them to participate. The website link and the QR code were also emailed to the staff and students of the project institution (Victoria University) to request eligible persons and their contacts residing in the Point Cook area to participate in the survey. A mobile iteration of Qualtrics XM, accessed offline on mobile digital devices such as smartphones and tablets, was used to collect survey data at public locations such as shopping centres, train stations, bus stops, and the local library. Data collection was carried out on weekdays and weekends at predetermined intervals for a duration of over one month, specifically from 2 September to 2 October 2022. By employing these data collection techniques, a diverse range of people, including residents, workers, students, and shoppers, were recruited to answer the survey questionnaire.

Participation in this study was limited to persons aged 18 and above to uphold ethical standards. To guarantee the collection of authentic responses, a screening question was incorporated at the beginning of the questionnaire. The first question inside the questionnaire was, "Kindly indicate your age". If the respondent chose the option "below 18 years", the survey was promptly terminated. Each survey had an estimated completion time of 10–15 min. The mean duration over which participants completed the survey was 2 min and 14 s. Throughout the survey period, 352 visitors responded to the survey questionnaire, resulting in 113 distinct users of the Social Pinpoint online survey. Of these, 62 comments and recommendations were received, and 100 attempts were made to complete the survey. Among these, 62 comments and suggestions were deemed valid, and 37 surveys were completed. The responses were gathered for data analysis.

Table 2 presents the demographic profiles of the survey respondents. More than half of the respondents were female (54.1%). The majority of the respondents lived in Point Cook (81.1%) and were middle-aged: 35–44 (40.5%) and 45–54 (29.7%). Over 44% of the respondents drove their cars as their usual mode of transport. Comparatively, only 15.4% and 12.3% of respondents walked and cycled respectively as their usual modes of transport.

Table 2. Demographic profile of the survey respondents.

Variable	Subgroup	Numbers (%)
Gender	Male	17 (45.9%)
	Female	20 (54.1%)
Age (years)	18–24	1 (2.7%)
	25–34	5 (13.5%)
	35–44	15 (40.5%)
	45–54	11 (29.7%)
	55–64	2 (5.4%)
	65+	2 (5.4%)
	Not disclosed	1 (2.7%)
Home location	Point Cook	30 (81.1%)
	Another area within Wyndham	3 (8.1%)
	Outside Wyndham	4 (10.8%)
Highest level of education	TAFE/college	11 (29.7%)
	Bachelor's degree	11 (29.7%)
	Master's degree or higher	15 (40.5%)
Usual mode(s) of transport	By car (as a driver)	29 (44.6%)
	By car (as a passenger)	8 (12.3%)
	By bus	4 (6.2%)
	By train	6 (9.2%)
	By bicycle	8 (12.3%)
	On foot	10 (15.4%)

Using the concept of 20-min neighbourhoods, the key destinations that residents walk to within their 800-metre radius (catchment areas) were identified. Figure 3 shows the key locations within Point Cook, which residents identified as major destinations.

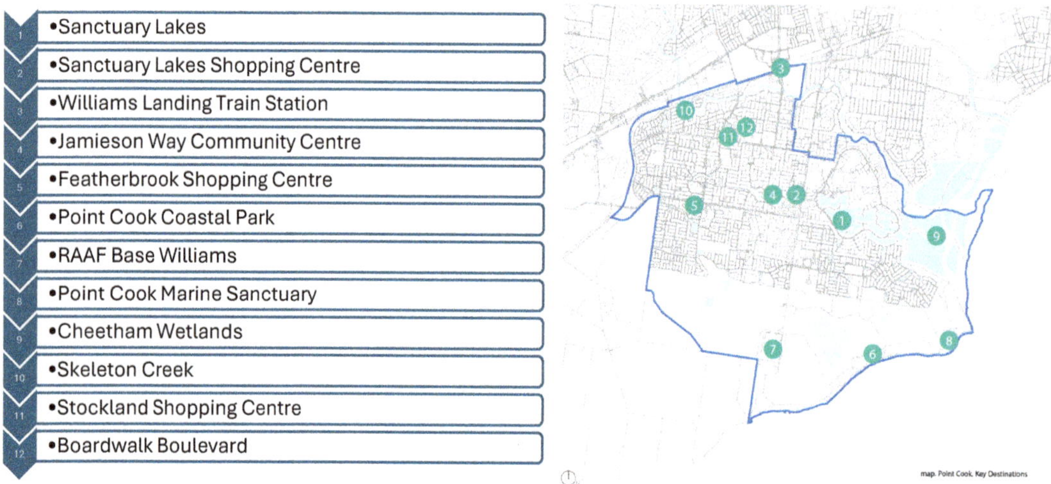

Figure 3. Major landmarks identified as key destinations within Point Cook based on survey outcomes from the respondents.

The survey also required respondents to identify areas where walking and cycling are challenging, and to provide feedback on the existing conditions of walking and cycling infrastructure.

The study employed a quantitative method for data analysis using Statistical Package for the Social Sciences (SPSS). The information on the demographics of residents and their opportunities to walk and cycle to the key destinations in Point Cook were analysed using SPSS version 29. Using the Social Pinpoint online survey interactive platform, active transport patterns for Point Cook residents were identified, and feedback on active transport around Point Cook was collected. The results of the survey were classified based on four themes listed in Table 3.

Table 3. Classification of feedback from survey respondents.

Themes	Explanation
Preferred Mode of Travel in Point Cook	This refers to what types of modes of transportation Point Cook residents prefer in terms of comfortability, financial perceptions, and safety.
Destination-based findings	This refers to how specific destinations impact residents' choices to walk or cycle. The study analyses accessibility, location, and urban design to see how these factors influence non-motorised travel. Results show that well-designed, accessible destinations can encourage residents to choose active transportation over car use. This emphasises the role of urban planning in shaping transportation behaviours, advocating for walkable and bike-friendly environments. Such planning can promote healthier and more sustainable lifestyles, encouraging a community shift towards active, non-motorised transport options.

Table 3. *Cont.*

Themes	Explanation
Walking-based findings	This refers to the factors that shape walking as the favoured method of active transportation in Point Cook. This study also investigated the impacts of environmental, socio-logical, and infrastructural factors on the preference of individuals to walk for transportation instead of utilising motorised means in Point Cook. Walking distances and durations to and from public transit vary greatly depending on exact circumstances and are affected by criteria such as personal characteristics, physical fitness-related environmental conditions, and the nature of the trip [49]. Utilising walking and cycling as means of travel offers significant health advantages through heightened physical activity and reduces the potential hazards of car accidents and exposure to air pollution [50].
Cycling in Point Cook	The findings from the survey showed that cycling at Point Cook is influenced by many interconnected elements including infrastructure, safety, proximity to destinations, and government policies. The suburbs' predominantly car-centric layout poses obstacles to promoting cycling as a feasible means of transport. However, the survey showed that if specific infrastructure improvements, safety enhancements, and cultural changes are implemented, cycling has the potential to become a more prevalent mode of transport for the residents of Point Cook. The findings from the survey mentioned that allocating resources to construct more connected cycling networks, enhancing intersection safety, and incorporating riding with public transit might greatly enhance cycling engagement in Point Cook, thus benefiting public health, the environment, and general quality of life.

3. Results

3.1. Point Cook Residents Preferred Modes for Mobility and Destination-Based Findings

According to the survey, the majority of respondents (57%) travel by vehicle, either as drivers (45%) or passengers (12%). A total of 15% and 12% of the respondents preferred walking or cycling, respectively. Public transport users comprised only 16% of the respondents (9% by train and 6% by bus). This finding reflects the lack of active transport users, primarily due to the inadequate or poorly connected active transport infrastructure in the Point Cook area. This is in line with previous studies [49–51], which showed that a lack of adequate infrastructure to support active transportation can lead to a reduction in its usage and affect the use of public transportation. In summary, these results showed that the concept of a 20-min neighbourhood is not yet feasible in the Point Cook area. This finding is supported by previous research conducted by [52–54], in which it was stated that an increase in private vehicle ownership will reduce the usage of active transportation modes (e.g., cycling and walking).

The majority of residents indicated that there was a lack of accessibility for disabled people in certain areas of Point Cook. Some respondents also expressed that several pathways in Point Cook have inaccessible elevated steps as a significant hindrance to people with disabilities and people who prefer to use any form of micromobility, such as e-bikes or e-scooters. These responses highlight the importance of the ease of use of e-micromobility in the community, which enables a sustainable mode of transportation [55,56].

Inadequate modes of active transportation in a community are a significant roadblock to the integration of people with disabilities into public transportation. People with disabilities face several obstacles when attempting to use transportation, including physical, attitudinal, and infrastructural obstacles. These obstacles decrease the frequency of their trips and restrict their transport modes options, thereby affecting their education. There is a need to prioritise the installation of suitable ramps and pathways to ensure that all residents, regardless of their mobility, can have safe access to different modes of active transportation in the community. This finding is consistent with previous studies by Kapsalis et al. and Rosenberg et al. [57,58], which stated that the use of universal design principles and adherence to standards for ramps and pathways are crucial for guaranteeing safe access to various forms of active transportation for all residents.

The findings of this study also support the Idea that Increasing public awareness of the importance of accessibility and promoting inclusive design can promote inclusive and equitable modes of public and active transport for communities. This finding corroborates the previous studies of Allen and Farber, Curtis and Scheurer, and Lättman et al. [59–61], which conclude that increasing awareness among the public and implementing inclusive design principles in transportation can result in fairer and more comprehensive mobility solutions. The effective development of transport systems that serve all community members requires the use of tools and policies that assess and improve accessibility by considering user experiences and addressing social issues.

Another important finding from the survey highlighted the need to construct more paths that are more than 50 m long to support more people using different modes of active transportation in the Point Cook area, which will reduce overreliance on private vehicles and public transport [62,63]. This finding aligns with a series of previous studies [37,39,45,64,65] for reiterating the importance of the 20-min neighbourhood in establishing sustainable, habitable, and resilient communities. Nevertheless, the effective execution of this concept requires resolving obstacles to policy funding, fairness, design density, and well-defined operational protocols. By addressing these obstacles, cities can improve their performance by encouraging active transportation, improving public health, and cultivating social unity.

3.2. *Walking-Based Findings*

The responses suggested that only approximately one-third of the participants walked frequently (making a walking trip once a day to once every two days). Approximately one-third of the participants made walking trips once or twice a week and approximately one-third made walking trips less than once a week. However, this response contradicts the research conducted by Van Soest et al., Lanchapelle and Pinto [66,67], which stated that walking is a fundamental part of active transportation, providing a multitude of health advantages, promoting the use of public transit, and enhancing environmental sustainability. Majority of survey participants i.e (89%) (majority of the participants) were interested in walking to varying degrees and participated in various types of walking trips as illustrated in Figure 4.

The survey findings show that 27% of residents walking to gardens, parks, and waterfronts; 19% prefer walking to local shops; 18% visit places with family/ friends; 9% walking to major shops; and 8% walking to visit family/ friends. According to Marquet and Miralles-Guasch and Chan et al. [68,69], walking trips significantly improve the viability of a 20-min neighbourhood by improving local accessibility, encouraging active ageing, and bolstering community involvement. The key areas often visited by walking in Point Cook included parks and wetlands. Other key destinations were shops and shopping centres. The findings also showed that 23% of respondents needed wider footpaths that were easily accessible, particularly to accommodate disabled people to walk more often to key destinations. Another important finding was that 10% of these participants believed that priority should be given to pedestrians and e-micromobility users (vulnerable road users) when crossing signalised or unsignalised road intersections. The findings from the survey also showed that the majority of the respondents preferred going to parks and wetlands because of the high level of security along walkways and bike lanes. This finding corroborates the research conducted by Loukaitou-Sideris, Rothman et al, and Bennett et al. [70–72], which stated that neighbourhood safety and security factors significantly influence physical activity, and design and policy interventions can help create environments more conducive to walking.

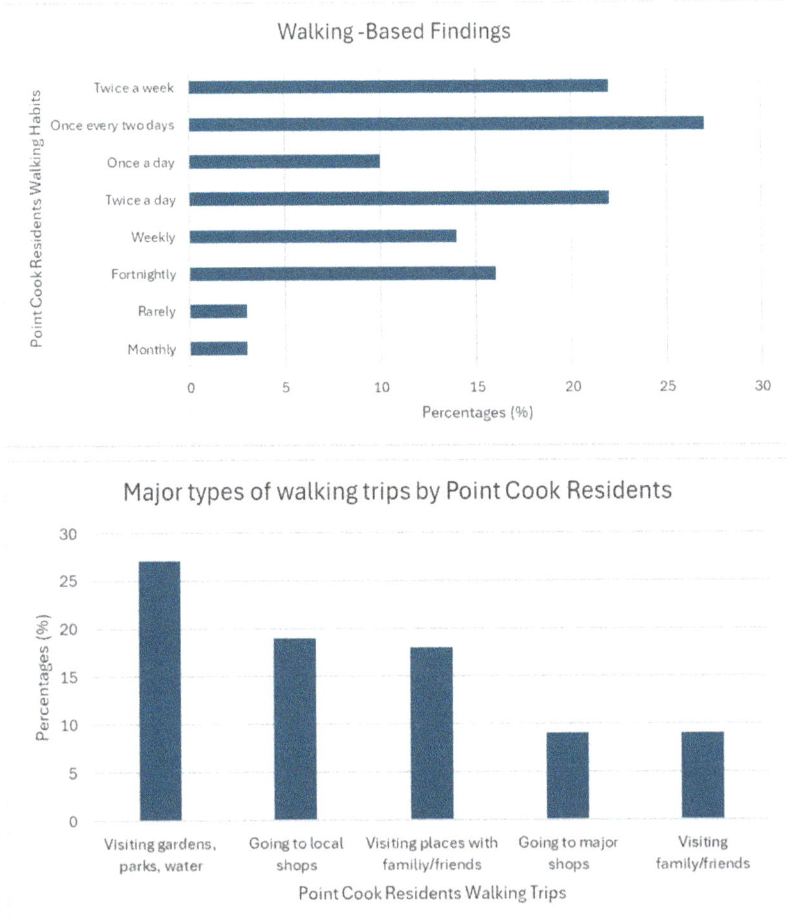

Figure 4. Analysis of how often Point Cook residents take a walking trip.

3.3. Cycling in Point Cook

The findings of the survey showed that, while 70% of respondents had their own bicycles, approximately half of the bicycle owners used their bicycle less than once a month. They were mainly concerned with safety issues along the road as well as the quality of cycling infrastructure. This corroborates the findings of Lusk et al., Chieng et al., and Sanders [73–75], in which the authors suggest that communities will perceive cycling on the road as dangerous if there is a lack of cycling infrastructure. According to the survey findings, over half of the non-bicycle owners (30%) were interested in cycling, while 14% of the total respondents did not show any interest in cycling. The primary reasons people cycled in Point Cook were health-related (18% of respondents), scenery (11% of respondents), frustration with traffic while driving (7%), frustration with public transport (6%), and encouragement to ride by friends or family (5%). This finding supports previous research conducted by De Hartog et al. [76], which stated that elevated physical activity by cycling yields considerable health advantages, including a significant rise in life expectancy, outweighing the hazards linked to heightened exposure to air pollution and traffic accidents.

Another significant survey finding is that more than 32% of respondents preferred not to cycle on the road, and 16% considered that the travel distance and the time required to cycle were both major hinderances to cycling. However, 9% of respondents did not own bicycles, and another 9% lacked the confidence to learn how to cycle (Figure 5). These results are consistent with the studies by Sanders and Gössling et al. [75,77]. According to these studies, riding a bike on the road increases the likelihood of being hurt or killed by other drivers. It also takes time to determine whether one is going to avoid dangerous situations.

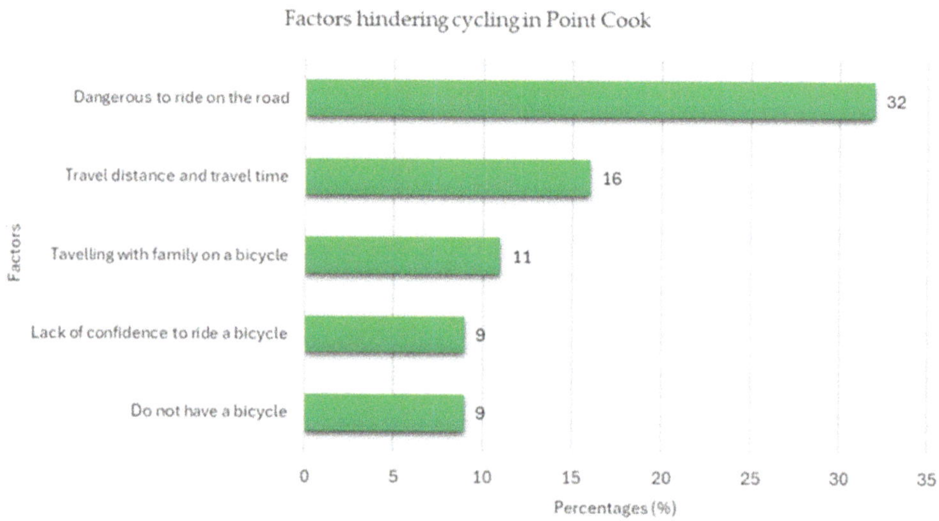

Figure 5. Reasons why Point Cook residents do not ride bicycles.

Almost half of the participants believed that the addition of quality bicycle paths could improve biking uptake and contribute to a more accessible cycling network. These findings agree with the research conducted by Panter et al., Standen et al., Fraser and Lock [29,78,79], which suggested that improving accessibility and connection, together with improving traffic and personal safety, are essential measures to encourage walking and cycling. Other key factors for accessibility were increasing driver awareness of bicycle safety (15%), the availability of bicycle parking and bicycle lockers (15%), increased individual knowledge of road regulations (9%), and access to electric bikes (9%). However, more than 29% of the respondents suggested that it is important to separate bike paths from car traffic. These results corroborate the ideas of Schimek, Nanayakkara et al., and Cicchino et al. [80–82], which state that bike paths that are separate from car traffic usually enable cyclists to feel safer and are preferred by both cyclists and drivers. However, intersections and changes between separate- and mixed-traffic areas must be carefully planned to avoid traffic conflicts.

4. Discussion and Future Recommendations

Informed by the survey results and feedback received from the Point Cook community, four policies and recommendations were formulated for the future of active transportation in Point Cook. These policies and recommendations were developed based on the findings of this study. Each policy/recommendation is followed by detailed actions to articulate infrastructure and accessibility improvements specific to locations within the Point Cook catchment (20-min catchment) and the surrounding environment.

4.1. Connection of Active and Public Transport

Road and cycle pathways surrounding train stations are often heavier with vehicle traffic and more catered towards motor vehicles. More pedestrian-focused streets and cycleways should be provided leading to stations to improve key walking/ cycling corridors and encourage walking and cycling as part of commuting/travel. This aligns with the studies by Welle and Xie et al. [83,84], which suggest that making important bike and walking paths better for trips to stations can improve connectivity with safety and encourage more people walking/ cycling towards station areas.

Opportunities exist for increased signage for wayfinding at train stations and key junctions to inform community members and visitors of journey times, nearby destinations, and possible route options, in combination with the use of public transport. Such information is crucial for rail transportation, especially when train disruptions lead to bus replacements. Researchers, such as Tardieu et al. and Li and Morita, suggest that improving wayfinding in train stations can be accomplished by uniform and explicit signage, efficient audio signals, and thoughtfully crafted maps [85,86].

Improved bicycle storage at key public transport nodes for both short- and long-term usage is useful to promote active transport. Increased cycle hoops and enclosed long-term bicycle storage facilities near public transport nodes encourage people to cycle more often and ensure that their bicycles are safe and secure away from outdoor weather influence. This aligns with the studies by Hochmair and Arbis et al. [87,88], which found that improved bicycle storage at critical public transport hubs can elevate service quality, boost bicycle parking usage, and more effectively address evolving user requirements through predictive modelling, infrastructure development, and optimisation strategies.

The provision of spaces for rest and shelter is useful for community members with limited mobility during various weather conditions. Sustainable and effective urban mobility systems require active transportation, such as walking and cycling, to be integrated with public transportation. However, bad weather prevents seamless communication among various modalities. Extreme heat, rain, or cold prevent people from using active transportation to reach public transit stations. Shelters at key locations enable multimodal travel more comfortable, reliable, and accessible year-round by simplifying the transition from active transportation to public transportation. Shelters near bus stops, train stations, and bike-sharing docks serve various purposes. Commuters can wait in dry, shaded, or even air-conditioned locations to protect them from harsh weather [89]. These facilities are helpful to passengers while they are waiting especially when public transit is overcrowded during peak commute hours. Waiting spaces offer comfort and safety and encourage more people to use active transportation, which can lead to reduced car use and less carbon emissions [90].

Integrating cycle pathways with significant bus lines improves urban mobility by facilitating seamless multimodal commuting alternatives that reduce the reliance on vehicles. Such connectivity enables commuters to cycle a portion of their journey while using buses for extended durations, thereby enhancing their flexibility and accessibility. By connecting bike pathways to transit hubs and bus stations, cities facilitate public transit access for residents in suburban and semi-urban regions, thereby providing cost-effective alternatives to vehicle ownership and fostering inclusion. Improved safety features, such as designated bike lanes and improved illumination, incentivise greater participation in riding, thereby minimisingaccident risks and rendering cycling a more appealing choice [91].

Well-interconnected cycling pathways promote environmental sustainability by decreasing emissions and mitigating urban air pollution due to reduced reliance on vehicles for short journeys. This integration establishes dynamic transit hubs that benefit local communities and improve liveability by promoting cleaner air, mitigating noise, and augmenting pedestrian traffic for businesses. Connecting cycle lanes with essential bus lines is a pragmatic measure for developing resilient, sustainable cities that encourage active travel and advance climate and public health objectives [79].

Incorporating electric bicycle renting facilities atmajor public transport hubs offers a substantial opportunity to improve active transportation systems and augment accessibility. Electric bicycles (e-bikes) are a viable option for those who struggle with conventional cycling over extended distances or steep inclines. Implementing e-bike rental services near transit hubs enables cities to improve their first- and last-mile connectivity, thereby streamlining commuter access to their ultimate destinations [92]. This ease expands the attractiveness of cycling to a broader demographic, including older adults and anyone with restricted physical capabilities, while simultaneously promoting a sustainable, low-emission transportation alternative.

The installation of charging stations near public transport hubs facilitates both private e-bike owners' travels and shared e-bike services, promoting the increased usage of e-bikes for daily travel. The charging infrastructure guarantees that users can maintain their e-bike power, enabling prolonged trips and alleviating range anxiety. This corresponds to urban sustainability objectives, as e-bikes reduce emissions and mitigate congestion by decreasing car dependence for short journeys. Furthermore, the availability of charging stations and e-bike rentals at strategic locations underscores a city's dedication to contemporary, eco-friendly transportation alternatives, likely drawing more environmentally conscious residents and visitors while establishing e-bike infrastructure as a fundamental component of a sustainable, multimodal transit network.

4.2. Creating Active Transport Hubs and Nodes

Establishing active transport hubs and nodes is pivotal for the development of integrated multimodal urban transport systems. These hubs, strategically positioned at critical transit locations, serve as centralised areas where various forms of active transportation (such as cycling, walking) and public transit interconnect. Active transport hubs improve the transition between transport modes by integrating amenities such as bicycle parking, repair stations, pedestrian walkways, and e-bike or scooter rental services, thereby promoting the convenience of sustainable alternatives to personal vehicles. This degree of connectedness is particularly advantageous in closing the first- and last-mile gaps, facilitating a smooth commuting experience for people requiring short-distance travel from transit stops to their destinations. This finding corroborates the ideas of Cai et al. [93], which suggested that establishing active transport hubs and nodes substantially improves journey time, service quality, and the transition from using private vehicles to public transport, while fostering sustainable mobility and optimising transportation network efficiency.

Active transit hubs not only boost functionality but also improve urban liveability and bolster local economies. These hubs can function as dynamic communal places that attract pedestrians, cyclists, and transit users, thereby generating foot circulation, which is advantageous for local businesses nearby. By incorporating facilities such as covered lounging areas, green spaces, coffee shops and retail establishments, these hubs can be transformed into vibrant urban centres that promote social interactions and enhance community participation. Moreover, active transport hubs significantly mitigate urban congestion and pollution by facilitating low-emission commutes and reducing car dependence. Active transport hubs and nodes collectively promote healthier, more sustainable, and interconnected urban settings that benefit both residents and broader urban landscapes. Previous studies by Nieuwenhuijsen and Maltese et al. have demonstrated that active transport hubs and nodes jointly improve public health, reduce car dependency, improve accessibility, and promote equitable and efficient multimodal transportation, thereby fostering healthier, more sustainable, and integrated urban environments [94,95].

4.3. Safer, More Comfortable Journerys

Key cycling routes passing through Point Cook hold enormous value for cyclists and pedestrians in the area and should be properly maintained. To facilitate their preservation and improvement, connections between these valuable assets should be promoted. This connection extends to linking other walking and cycling infrastructures in the area. According to the work by Meisel and Kopfer [96], it is necessary to synchronise the operations of passive and active transport principles to optimise transport routes. Minor upgrades, such as adding reflective strips and/or providing physical barriers along bike lanes, enable cyclists to feel safer and more comfortable during their riding journeys.

During survey collection, it was common for respondents to share their concerns with the absence of safe, facilitated crossing points. To facilitate accessible and comfortable journeys, additional safeguards must be placed at key intersections (mostly along busy roads and streets) to make the community feel safe and secure at road crossings. The survey and community engagement process revealed that a lack of space for rest along walking routes discouraged community members from making additional journeys. These spaces include a mixture of seating spaces, shade, drinking fountains and additional lighting, which promote walking as a mode of active transport and make existing pathways more accessible to those with limited mobility.

Key pedestrian routes were identified through an analysis of existing mapping data and community survey responses. Slower vehicle speeds have been proven to reduce the chances of accidents and fatalities. Therefore, the identified key pedestrian routes would benefit from slower traffic movement to enhance pedestrian safety. Signage and wayfinding also serves as useful interpretive tools to communicate the historical and cultural backgrounds of an area to new visitors.

4.4. Promoting Mobility for Leisure and Well-Being

Advancing mobility for recreation and wellness within active transportation networks is essential for cultivating healthier, more involved communities. Nonetheless, it encounters considerable obstacles that affect execution and utilisation. Establishing pedestrian and cycling networks that facilitate leisure movements requires meticulous design considerations, including connectivity to green spaces, picturesque routes, and sites of cultural or recreational significance. Nonetheless, the physical infrastructure is inadequate if these networks lack accessibility, proper maintenance, and perceived safety. Obstacles such as insufficient path continuity, the absence of protective measures against traffic, and inadequate lighting reduce the attractiveness of these networks for recreational purposes, especially for vulnerable populations, including children, older adults, and those with mobility impairments. Overcoming these obstacles requires dedicated efforts by urban designers and planners to create active transport networks that are inclusive and facilitate diverse forms of mobility beyond mere commuting.

Advocating for leisure mobility encounters obstacles associated with social perceptions and cultural norms, as active transportation is frequently regarded as a utilitarian mode of travel rather than a way for relaxation and well-being. Cities can mitigate these obstacles by promoting leisure mobility through the integration of pedestrian and cycling networks with public spaces, including parks and waterfronts, while offering amenities such as benches, shaded places, and wayfinding signage to enrich leisure experiences. Moreover, public campaigns highlighting the health advantages and pleasures of walking and cycling for recreation may facilitate changes in perceptions and increase usage of active transport networks. By addressing both physical and social impediments to leisure mobility, cities can establish active transport networks that enhance connectivity and promote well-being, rendering active transport a comprehensive solution that fosters physical health, mental well-being, and community resilience. These results agree with the findings of studies by Buliung et al., Allen and Nolmark, Prince et al., and Alessio et al. [97–100], which state that encouraging active transportation, such as walking and cycling, can improve physical activity levels, health, and well-being, with environmental and socioeconomic advantages.

5. Conclusions and Limitations of the Study

This study aims to first conduct an analysis of the existing conditions of cycling and walking infrastructure by identifying disconnected paths in one of the fastest-growing suburbs in Melbourne, using the 20-min neighbourhood principle. The present study also aims to determine the segments of pedestrian and cycling networks that are inappropriate for active transport, especially for people with disabilities and older adults, using Point Cook in Melbourne's west as a case study. An online interactive survey platform, "Social Pinpoint", was used to conduct the surveys at Point Cook. The following conclusions were drawn from this study:

- In this study, the concept of a 20-min neighbourhood was used to assess the accessibility of key destinations in which 800 metres was adopted as a guide for measuring the spatial accessibility of a walkable neighbourhood. This highlights the significance of active transport in Point Cook. An efficiently planned 20-min neighbourhood ensures that residents can easily accessthe necessary services and facilities by walking or cycling, thus minimising their reliance on vehicles for transportation. However, the implementation of this idea at Point Cook is impeded by the fragmented state of the current active transport system. By rectifying these deficiencies and establishing a unified and integrated system of walkways and bicycle lanes, the suburb of Point Cook may have substantial progress towards achieving its goal of becoming an easily accessible and well-integrated area.
- Our investigation of active transport networks in the suburb of Point Cook uncovered significant shortcomings in the connectivity between pedestrian and cycling routes. These shortcomings are due to dependence on private vehicles that are deeply rooted in inadequate, insecure, and lack of seamlessly integrated modes of transportation. The results of this research highlight the urgent need to address these shortcomings, especially under the rapid expansion of the suburb and the continuous rise in population density. By addressing these shortcomings, the development of active transport in Point Cook can have a crucial impact on reducing the overreliance on private vehicles.
- An important finding of this study is the identification of segments of pedestrian and cycling networks that are inaccessible to disabled people and older adults. The lack of accessibility not only restricts mobility choices for vulnerable people but also reduces the overall efficacy of the transportation system in encouraging active transport. The survey response data from Point Cook residents using the Social Pinpoint interactive platform, which involved both online interactive questionnaires and face-to-face interviews, revealed a significant desire among Point Cook residents to improve their involvement in active transport. Approximately 89% of the participants showed different degrees of interest in increasing their frequency of walking; however, the current transportation infrastructure failed to sufficiently accommodate this prospective change. To take advantage of this enthusiasm, it is crucial that the city's transport

planning prioritises the establishment of pedestrian and cycling infrastructure that is inclusive and easily accessible to all inhabitants, irrespective of their age or physical capabilities.

- Improving pedestrian and bicycle infrastructure in Point Cook will not only ease traffic congestion but also contribute to the wider sustainability objectives of the suburb of Point Cook. Promoting the adoption of active modes of transportation can result in a reduction in greenhouse gas emissions, lower levels of air pollution, and a minimised overall environmental impact of urban transportation. These advantages are particularly crucial in the context of climate change and the necessity for cities to embrace more sustainable practices in every aspect of the urban life.
- The results of this study provide strong evidence supporting the implementation of active transport in Point Cook as a crucial approach to alleviate traffic congestion and facilitate progress towards the development of 20-min neighbourhoods. Point Cook can enhance the resilience and sustainability of its urban environment by addressing gaps in its transport network in terms of accessibility and connectivity. To maintain the liveability and sustainability of Point Cook as it expands, it is crucial to provide a strong, easily accessible, and well-integrated network for both pedestrians and cyclists.
- By prioritising active transport, the city can minimise its carbon emissions, improve public health, and foster a livelier and more interconnected society. Promoting active transport in Point Cook is not only a remedy for existing traffic congestion, but also a proactive approach that will enhance the long-term sustainability and quality of life in the suburb of Point Cook. The results of this study establish a strong basis for future initiatives and encourage active travel in Point Cook. This study highlights the significance of strategic planning and policy formulation in realising the goal of creating 20-min neighbourhoods.

This study has also highlighted the following areas for further investigation:

- Further research should be conducted to investigate ways to improve the pedestrian and cycling network within and between municipalities in Australia and contribute to the development of 20-min neighbourhoods at both local and global scales. Prospective researchers might employ gravity models to quantify the attraction between residential areas and key destinations (such as schools, shopping centres, and parks). This model would consider factors like distance decay and the significance of various destinations, offering a mathematical representation of how accessibility influences active transport patterns.
- A logical progression of this work would be to implement agent-based models to simulate individual and group behaviour within a 20-min neighbourhood. By modelling residents as agents with distinct preferences and constraints, ABM (Agent based modelling) could provide deeper insights into how network connectivity and socio-demographic variables influence active transport choices over time.

The findings of this study are subject to the following two main limitations:

- This study may be limited by the accessibility of data regarding pedestrian and cycling network connectivity in the suburb of Point Cook in Melbourne's west. The lack of comprehensive data during our examination of feedback from Point Cook residents may have restricted the depth of this study.
- Another limitation of this study is the focus on Melbourne's west, which may have limited the findings to a localised setting, thus failing to include the barriers or opportunities in advocating for active transport in other areas of Victoria, other states in Australia, and even other countries, particularly in developing countries.

Despite these limitations, this study significantly advances active transport by analysing pedestrian and cycling network connectivity barriers in Melbourne's west. These findings can shape local active transport strategies. The results also highlight broader issues that may apply to other developed countries, influencing future research and policy in both developed and developing countries.

Author Contributions: Conceptualization, E.J., I.O.O. and H.-W.C.; methodology, E.J. and H.-W.C.; software, E.J. and H.-W.C.; validation, E.J., I.O.O. and H.-W.C.; formal analysis, E.J., I.O.O. and H.-W.C.; investigation, E.J. and H.-W.C.; resources, E.J. and H.-W.C.; data curation, I.O.O.; writing—original draft preparation, I.O.O.; writing—review and editing, E.J. and H.-W.C.; visualization, I.O.O.; supervision, E.J. and H.-W.C.; project administration, E.J. and H.-W.C.; funding acquisition, E.J. All authors have read and agreed to the published version of the manuscript.

Funding: This research was supported by the project "Social Value Creation for Transport and Infrastructure", funded by the Victorian Higher Education State Investment Fund, Australia (2021–2022).

Institutional Review Board Statement: The study was approved by the University Human Research Ethics Committee of Victoria University (reference number: HRE22-108, and the date of approval was 1 August 2022.

Data Availability Statement: The original contributions presented in the study are included in the article, further inquiries can be directed to the corresponding author.

Acknowledgments: The authors are grateful to the reviewers for reviewing this manuscript and providing valuable feedback.

Conflicts of Interest: The authors declare no conflicts of interest.

References

1. Australian Institute of Health and Welfare. *Australian Burden of Disease Study: Impct and Causes of Illnessand Death in Australia 2015*; Australian Institute of Health and Welfare: Canberra, Australia, 2019.
2. Australian Institute of Health and Welfare. *Pedal Cyclist Injury Deaths and Hospitalisations 1999–00 to 2015–16*; Australian Institute of Health and Welfare: Canberra, Australia, 2019.
3. Henley, G.; Harrison, J.E. *Trends in Serious Injury Due to Road Vehicle Traffic Crashes, Australia 2001 to 2010*; Australian Institute of Health and Welfare: Canberra, Australia, 2015.
4. Beck, B.; Cameron, P.A.; Fitzgerald, M.C.; Judson, R.T.; Teague, W.; Lyons, R.A.; Gabbe, B.J. Road safety: Serious injuries remain a major unsolved problem. *Med. J. Aust.* **2017**, *207*, 244–249. [CrossRef] [PubMed]
5. Giles-Corti, B.; Vernez-Moudon, A.; Reis, R.; Turrell, G.; Dannenberg, A.L.; Badland, H.; Foster, S.; Lowe, M.; Sallis, J.F.; Stevenson, M. City planning and population health: A global challenge. *Lancet* **2016**, *388*, 2912–2924. [CrossRef] [PubMed]
6. Australian Bureau of Statistics. 2071.0.55.001—Census of Population and Housing: Commuting to Work—More Stories from the Census, 2016. Available online: https://www.abs.gov.au/ausstats/abs@.nsf/Lookup/by%20Subject/2071.0.55.001~2016~Main%20Features~Feature%20Article:%20Journey%20to%20Work%20in%20Australia~40 (accessed on 30 September 2024).
7. Zapata-Diomedi, B.; Veerman, J.L. A Shift from Motorised Travel to Walking, Cycling and Public Transport: What Are the Potential Health Gains for an Australian City? *J. Transp. Health* **2017**, *5*, S5. [CrossRef]
8. Brown, V.; Diomedi, B.Z.; Moodie, M.; Veerman, J.L.; Carter, R. A systematic review of economic analyses of active transport interventions that include physical activity benefits. *Transp. Policy* **2016**, *45*, 190–208. [CrossRef]
9. Cirianni, F.; Monterosso, C.; Panuccio, P.; Rindone, C. A review methodology of sustainable urban mobility plans: Objectives and actions to promote cycling and pedestrian mobility. In *Smart and Sustainable Planning for Cities and Regions: Results of SSPCR 2017 2*; Springer: Cham, Switzerland, 2018; pp. 685–697.
10. Feltran, F.R. Instrumentos de Efetivação do Direito à Cidade a Lei 12.587/2012-Política Nacional de Mobilidade Urbana como Instrumento de Efetivação do Direito à Cidade. *Rev. Direito Dialogicidade* **2016**, *7*, 86–104.
11. Mugion, R.G.; Toni, M.; Raharjo, H.; Di Pietro, L.; Sebathu, S.P. Does the service quality of urban public transport enhance sustainable mobility? *J. Clean. Prod.* **2018**, *174*, 1566–1587. [CrossRef]
12. United Nations. Mobilizing for Development. Analysis and Policy Recommendations from the United Nations Secretary-General's High-Level Advisory Group on Sustainable Transport. 2016. Available online: https://sustainabledevelopment.un.org/index.php?page=view&type=400&nr=2375&menu=35 (accessed on 11 July 2024).

13. Brown, V.; Moodie, M.; Cobiac, L.; Mantilla, H.; Carter, R. Obesity-related health impacts of active transport policies in Australia—A policy review and health impact modelling study. *Aust. N. Z. J. Public Health* **2017**, *41*, 611–616. [CrossRef]
14. Liu, Y.; Wang, S.; Xie, B. Evaluating the effects of public transport fare policy change together with built and non-built environment features on ridership: The case in South East Queensland, Australia. *Transp. Policy* **2019**, *76*, 78–89. [CrossRef]
15. Taylor, M.; Thompson, S. An analysis of active transport in Melbourne: Baseline activity for assessment of low carbon mobility interventions. *Urban Policy Res.* **2019**, *37*, 62–81. [CrossRef]
16. Spandonide, B. Sustainable remote Australian transport for living on country and going out bush. *Int. J. Reg. Rural Remote Law Policy* **2017**, 1–19. [CrossRef]
17. Council, A.B. *National Cycling Strategy 2011–2016*; Austroads: Canberra, Australia, 2010.
18. Beattie, L.; Hayes, O. City of Melbourne transport strategy 2030. *Plan. News* **2021**, *47*, 26–27.
19. Government, Q. *Queensland Cycling Strategy 2017–2027*; Department of Transport and Main Roads: Brisbane, Australia, 2020.
20. Victoria, T.F. *Victorian Cycling Strategy 2018–28*; Victorian Government: Melbourne, Australia, 2017.
21. Pooley, C.G.; Horton, D.; Scheldeman, G.; Mullen, C.; Jones, T.; Tight, M.; Jopson, A.; Chisholm, A. Policies for promoting walking and cycling in England: A view from the street. *Transp. Policy* **2013**, *27*, 66–72. [CrossRef]
22. Pucher, J.; Dill, J.; Handy, S. Infrastructure, programs, and policies to increase bicycling: An international review. *Prev. Med.* **2010**, *50*, S106–S125. [CrossRef] [PubMed]
23. Olayode, I.O.; Severino, A.; Alex, F.J.; Macioszek, E.; Tartibu, L.K. Systematic review on the evaluation of the effects of ride-hailing services on public road transportation. *Transp. Res. Interdiscip. Perspect.* **2023**, *22*, 100943. [CrossRef]
24. Alex, F.J.; Tan, G.; Kyei, S.K.; Ansah, P.O.; Agyeman, P.K.; Fayzullayevich, J.V.; Olayode, I.O. Transmission of viruses and other pathogenic microorganisms via road dust: Emissions, characterization, health risks, and mitigation measures. *Atmos. Pollut. Res.* **2022**, *14*, 101642. [CrossRef]
25. Shannon, T.; Giles-Corti, B.; Pikora, T.; Bulsara, M.; Shilton, T.; Bull, F. Active commuting in a university setting: Assessing commuting habits and potential for modal change. *Transp. Policy* **2006**, *13*, 240–253. [CrossRef]
26. Rietveld, P.; Daniel, V. Determinants of bicycle use: Do municipal policies matter? *Transp. Res. Part A Policy Pract.* **2004**, *38*, 531–550. [CrossRef]
27. Heinen, E.; Buehler, R. Bicycle parking: A systematic review of scientific literature on parking behaviour, parking preferences, and their influence on cycling and travel behaviour. *Transp. Rev.* **2019**, *39*, 630–656. [CrossRef]
28. Mölenberg, F.J.; Panter, J.; Burdorf, A.; van Lenthe, F.J. A systematic review of the effect of infrastructural interventions to promote cycling: Strengthening causal inference from observational data. *Int. J. Behav. Nutr. Phys. Act.* **2019**, *16*, 93. [CrossRef]
29. Panter, J.; Guell, C.; Humphreys, D.; Ogilvie, D. Can changing the physical environment promote walking and cycling? A systematic review of what works and how. *Health Place* **2019**, *58*, 102161. [CrossRef] [PubMed]
30. Stewart, G.; Anokye, N.K.; Pokhrel, S. What interventions increase commuter cycling? A systematic review. *BMJ Open* **2015**, *5*, e007945. [CrossRef]
31. Kelly, P.; Williamson, C.; Baker, G.; Davis, A.; Broadfield, S.; Coles, A.; Connell, H.; Logan, G.; Pell, J.P.; Gray, C.M. Beyond cycle lanes and large-scale infrastructure: A scoping review of initiatives that groups and organisations can implement to promote cycling for the Cycle Nation Project. *Br. J. Sports Med.* **2020**, *54*, 1405–1415. [CrossRef] [PubMed]
32. Ton, D.; Duives, D.C.; Cats, O.; Hoogendoorn-Lanser, S.; Hoogendoorn, S.P. Cycling or walking? Determinants of mode choice in the Netherlands. *Transp. Res. Part A Policy Pract.* **2019**, *123*, 7–23. [CrossRef]
33. Bruzzone, F.; Scorrano, M.; Nocera, S. The combination of e-bike-sharing and demand-responsive transport systems in rural areas: A case study of Velenje. *Res. Transp. Bus. Manag.* **2021**, *40*, 100570. [CrossRef]
34. Cobbold, A.; Standen, C.; Shepherd, L.; Greaves, S.; Crane, M. Multimodal trips, quality of life and wellbeing: An exploratory analysis. *J. Transp. Health* **2022**, *24*, 101330. [CrossRef]
35. Scheepers, C.; Wendel-Vos, G.; Den Broeder, J.; Van Kempen, E.; Van Wesemael, P.; Schuit, A. Shifting from car to active transport: A systematic review of the effectiveness of interventions. *Transp. Res. Part A Policy Pract.* **2014**, *70*, 264–280. [CrossRef]
36. Tuominen, A.; Sundqvist-Andberg, H.; Aittasalo, M.; Silonsaari, J.; Kiviluoto, K.; Tapio, P. Building transformative capacity towards active sustainable transport in urban areas–Experiences from local actions in Finland. *Case Stud. Transp. Policy* **2022**, *10*, 1034–1044. [CrossRef]
37. Kamruzzaman, L. Typologies of 20-minute neighbourhoods, active transport use, and spatial spillovers. *Findings* **2022**. [CrossRef]
38. AlWaer, H.; Cooper, I. Unpacking the concept of 20-minute neighbourhoods: Disentangling "desired outcomes" from the "means" available for achieving them. *Open House Int.* **2023**, *48*, 704–728. [CrossRef]
39. Calafiore, A.; Dunning, R.; Nurse, A.; Singleton, A. The 20-minute city: An equity analysis of Liverpool City Region. *Transp. Res. Part D Transp. Environ.* **2022**, *102*, 103111. [CrossRef]
40. Government of Victoria. Plan Melbourne 2017–2050. Available online: https://www.planning.vic.gov.au/guides-and-resources/strategies-and-initiatives/plan-melbourne (accessed on 27 August 2024).

41. Government of Victoria. Plan Melbourne: Metropolitan Planning Strategy. Available online: https://www.planning.vic.gov.au/guides-and-resources/strategies-and-initiatives/melbournes-strategic-planning-history/plan-melbourne-2014 (accessed on 27 August 2024).
42. Dunning, R.; Calafiore, A.; Nurse, A. 20-minute neighbourhood or 15-minute city? *Town Ctry. Plan.* **2021**, *16*. Available online: https://livrepository.liverpool.ac.uk/id/eprint/3127722 (accessed on 27 August 2024).
43. Wyndham City Council. *2023 Wyndham Integrated Transport Strategy*; Wyndham City Council: Werribee, Australia, 2023.
44. Moreno, C.; Allam, Z.; Chabaud, D.; Gall, C.; Pratlong, F. Introducing the "15-Minute City": Sustainability, resilience and place identity in future post-pandemic cities. *Smart Cities* **2021**, *4*, 93–111. [CrossRef]
45. O'Gorman, S.; Dillon-Robinson, R. *(Ramboll). 20 Minute Neighbourhoods in a Scottish Context*; Ramboll: Mumbai, India, 2021.
46. Grodach, C.; Kamruzzaman, L.; Harper, L. *20-Minute Neighbourhood: Living Locally Research*; Department of Environment, Land, Water and Planning—State Government of Victoria: Melbourne, Australia, 2019.
47. Wyndham City Council. Victorian Population Health Survey 2017. Available online: https://www.wyndham.vic.gov.au/sites/default/files/2019-10/Victorian%20Population%20Health%20Survey%20(VPHS)%202017%20-%20Wyndham%20results%20-%20October%202019.pdf (accessed on 28 August 2024).
48. Chau, H.-W.; Gaisie, E.; Jamei, E.; Chan, M.; Lättman, K. Perceived Accessibility: Impact of Social Factors and Travel Modes in Melbourne's West. *Appl. Sci.* **2024**, *14*, 6399. [CrossRef]
49. Le Thi, H. Improving Transport-Related Health Impacts by Promoting Active Transport and Public Transport. Ph.D. Thesis, Technische Universität Darmstadt, Darmstadt, Germany, 2021.
50. Mueller, N.; Rojas-Rueda, D.; Cole-Hunter, T.; De Nazelle, A.; Dons, E.; Gerike, R.; Götschi, T.; Panis, L.I.; Kahlmeier, S.; Nieuwenhuijsen, M. Health impact assessment of active transportation: A systematic review. *Prev. Med.* **2015**, *76*, 103–114. [CrossRef] [PubMed]
51. Leger, S.J.; Dean, J.L.; Edge, S.; Casello, J.M. "If I had a regular bicycle, I wouldn't be out riding anymore": Perspectives on the potential of e-bikes to support active living and independent mobility among older adults in Waterloo, Canada. *Transp. Res. Part A Policy Pract.* **2019**, *123*, 240–254. [CrossRef]
52. Olabarria, M.; Pérez, K.; Santamariña-Rubio, E.; Novoa, A.M.; Racioppi, F. Effect of neighbourhood motorization rates on walking levels. *Eur. J. Public Health* **2015**, *25*, 740–747. [CrossRef]
53. Maibach, E.; Steg, L.; Anable, J. Promoting physical activity and reducing climate change: Opportunities to replace short car trips with active transportation. *Prev. Med.* **2009**, *49*, 326–327. [CrossRef] [PubMed]
54. Mahmoud, M.; El-Assi, W.; Habib, K.; Shalaby, A. How Active Modes Compete with Motorized Modes in High-Density Areas: A Case Study of Downtown Toronto. In *Canadian Transportation Research Forum*; University of Toronto: Toronto, ON, Canada, 2015.
55. Oeschger, G.; Carroll, P.; Caulfield, B. Micromobility and public transport integration: The current state of knowledge. *Transp. Res. Part D Transp. Environ.* **2020**, *89*, 102628. [CrossRef]
56. Fazio, M.; Giuffrida, N.; Le Pira, M.; Inturri, G.; Ignaccolo, M. Planning suitable transport networks for e-scooters to foster micromobility spreading. *Sustainability* **2021**, *13*, 11422. [CrossRef]
57. Kapsalis, E.; Jaeger, N.; Hale, J. Disabled-by-design: Effects of inaccessible urban public spaces on users of mobility assistive devices–a systematic review. *Disabil. Rehabil. Assist. Technol.* **2024**, *19*, 604–622. [CrossRef]
58. Rosenberg, D.E.; Huang, D.L.; Simonovich, S.D.; Belza, B. Outdoor built environment barriers and facilitators to activity among midlife and older adults with mobility disabilities. *Gerontologist* **2013**, *53*, 268–279. [CrossRef] [PubMed]
59. Allen, J.; Farber, S. Planning transport for social inclusion: An accessibility-activity participation approach. *Transp. Res. Part D Transp. Environ.* **2020**, *78*, 102212. [CrossRef]
60. Lättman, K.; Friman, M.; Olsson, L.E. Perceived accessibility of public transport as a potential indicator of social inclusion. *Soc. Incl.* **2016**, *4*, 36–45. [CrossRef]
61. Curtis, C.; Scheurer, J. Planning for sustainable accessibility: Developing tools to aid discussion and decision-making. *Prog. Plan.* **2010**, *74*, 53–106. [CrossRef]
62. Woodcock, J.; Edwards, P.; Tonne, C.; Armstrong, B.G.; Ashiru, O.; Banister, D.; Beevers, S.; Chalabi, Z.; Chowdhury, Z.; Cohen, A. Public health benefits of strategies to reduce greenhouse-gas emissions: Urban land transport. *Lancet* **2009**, *374*, 1930–1943. [CrossRef] [PubMed]
63. Möller, H.; Haigh, F.; Hayek, R.; Veerman, L. What is the best practice method for quantifying the health and economic benefits of active transport? *Int. J. Environ. Res. Public Health* **2020**, *17*, 6186. [CrossRef]
64. Gower, A.; Grodach, C. Planning innovation or city branding? Exploring how cities operationalise the 20-minute neighbourhood concept. *Urban Policy Res.* **2022**, *40*, 36–52. [CrossRef]
65. Chau, H.-W.; Gilzean, I.; Jamei, E.; Lesley, P.; Preece, T.; Quirke, M. Comparative analysis of 20-minute neighbourhood policies and practices in Melbourne and Scotland. *Urban Plan.* **2022**, *7*, 13–24. [CrossRef]
66. Van Soest, D.; Tight, M.R.; Rogers, C.D. Exploring the distances people walk to access public transport. *Transp. Rev.* **2020**, *40*, 160–182. [CrossRef]

67. Lachapelle, U.; Pinto, D.G. Longer or more frequent walks: Examining the relationship between transit use and active transportation in Canada. *J. Transp. Health* **2016**, *3*, 173–180. [CrossRef]
68. Chan, E.T.; Schwanen, T.; Banister, D. The role of perceived environment, neighbourhood characteristics, and attitudes in walking behaviour: Evidence from a rapidly developing city in China. *Transportation* **2021**, *48*, 431–454. [CrossRef]
69. Marquet, O.; Miralles-Guasch, C. Neighbourhood vitality and physical activity among the elderly: The role of walkable environments on active ageing in Barcelona, Spain. *Soc. Sci. Med.* **2015**, *135*, 24–30. [CrossRef] [PubMed]
70. Loukaitou-Sideris, A. Is it safe to walk? 1 neighborhood safety and security considerations and their effects on walking. *J. Plan. Lit.* **2006**, *20*, 219–232. [CrossRef]
71. Rothman, L.; Buliung, R.; Macarthur, C.; To, T.; Howard, A. Walking and child pedestrian injury: A systematic review of built environment correlates of safe walking. *Inj. Prev.* **2014**, *20*, 41–49. [CrossRef] [PubMed]
72. Bennett, G.G.; McNeill, L.H.; Wolin, K.Y.; Duncan, D.T.; Puleo, E.; Emmons, K.M. Safe to walk? Neighborhood safety and physical activity among public housing residents. *PLoS Med.* **2007**, *4*, e306. [CrossRef] [PubMed]
73. Lusk, A.C.; Furth, P.G.; Morency, P.; Miranda-Moreno, L.F.; Willett, W.C.; Dennerlein, J.T. Risk of injury for bicycling on cycle tracks versus in the street. *Inj. Prev.* **2011**, *17*, 131–135. [CrossRef] [PubMed]
74. Chieng, M.; Lai, H.; Woodward, A. How dangerous is cycling in New Zealand? *J. Transp. Health* **2017**, *6*, 23–28. [CrossRef]
75. Sanders, R.L. Perceived traffic risk for cyclists: The impact of near miss and collision experiences. *Accid. Anal. Prev.* **2015**, *75*, 26–34. [CrossRef]
76. De Hartog, J.J.; Boogaard, H.; Nijland, H.; Hoek, G. Do the health benefits of cycling outweigh the risks? *Environ. Health Perspect.* **2010**, *118*, 1109–1116. [CrossRef]
77. Gössling, S.; Humpe, A.; Litman, T.; Metzler, D. Effects of perceived traffic risks, noise, and exhaust smells on bicyclist behaviour: An economic evaluation. *Sustainability* **2019**, *11*, 408. [CrossRef]
78. Standen, C.; Crane, M.; Greaves, S.; Collins, A.T.; Rissel, C. How equitable are the distributions of the physical activity and accessibility benefits of bicycle infrastructure? *Int. J. Equity Health* **2021**, *20*, 208. [CrossRef]
79. Fraser, S.D.; Lock, K. Cycling for transport and public health: A systematic review of the effect of the environment on cycling. *Eur. J. Public Health* **2011**, *21*, 738–743. [CrossRef]
80. Schimek, P. Cycle track safety remains unproven. *Am. J. Public Health* **2013**, *103*, e6–e7. [CrossRef] [PubMed]
81. Nanayakkara, P.K.; Langenheim, N.; Moser, I.; White, M. Do safe bike lanes really slow down cars? A simulation-based approach to investigate the effect of retrofitting safe cycling lanes on vehicular traffic. *Int. J. Environ. Res. Public Health* **2022**, *19*, 3818. [CrossRef] [PubMed]
82. Cicchino, J.B.; McCarthy, M.L.; Newgard, C.D.; Wall, S.P.; DiMaggio, C.J.; Kulie, P.E.; Arnold, B.N.; Zuby, D.S. Not all protected bike lanes are the same: Infrastructure and risk of cyclist collisions and falls leading to emergency department visits in three US cities. *Accid. Anal. Prev.* **2020**, *141*, 105490. [CrossRef] [PubMed]
83. Welle, B. Health Impacts and Assessment from BRT and Complete Streets: Mexico City's Line 5 Metrobus Corridor. *J. Transp. Health* **2017**, *5*, S66. [CrossRef]
84. Xie, K.; Liang, B.; Song, Y.; Dong, X. Analysis of walking-edge effect in train station evacuation scenarios: A sustainable transportation perspective. *Sustainability* **2019**, *11*, 7188. [CrossRef]
85. Tardieu, J.; Susini, P.; Poisson, F.; Kawakami, H.; McAdams, S. The design and evaluation of an auditory way-finding system in a train station. *Appl. Acoust.* **2009**, *70*, 1183–1193. [CrossRef]
86. Li, Z.; Morita, Y. The Consistency of Information based on Observations of Users' Behavior Regarding the Wayfinding Transfer between Trains and City Buses at Hakata Station in Fukuoka City. *Arch. Des. Res.* **2016**, *29*, 23–45. [CrossRef]
87. Hochmair, H.H. Assessment of bicycle service areas around transit stations. *Int. J. Sustain. Transp.* **2015**, *9*, 15–29. [CrossRef]
88. Arbis, D.; Rashidi, T.H.; Dixit, V.V.; Vandebona, U. Analysis and planning of bicycle parking for public transport stations. *Int. J. Sustain. Transp.* **2016**, *10*, 495–504. [CrossRef]
89. Miao, Q.; Welch, E.W.; Sriraj, P. Extreme weather, public transport ridership and moderating effect of bus stop shelters. *J. Transp. Geogr.* **2019**, *74*, 125–133. [CrossRef]
90. Lanza, K.; Durand, C.P. Heat-moderating effects of bus stop shelters and tree shade on public transport ridership. *Int. J. Environ. Res. Public Health* **2021**, *18*, 463. [CrossRef] [PubMed]
91. Prins, R.; Panter, J.; Heinen, E.; Griffin, S.; Ogilvie, D. Causal pathways linking environmental change with health behaviour change: Natural experimental study of new transport infrastructure and cycling to work. *Prev. Med.* **2016**, *87*, 175–182. [CrossRef] [PubMed]
92. Bieliński, T.; Dopierała, Ł.; Tarkowski, M.; Ważna, A. Lessons from implementing a metropolitan electric bike sharing system. *Energies* **2020**, *13*, 6240. [CrossRef]
93. Cai, Y.; Chen, J.; Lei, D.; Yu, J. The integration of multimodal networks: The generalized modal split and collaborative optimization of transportation hubs. *J. Adv. Transp.* **2022**, *2022*, 3442921. [CrossRef]

94. Nieuwenhuijsen, M.J. Urban and transport planning pathways to carbon neutral, liveable and healthy cities; A review of the current evidence. *Environ. Int.* **2020**, *140*, 105661. [CrossRef] [PubMed]
95. Maltese, I.; Gatta, V.; Marcucci, E. Active travel in sustainable urban mobility plans. An Italian overview. *Res. Transp. Bus. Manag.* **2021**, *40*, 100621. [CrossRef]
96. Meisel, F.; Kopfer, H. Synchronized routing of active and passive means of transport. *OR Spectr.* **2014**, *36*, 297–322. [CrossRef]
97. Buliung, R.N.; Mitra, R.; Faulkner, G. Active school transportation in the Greater Toronto Area, Canada: An exploration of trends in space and time (1986–2006). *Prev. Med.* **2009**, *48*, 507–512. [CrossRef]
98. Allen, H.; Nolmark, H. Active Transportation, the Ultimate Low Carbon Way to Travel—A Review of International Research and Education. *Front. Sustain. Cities* **2022**, *4*, 824909. [CrossRef]
99. Prince, S.A.; Lancione, S.; Lang, J.J.; Amankwah, N.; de Groh, M.; Garcia, A.J.; Merucci, K.; Geneau, R. Are people who use active modes of transportation more physically active? An overview of reviews across the life course. *Transp. Rev.* **2022**, *42*, 645–671. [CrossRef]
100. Alessio, H.M.; Bassett, D.R.; Bopp, M.J.; Parr, B.B.; Patch, G.S.; Rankin, J.W.; Rojas-Rueda, D.; Roti, M.W.; Wojcik, J.R. Climate change, air pollution, and physical inactivity: Is active transportation part of the solution? *Med. Sci. Sports Exerc.* **2021**, *53*, 1170–1178. [CrossRef] [PubMed]

land

MDPI

Article

Reconstructing the Silk Road Network: Insights from Spatiotemporal Patterning of UNESCO World Heritage Sites

Yingning Shen [1,†], Junmin Liu [1,†], Jianan Han [2] and Xiang Wan [3,*]

1 School of Cultural Heritage, Northwest University, Xi'an 710127, China; shenyingning@stumail.nwu.edu.cn (Y.S.); ljm@nwu.edu.cn (J.L.)
2 Institute of Silk Road Studies, Northwest University, Xi'an 710127, China; hanjianan@stumail.nwu.edu.cn
3 Collaborative Research Centre for Archaeology of the Silk Roads, Northwest University, Xi'an 710127, China
* Correspondence: wanxiang@nwu.edu.cn; Tel.: +86-186-9152-3093
† These authors contributed equally to this work.

Abstract: Building on the observation of gaps in current research, this study provides a comprehensive analysis of the spatial patterns of heritage sites along the Silk Road, focusing on how historical trade routes shaped what are now recognized as heritage sites. Using data from the United Nations Educational, Scientific and Cultural Organization (UNESCO) World Heritage List, the research examines heritage sites across Eurasia and North Africa, with a specific emphasis on the Silk Road corridors. This study employs a spatiotemporal approach, categorizing sites into northern overland routes and southern maritime routes to highlight regional variations in network development. The key findings of this study reveal the significant influence of historical trade routes on the development of settlements, cities, and cultural landmarks along the Silk Road. These findings identify clear trends in the Silk Road network's evolution over time, illustrating a shift in its spatial focus across different historical periods. Initially, the network was centered in the eastern Mediterranean during the Classical Period. In the medieval period, this focus expanded to include a dual core area in both the eastern Mediterranean and Central Asia. By the late Medieval period, the network had shifted again, with a new core emerging in Europe. This chronological and spatial analysis allows for a detailed examination of the Silk Road network's heritage landscape evolution. The study underscores the interconnectedness of heritage sites across these regions, contributing to a deeper understanding of how landscape connectivity and trade network dynamics evolved over time. Furthermore, by identifying patterns of network development and shifts in centrality and density, this research offers valuable insights for the conservation and management of heritage landscapes. These findings are particularly crucial for preserving the historical and cultural integrity of Silk Road heritage sites.

Keywords: Silk Road network; kernel density estimation; spatial cluster analysis; regional connectivity; heritage landscapes conservation

1. Introduction

Systematic efforts are required for precise a definition and understanding of the vicissitudes of the Silk Road for researchers on this concept. The earliest definition of the Silk Road comes from the German geographer Ferdinand von Richthofen in 1877, although evidence shows that the concept was already emerging in the works of the German geographer Carl Ritter as early as 1838. It was first employed to refer to the ancient routes of transportation and trade between the three continents of the Eastern Hemisphere, particularly between Europe and Asia, and from the beginning, it included both the overland and maritime Silk Roads [1]. Richthofen's description of the Silk Road marks the establishment of official interactions between China and the West in the 2nd century B.C.E., and his research made the concept widely known. Subsequent scholars further refined the description of the Silk Road, defining its eastern starting point in the middle and lower reaches of the Yellow River in China, while the western starting point extended from the eastern Mediterranean to Italy. They also proposed various perspectives on the trade patterns and historical significance of the Silk Road [2–8].

Based on the intuitive impressions formed by existing research on the Silk Road, the official description provided by the United Nations Educational, Scientific and Cultural Organization (UNESCO) offers a reference for the definition in this study. Here, the Silk Roads (*sic*) are described as "a diversity of routes and cargos" and "routes of dialogue", acknowledging them as not only a trade network but also a network of culture, knowledge, and religion [9,10]. This provides a basis for defining the Silk Road beyond just trade. The Silk Road network (hereafter SRN) can be defined as the total network of routes connecting most cultural areas of the Old World, including the continents of Africa, Asia, and Europe. Precisely speaking, after the historically attested opening of the Silk Road by General Zhang Qian in the latter half of the 2nd century B.C.E., the SRN began to function as a system of contact, communication, and exchange across the three continents [11]. This network remained systematically integrated and consistent until Christopher Columbus's first navigation to the Americas in 1492 [12]. Columbus's navigation to the Americas initiated a new open system of contact among the five continents, and subsequent navigation of the Oceanic islands extended the reach of European culture into the Americas and Oceania. Since then, the route network has no longer been enclosed, thus incorporating a large portion of the global population and resources into the global route network (hereafter GRN) of communication and exchange. The marked difference in the range between the SRN and the GRN leads us to conclude that the study of the SRN as a synchronic system must end in the 15th century, as diachronically, this system evolved into a completely different GRN.

In fact, studies of the SRN based on documentary sources can be so precise that most of the connecting routes between major cities can be illustrated on maps. However, such studies face several issues. Firstly, the recorded distances in historical texts are not always accurate. The distance measurement units used by ancient civilizations often differ significantly from today's precise units, and historical texts do not provide conversions between these ancient and modern units [13]. Furthermore, the locations mentioned in historical texts may differ from those of the same name today, due to differences in site selection [14]. Additionally, historical texts may mention many specific routes of interaction, but the extent to which these routes were used varies greatly [15]. Some routes were rarely used but received greater attention in SRN research due to their reference in texts [16]. In the absence of precise statistical data on ancient city populations and trade volumes, the weights of various routes and relevant nodes can only be estimated based on experience [17].

The challenges associated with documentary-based studies underscore the need for alternative methods to reflect the SRN more accurately. Recognizing the importance of comprehensive approaches to understanding the SRN, international organizations have taken significant steps to address these challenges. Recognizing the importance of diverse cultural exchanges along the Silk Roads, UNESCO in 1994 launched the "Global Strategy for a Representative, Balanced and Credible World Heritage List" [18]. This initiative aimed to ensure that the world's cultural and natural diversity was fully represented. Central Asia, identified as one of the most under-represented regions, became a focal point for efforts to fill this gap in global heritage representation. UNESCO had earlier acknowledged the significance of the Silk Roads for intercultural exchange and the wealth of cultural heritage sites it encompassed. Consequently, UNESCO proposed a trans-national nomination project for the Silk Roads as part of their strategy, fostering international collaboration and addressing the lack of representation on the World Heritage List, such as the UNESCO/Japanese Funds-in-Trust project: Support for Silk Roads World Heritage Sites in Central Asia (Phase II) [19].

Given the existing issues in estimating the SRN based on documentary sources and these international efforts, a supplementary method is proposed: using the heritage site databases of the three continents to estimate and reconstruct SRN routes. The most concentrated and comprehensive information on World Heritage Sites (hereafter, WHSs) comes from the UNESCO's World Heritage List. The downloadable WHS database provides the latitude and longitude information of heritage sites, and the UNESCO WHS List website provides detailed information on the establishment and abandonment dates of these sites [20–22]. By logging these heritage sites, this study has established a geographic information database reflecting the emergence, prosperity, and abandonment of heritage sites over different periods [23].

Using the geographic information database of heritage sites to reflect the evolution of the SRN is based on the following assumptions:

1. The SRN, the system of interactions among Eurasia and Africa, is an enclosed and stable system.
2. Heritage sites that once existed and prospered can reflect the prosperity of the aforementioned interaction system.
3. The network density between heritage sites reflects the network density of the SRN.

Using heritage sites to reflect the SRN offers greater precision compared to documentary sources, as the geographic locations of the archeological sites and cultural landscapes represented by these heritage sites are accurate and do not require textual verification. The distances between heritage sites are precise. The issues of distance, location, and unit conversion that reduce precision in the textual verification of documentary sources do not arise when using heritage sites to reflect the SRN. Due to the lack of exact route markings, only straight-line measurements can be calculated. Thus, this study adopts spatial clustering methods.

Based on this assumption, this study first introduces the research methods and then demonstrate how heritage sites reflect the SRN. The network relationships between heritage sites indicate the network density of the SRN. The resulting network will exhibit varying levels of route usage. In areas with high network density between heritage sites, the density will be significantly higher than in areas with low density, forming a heterogeneous network in terms of usage levels. This situation contrasts with the uniform recording of multiple routes based on textual verification of documentary sources. Both methods have distinct focuses: faithfully maintaining textual records maximizes the representation of all possible routes in the SRN. All possible routes are uniformly recorded based on historical records. Routes formed between heritage sites give higher weight to those with higher network density, accurately reflecting the weights of various network parts and nodes, though influenced by existing heritage site registrations. This study integrates both methods to construct an evolution map of the SRN based on documented routes and heritage site weights, offering a comprehensive research framework. Finally, the study analyzes and discusses related factors influencing the changes in the SRN. Based on these, the study provides recommendations for the future preservation of Silk Road cultural heritage.

2. Materials and Methods

This study analyses the distribution of the UNESCO World Heritage Site Database heritage sites along the SRN. As a foundational step, our methodological framework, depicted in Figure 1, outlines the comprehensive approach taken from data collection through to spatial analysis. More specific, to achieve the aim and objectives of this research, which involves using heritage site databases from three continents to estimate and reconstruct SRN routes, three primary analytical methods are employed: spatial distribution analysis, proximity analysis, and Kernel Density Estimation. Firstly, spatial distribution analysis integrates multiple methods to examine the geographical patterns and spatial structures of heritage sites along the Silk Road. Spatial cluster analysis is employed to detect clusters of heritage sites, revealing areas of significant cultural and economic activity. This analysis is complemented by the chi-square test, which assesses the distribution of heritage sites along the northern and southern routes of the SRN, identifying periods of significant transformation in regional connectivity and land use. Additionally, the study utilizes spatial autocorrelation functions (Ripley's K, L, and G) to quantify the degree of clustering or dispersion within the distribution of heritage sites. Ripley's K function is particularly useful for identifying the extent of clustering compared to a random distribution, while Ripley's L and G functions offer more detailed insights into the spatial processes influencing site distribution. Secondly, proximity analysis, primarily through the nearest neighbor distance (NND) method, provides critical insights into the spatial relationships between heritage sites over time. By calculating the distance between each site and its nearest neighbor, this method evaluates the concentration and dispersion of heritage sites along the SRN. The study also incorporates the DBSCAN (Density-Based Spatial Clustering of Applications with Noise) algorithm, which identifies clusters based on density and effectively manages noise and outliers. DBSCAN is particularly valuable in archeological research because it does not require a predetermined number of clusters, allowing for the identification of geographically significant differences between sites. Moreover, the silhouette coefficient is calculated to assess the quality of clusters identified by DBSCAN, providing a measure of cohesion within clusters and separation between different clusters. Finally, Kernel Density Estimation (KDE) is used to analyze the spatial distribution of heritage sites across various centuries, providing both statistical and visual representations of site density. KDE smooths spatial data to produce a continuous surface that highlights areas of high cultural and economic activity along the Silk Road. This method effectively demonstrates how the distribution of heritage sites evolved over time, reflecting changes in settlement patterns and land use. The KDE analysis is complemented by distance matrix construction and proximity analysis, which further explore the spatial relationships between sites. The results from KDE, combined with these other methods, provide a detailed understanding of the spatial dynamics and the role of the SRN in shaping connectivity and landscape evolution across Eurasia and North Africa.

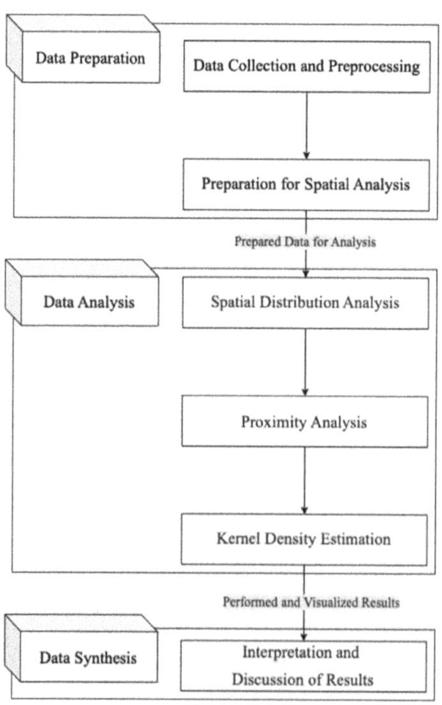

Figure 1. Methodological framework.

2.1. Material Collection

The cultural heritage data used in this study were sourced from the UNESCO World Heritage site (www.unesco.org) official platform of the United Nations Educational, Scientific and Cultural Organization. The data and information provided are sourced from UNESCO and its partner organizations, and are subject to authoritative verification and approval, exerting a significant influence in the field of cultural heritage. Details such as the name of the heritage site, the country it is in, its coordinates, the region, and the dates of establishment and abandonment were obtained by accessing the descriptive pages of cultural heritage entries across Asia, Europe, and North Africa. The names of the heritage sites are derived from the entries themselves, while the country and coordinates are from the basic information section, and the region along with the dates are from the descriptive section of the entries.

The data are categorized chronologically, sorted by the date of establishment, and recorded in centuries from the time of their establishment to their abandonment. The temporal scope for the cultural heritage sites in this study is designated from the 2nd century B.C.E. to the 15th century. This timeframe was specifically chosen to align with research on the SRN, beginning with the historical mission of Zhang Qian to the Western Regions. Zhang Qian's journey not only opened up routes to these regions but also marked the beginning of cultural and commercial exchanges between the East and the West, as detailed in the *Records of the Grand Historian: Treatise on the Dayuan* [24]. Over time, the prominence of the terrestrial Silk Road gradually diminished, supplanted by the maritime periods, particularly by the 15th century—a period notable for the Age of Discovery in Europe. During this era, European explorers, emboldened by advancements in Portuguese navigational techniques that refined traditional Mediterranean maritime skills to meet the demands of oceanic voyages, ventured into the New World, establishing new sea routes. This period is marked by the navigational achievements of explorers, most notably Christopher Columbus, whose voyages from 1492 to 1493 were meticulously planned and executed based on his exceptional navigational skills as documented in his logs [25], with the achievement of the discovery of maritime routes to the Americas. Italian navigators also played a significant role in these maritime explorations, though it was undoubtedly the golden age of Iberian power that first propelled European maritime expansion [26]. Concurrently, the overland Silk Road began to decline as these new sea routes emerged, allowing Europe to trade directly with Asia and the Americas via the newly discovered maritime routes, thereby reducing reliance on the traditional Silk Road, the nexus of the Old World [27]. The selected period from the 2nd century B.C.E. to the 15th century effectively reflects the changes in the Silk Road's cultural heritage sites over its historical phases, thereby providing robust support for further analysis. Once the timeframe was established, the data were then systematically categorized by century.

In terms of data preprocessing for this study, cultural heritage information was entered in the sequence of heritage name, country, latitude, longitude, region, time of establishment, and time of abandonment. During the selection process, entries without specific establishment dates were removed, and smaller sub-entries under major entries were included to better highlight the Silk Road's evolutionary process. For data transformation, the original geographical coordinates, originally in degrees, minutes, and seconds, were converted to decimal degrees format. This conversion facilitates the eventual integration of the data into illustrations produced by JupyterLab platform of Python (3.11.5) language and Rstudio platform of R (4.3.3) language. For heritage sites that remain to this day without a recorded date of abandonment, no abandonment time was entered; however, these sites are included in the dataset up to the 15th century, placed within the time frame from the 2nd century B.C.E. to the 15th century.

2.2. Research Methods

2.2.1. Spatial Distribution Analysis

In pursuit of clarifying the spatial clusters of SRN sites through 2nd century B.C.E. to the 15th century mentioned in the previous section, this study advances into the second pivotal phase: spatial cluster analysis. This analysis aims to unearth the underlying spatial structures and tendencies inherent within the geographic distribution of the sites. To achieve this objective, this study embarks on an application of spatial statistical techniques. To investigate the relationship between the establishment centuries of SRN sites and their geographical coordinates, this study implements the chi-square test, a statistical assay for its efficacy in discerning and comparing the distributions of categorical variables across distinct groups. The chi-square test is defined by the following formula:

$$\chi^2 = \sum \frac{(O_i - E_i)^2}{E_i} \tag{1}$$

where O_i represents the observed frequency of heritage sites in each period, and E_i is the expected frequency assuming no change over periods. This test enables us to quantitatively determine the homogeneity or heterogeneity in the geographical spread of sites through time. Meanwhile, this test serves to ascertain whether the distribution of heritage sites exhibits significant variation across several centuries, thereby permitting a quantified assessment of distributional homogeneity or disparity [28].

Complementing the chi-square analysis, this study invokes the use of spatial autocorrelation functions—namely Ripley's K, L, and G. These functions are instrumental in dissecting the spatial clustering of sites, providing a spectrum of scales at which any agglomeration might manifest. Ripley's K function is particularly adept at identifying the degree of clustering within a given radius compared to a random distribution. It is computed as

$$K(t) = \frac{1}{\lambda}\sum_{i \neq j} I(d_{ij} \leq t) \tag{2}$$

where λ is the intensity of points (number of points per unit area), d_{ij} is the distance between points i and j, and I is an indicator function that equals 1 if $d_{ij} \leq t$ and 0 otherwise. This function allows us to quantify the extent to which heritage sites are more clustered than would be expected in a random distribution.

In extension, Ripley's L function, a transformation of the K function, affords an expected distance metric that is intuitively more interpretable, delineating clustering or dispersion phenomena in relation to complete spatial randomness. It can be used as

$$L(t) = \sqrt{\frac{K(t)}{\pi}} - t \tag{3}$$

The G function further enhances research analysis by offering an empirical distribution function of the distances from a randomly selected site to its nearest neighbor, thereby allowing an evaluation of the spatial processes influencing the point patterns [29,30]. It is given by

$$G(t) = \frac{1}{n}\sum_{i=1}^{n} I(d_{i,\text{nearest}} \leq t) \tag{4}$$

where $d_{i,\text{nearest}}$ is the distance from point i to its nearest neighbor. This metric provides insights into the intensity of clustering by looking at the proximity of each site to its closest neighbor.

2.2.2. Proximity Analysis

In the Proximity Analysis section, this study constructs a distance matrix for various sites to assess their spatial relationships and interactions over different historical periods. The nearest neighbor distance (NND) method, essential for evaluating the concentration of archeological sites, is widely employed in the statistical analysis of archeological and historical studies. This method calculates the distance between a site and its nearest neighbor to deduce the spatial distribution characteristics of these sites.

In applying this method to this research, the first step involves identifying archeological sites that existed from the 2nd century B.C.E through the 15th century. Subsequently, the NNDs for each site are calculated. Assuming each site is denoted as i, the distance d_i to its closest heritage site is computed using the following formula:

$$d_i = \min_{j \neq i} d(s_i, s_j) \tag{5}$$

In this analysis, $d(s_i, s_j)$ represents the distance between heritage site s_i and heritage site s_j, where $j \neq i$ to ensure that the site does not measure the distance to itself.

Subsequently, this study calculates the average nearest neighbor distance for all heritage sites across various centuries. The formula is as follows:

$$\bar{d} = \frac{1}{n}\sum_{i=1}^{n} d_i \tag{6}$$

n represents the total number of heritage sites within a given century, while d_i refers to the nearest neighbor distance for the ith heritage site.

The spatial distribution characteristics across different historical periods can be revealed by calculating the average NND coefficient for each century. This approach enhances the understanding of the evolving spatial patterns over time and provides a comprehensive perspective on the spatial layout trends of each period. This exploration of both relative distances and statistical measures of spatial distribution enables a thorough understanding of how cultural and historical influences may have shaped the placement and significance of heritage sites.

This study utilized DBSCAN (Density-Based Spatial Clustering of Applications with Noise), a popular clustering algorithm for density-based grouping of points that can identify clusters of arbitrary shapes while managing noise and outliers [31]. DBSCAN's primary parameters are 'eps', which defines the neighborhood radius, and 'minPts', which specifies the minimum number of neighboring points required for a point to be considered a core point.

The analysis commenced with the computation of the k-distance for each heritage site. This k-distance, defined as the Euclidean distance to a site's k^{th} nearest neighbor, is crucial for 'eps' for the DBSCAN algorithm. The systematic calculation of these distances for each site within the dataset is performed as follows:

$$k\text{-distance}(p) = ||p - N_k(p)|| \tag{7}$$

where p denotes an individual heritage site, $N_k(p)$ signifies its kth nearest neighbor, and $||p - N_k(p)||$ represents the Euclidean distance between them. The selection of a k-value, often predicated on the dataset's intrinsic characteristics and the clustering objectives, leads to a collection of k-distances which, when ordered and visualized, elucidate a discernible 'elbow' or inflection point in their progression.

This study computed the k-distance to facilitate DBSCAN clustering. Unlike other clustering algorithms such as K-means, DBSCAN does not rely on a predetermined number of clusters but instead bases clustering on the density and proximity of points. This makes DBSCAN particularly sensitive to geographic differences between sites, a feature that is especially valuable in archeological research [31].

Subsequently, this study calculates the silhouette coefficient, a metric used to assess the quality of clustering. This coefficient integrates both cohesion—measured as the average distance of a point to others within the same cluster—and separation, or the average distance to points in the nearest different cluster. Each point is assigned a score ranging from -1 to 1. A silhouette coefficient close to 1 indicates well-defined clustering;

a coefficient near −1 suggests poor clustering quality; and a score approaching 0 may indicate overlapping or randomly distributed clusters.

For each heritage site as one point i, the calculation of the silhouette coefficient includes two crucial distances: $a(i)$, the mean distance to other points within the same cluster, reflecting the cluster's cohesion, and $b(i)$, the mean distance to points in the nearest different cluster, indicating the degree of separation. These distances are essential for determining the silhouette coefficient for each point, calculated as follows:

$$s(i) = \frac{b(i) - a(i)}{\max\{a(i), b(i)\}} \tag{8}$$

After calculating the silhouette values for each point, it is necessary to compute the average values for each century. This allows the degree of clustering for different centuries to be presented and determines whether the results are similar across these periods.

2.2.3. Kernel Density Estimation

This study conducts a Kernel Density Estimation (KDE) to analyze the distribution of archeological sites across various centuries. KDE is a method extensively used in archeological research to examine spatial point distributions. Mathematically, KDE can be represented as follows:

$$\hat{f}(x,y) = \frac{1}{n \cdot h^2} \sum_{i=1}^{n} K\left(\frac{x - x_i}{h}, \frac{y - y_i}{h}\right) \tag{9}$$

(x,y) represents the coordinates of the point on the evaluation grid. (x_i, y_i) represents the longitude and latitude coordinates of the i-th data point. n is the number of site data points. h is the bandwidth. K is the Gaussian kernel function.

KDE could assess the density of heritage sites along the Silk Road, demonstrating KDE's recognized effectiveness in spatial analysis. This study utilized the 'scipy.stats' module in Python 3.11.5, which automatically selects the bandwidth h and applies kernel smoothing to the data points, generating a smoothed density estimate across the spatial region. Visualizing these density estimates on a map provides an intuitive understanding of the distribution density of heritage sites for each century.

3. Results

This section applies the above method and divides into three sections to show the findings. Before delving into the specific analysis, a statistical examination of the WHS data was conducted. Figure 2a presents the ranking of the top 20 countries or regions by the number of heritage sites. China ranks first with the most heritage sites, followed by Italy and Spain. Germany, France, and India also rank prominently. The number of heritage sites in these countries underscores their importance in global cultural heritage preservation. This figure indicates that Asian and European countries dominate in heritage site numbers, reflecting their rich historical and cultural legacy. Figure 2b illustrates changes in the number of heritage sites over various periods, with a polynomial trend line highlighting the overall trend. The horizontal axis represents centuries, from the 2nd century BCE to the 15th century CE, while the vertical axis shows the number of heritage sites. The number of heritage sites has gradually increased over time. The polynomial trend line further shows that this growth has become more pronounced in recent centuries, particularly after the 10th century CE. This suggests a continued increase in the number of cultural heritage sites over time.

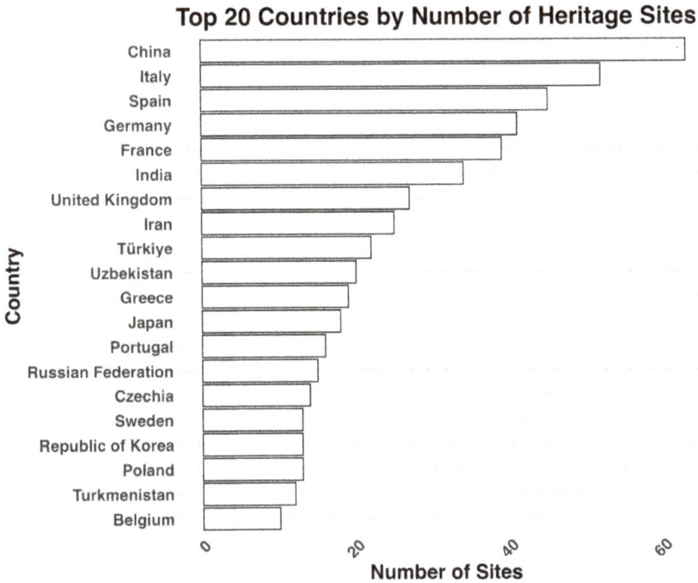

(a)

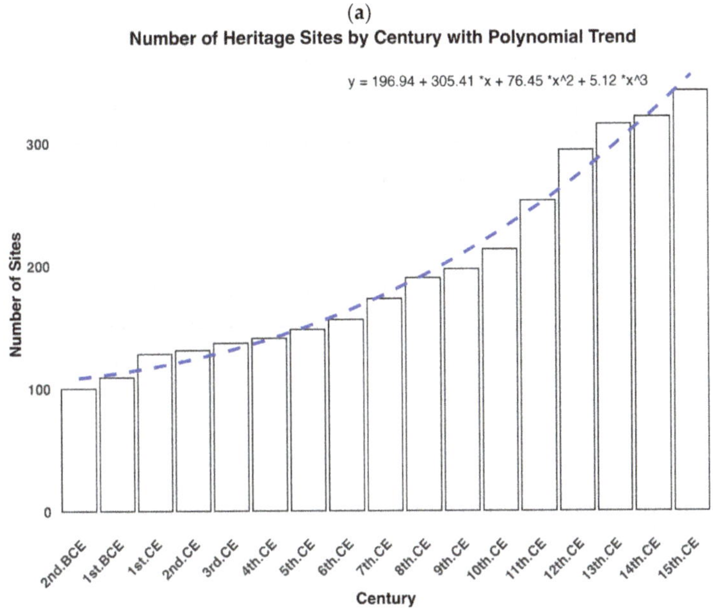

(b)

Figure 2. This figure presents the statistical data visualization by country and time. The figure contains two panels: (**a**) the top 20 countries by number of SRN sites and (**b**) the number of heritage sites by century with a polynomial trend.

3.1. Statistical Analysis

The chi-square test was employed to examine the association between the distribution of heritage sites along the terrestrial and maritime Silk Roads and their respective historical periods (shown in Figure 3). Using the average latitude of 33.4253° N, derived from cities on the UNESCO interactive map of the Silk Road, the SRN sites were divided into northern and southern regions. A contingency table was constructed to tally the number of sites across different centuries within these regions. The chi-square test yielded a value of 26.9515 with 16 degrees of freedom and a p-value of 0.04203. Since the p-value is below the significance level of 0.05, the results indicate a statistically significant association between the location of sites and their temporal classification. This suggests a notable, non-random pattern in the geographical placement of sites corresponding to their historical periods.

Figure 3. Major cities along the Silk Roads [32]; see Appendix B for detailed cities.

This does not, however, imply a causal relationship between the two variables. The following figures illustrate the spatial point patterns of a dataset as analyzed by Ripley's K, L, and G functions within the context of their respective significance bounds. The analysis using Ripley's K, L, and G functions reveals a clear pattern of non-randomness in the spatial distribution of the points studied. The consistent positioning of the observed functions above the theoretical lines and outside the significance envelopes for all three tests indicates a significant tendency towards clustering rather than randomness or regularity. These results should be interpreted as evidence of an underlying spatial process or interaction influencing the distribution of the points, which could be crucial for further research into the causative factors or mechanisms responsible for this pattern.

Figure 4 shows that the observed K function, represented by the solid black line, lies above the theoretical K function, indicated by the red dashed line, across the range of distances. This consistently higher value outside the grey confidence envelope suggests a significant clustering of points over the distance scale analyzed. This finding indicates that points are more densely located around each other than would be expected if they were randomly distributed, pointing to a potential underlying process that causes aggregation.

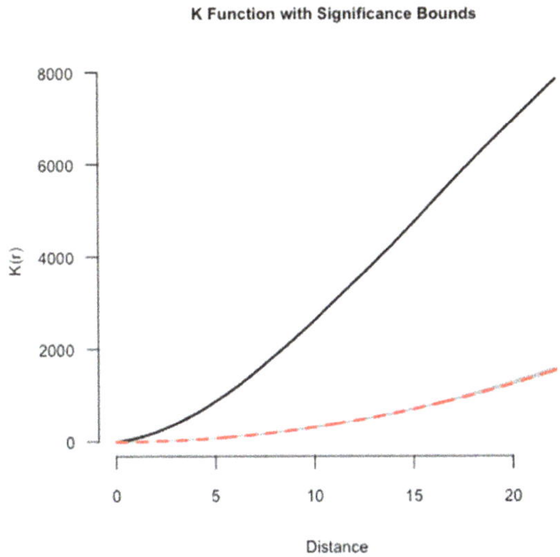

Figure 4. This figure displays the K function with significance bounds. The K function (represented by the solid black line) measures the degree of spatial clustering as a function of distance. The distance is plotted on the x-axis, while the K(r) values are on the y-axis. The red dashed line represents the significance bounds, providing a reference to assess whether the observed spatial clustering (black line) deviates significantly from a random distribution.

In Figure 5, the observed L-function, shown by the solid black line, is above the theoretical expectation (red dashed line) for all distances measured. Since the L-function is designed to have an expected value of zero under a random distribution, the observed positive values suggest clustering. The deviation of the L-function from zero beyond the grey significance bounds reinforces the evidence of spatial clustering at various scales.

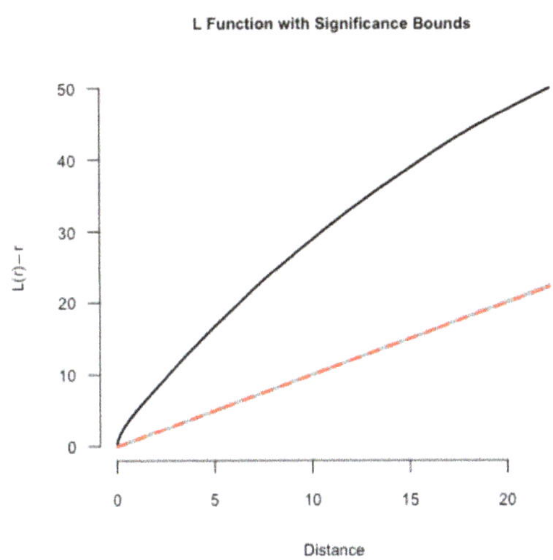

Figure 5. This figure illustrates the L function with significance bounds. The L function, represented by the solid black line, is a transformation of the K function used to detect spatial clustering or dispersion over distance. The x-axis denotes distance, while the y-axis shows L(r) − r, adjusting the L function to a zero baseline for easier interpretation. The solid black line depicts the observed spatial pattern, while the red dashed line represents the significance bounds.

Figure 6 depicts the G function. Here, the solid black line of the observed G function rises above the theoretical line (red dashed line), particularly at smaller distances, and remains outside the grey significance envelope. This pattern suggests that the nearest neighbor distances are generally shorter than expected under a random distribution, indicating a tendency toward clustering.

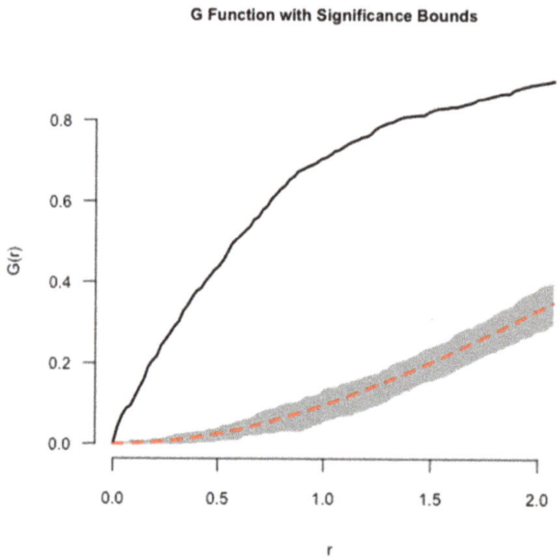

Figure 6. The G function with significance bounds. The solid black line represents the observed spatial distribution of points, while the red dashed line indicates the expected distribution under randomness. The grey shaded area marks the significance bounds. The sharp rise in the G function above these bounds suggests significant spatial clustering at shorter distances.

3.2. Spatial Clustering Analysis across Centuries

From the 2nd century B.C.E. to the 3rd century, the distribution of nearest neighbor coefficients (see Figure 7a) indicates that SRN sites were primarily concentrated in lower value ranges. The lighter color and smaller size of the bubbles indicate that the data points were closely spaced and relatively clustered. The silhouette scores (see Figure 7c) remained low and relatively stable at around 0.2, suggesting poor clustering performance with data points spatially dispersed or lacking significant aggregation. The clusters identified through the DBSCAN algorithm (see Figure 7b) show that most data points lie outside the red contour area, indicating they are relatively dispersed and likely classified as noise points or independent clusters. According to Figure 7d, from the 2nd to 1st century B.C.E., the clusters were relatively dispersed. This indicates that the sites did not form significant high-density aggregation areas, suggesting dispersed social activities or limited records. It also shows that during the 2nd and 3rd centuries CE, the clusters became more pronounced, indicating increased social activities or more the establishment of more heritage sites during this period.

From the 4th to the 8th century, the nearest neighbor coefficients gradually increased (see Figure 7a), with bubbles becoming darker and larger, indicating a more dispersed spatial distribution of data points. The silhouette scores (see Figure 7c) began to rise from the 6th century onwards, showing improved clustering performance and suggesting the emergence of spatial aggregation among sites. The DBSCAN-identified clusters (see Figure 7b) also became more distinct, with some data points located within high-density areas (inside the contour), while others remained more dispersed (outside the contour), indicating an increasing tendency toward clustering. According to Figure 7d, from the 4th to the 9th centuries, several clusters evenly appeared across these periods, indicating that sites had a certain degree of spatial aggregation.

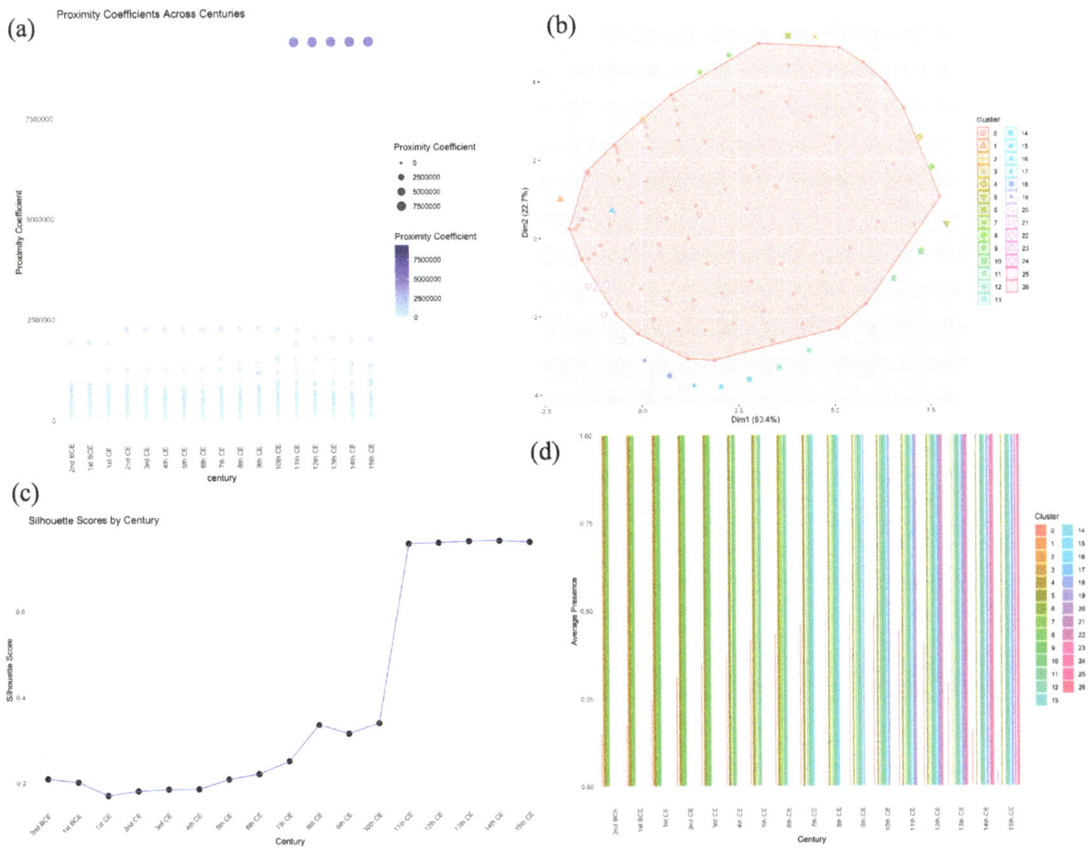

Figure 7. This is a figure illustrating the spatial and temporal analysis of heritage site distributions along the Silk Road network (SRN) using various statistical and clustering methods. The figure contains four panels: (**a**) the distribution of nearest neighbor coefficients (NNCs) over time; (**b**) the DBSCAN clustering results; (**c**) the silhouette scores over time; (**d**) the temporal distribution of clusters from the 2nd century B.C.E. to the 15th century C.E., showing the clusters' relative density and diversity over different periods.

From the 9th to the 13th century, the nearest neighbor coefficients significantly increased (see Figure 7a), especially during the 11th to 13th centuries, when some data points reached the highest values. The bubbles became darker and larger, indicating a highly dispersed spatial distribution of data points. The silhouette scores (see Figure 7c) rose sharply in the 10th century, reaching a peak, indicating significantly improved clustering performance. From the 11th century onwards, the silhouette scores remained high, close to 1.0, indicating very distinct clustering characteristics. The clusters identified through the DBSCAN algorithm (see Figure 7b) showed a marked increase in number and diversity, with several forming high-density areas, particularly during the 11th and 12th centuries. This indicates significant social, cultural, or environmental changes that led to concentrated site distributions. According to Figure 7d, from the 10th to the 15th century, the number and diversity of clusters significantly increased, indicating that the spatial distribution of sites became more complex and diverse. Particularly during the 11th and 12th centuries, several clusters occupied a large proportion, indicating significant social, cultural, or environmental changes that led to concentrated site distributions.

These findings illustrate the changes in the aggregation and dispersion of sites from the 2nd B.C.E. to the 13th century. During the early period (2nd B.C.E. to 9th century), site distribution was relatively dispersed with no significant high-density aggregation areas, possibly reflecting the dispersed nature of early societies or incomplete archeological records. During the middle period (10th to 12th century), site aggregation significantly increased, with multiple clusters forming high-density areas, possibly indicating a peak in social and cultural activities. During the late period (13th to 15th century), site distribution became complex and diverse again, with multiple clusters appearing at different times. This indicates diverse social activities and regional development, possibly reflecting significant social changes and environmental shifts.

This study presents intriguing findings where some areas exhibit low nearest neighbor coefficients but high clustering. This indicates that while the overall distribution may appear dispersed, certain localized areas have a dense concentration of heritage sites. The low nearest neighbor coefficients suggest that data points are closely spaced within these local regions, highlighting significant clustering at a smaller scale. Furthermore, high clustering observed at larger scales implies that within a broad area, several smaller regions have high point density, but the distances between these smaller regions are relatively large. This results in an overall low nearest neighbor coefficient. Archeological findings often concentrate in specific hotspot regions, likely centers of ancient civilization activities. Consequently, even though the distribution appears dispersed at a macro scale, the density of points within these hotspot regions is extremely high at a micro scale.

3.3. *Kernel Density Estimation Visualization*

To better explore the relationships between heritage site locations, the results of the Kernel Density Estimation (KDE) analysis are presented below.

3.3.1. Heatmap Visualization and Interpretation

This section uses heatmap visualization to present the spatial clustering of heritage sites across different centuries. KDE heatmaps are employed to estimate the probability density function of a random variable in continuous space. In the context of geographic data, the heatmaps visualize the density of data points (such as settlements or events) across a geographic area, highlighting where these points are concentrated. The heatmaps use color gradients to represent point density across different regions of the map, with warmer colors (e.g., red) indicating higher densities (more data points in each area) and cooler colors (e.g., blue) indicating lower densities. KDE heatmaps provide an intuitive visual summary of data point clustering within a geographical area.

From analyzing the Kernel Density heatmap, a consistent trend is identified. In the time span from the 2nd century B.C.E. to the 3rd century (shown in Figure 8), the heatmap displays a core concentration situated in the eastern Mediterranean region, expanding continuously in both eastern and western directions. The influence extends eastward through the Iranian Plateau into the Central Asian Transoxiana region, linking with the Yellow River Basin in China, and westward along the Mediterranean to the western Mediterranean region. Examining the cultural landscape depicted by the heatmap, the Greek–Roman cultural sphere, focalized in the eastern Mediterranean, demonstrated robust external connections during this era, notably interacting with the cultural zones of the Mesopotamian region and Central Asia [33]. Furthermore, the cultural realm of the distant Chinese region, positioned in the east, was impacted by this cultural interchange centered on the eastern Mediterranean, marking the nascent stages of cross-cultural exchanges among major empires.

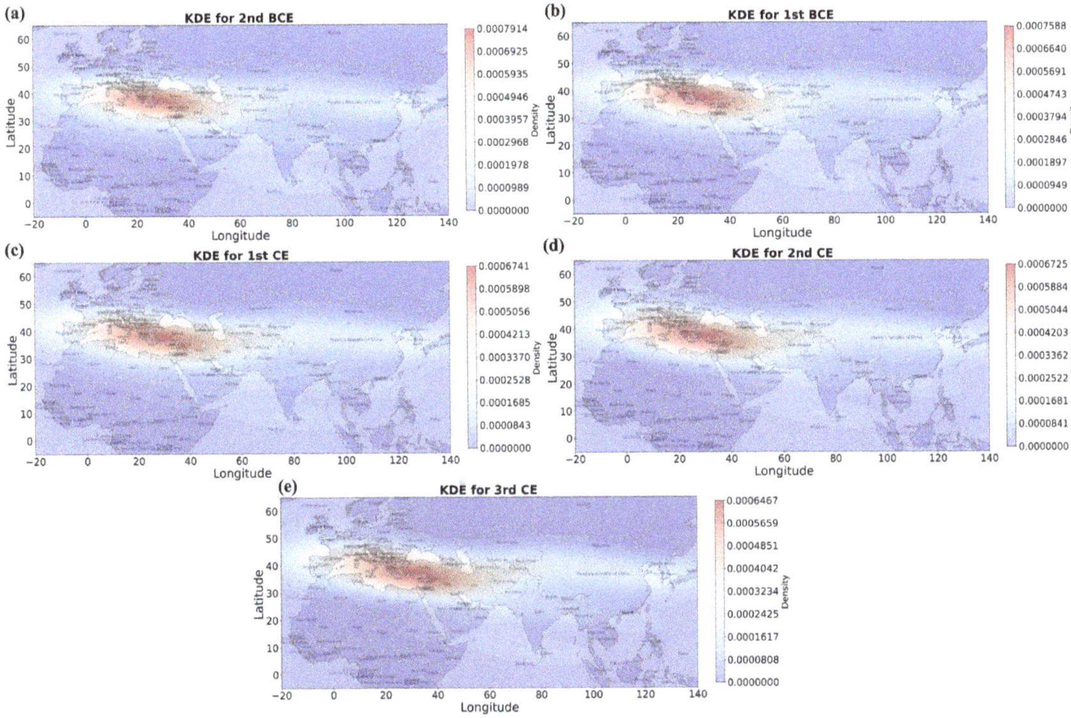

Figure 8. This is a figure illustrating the heatmap of Kernel Density Estimation (KDE) results for heritage sites along the Silk Road network (SRN). The figure contains five panels, each representing a different time period: (**a**) KDE for 2nd century B.C.E.; (**b**) KDE for 1st century B.C.E.; (**c**) KDE for 1st century; (**d**) KDE for 2nd century; (**e**) KDE for 3rd century C.E.

The Kernel Density heatmaps (shown in Figure 9) from the 4th to 8th centuries illustrate a transformative shift from a singular core initially centered on the eastern Mediterranean region to an extensive expansion in both easterly and westerly directions. This reconfiguration establishes a dual-core pattern encompassing the eastern Mediterranean–Mesopotamian region, and the Central Asian Transoxiana region, along with the Yellow River Basin in China emerging as a focal point, signifying a general shift towards the east. The former epitomizes the Near Eastern cultural realm, focusing on the eastern provinces of the Roman Empire and later the Byzantine Empire, situated amidst the Black Sea, Caspian Sea, Persian Gulf, Red Sea, and the Mediterranean—historically, a thriving hub and a key cultural center along the Silk Road network in classical antiquity [34]. The latter symbolizes the Central Asian cultural sphere, demonstrating remarkable prosperity and emphasizing the evolution of urban civilizations in the Central Asian Transoxiana region along the early medieval Silk Road, highlighting the crucial role of the Sogdian merchant network in facilitating East–West exchanges [35]. Notable bright areas in the heatmap extend eastward into the core of East Asia, establishing local hubs in the lower Yellow River region of China, showcasing the prosperity of Chinese culture and its interactions and amalgamations with other significant cultural spheres across the Eurasian continent [36].

The trend from the 9th to the 15th centuries indicates a notable decline in the Kernel Density heatmap of Central Asia and China (shown in Figure 10). This period can be segmented into two phases. Between the 9th and 11th centuries, the Silk Road continued its exchange patterns, showing cycles of prosperity and decline. The heat centers steadily shifted westward, with the focus in the Yellow River Basin of China weakening and shifting towards the southeast. Notably, during this time, the once significant heat centers in Syria and Iraq, along with the Central Asian Transoxiana region, experienced diminishing intensity and a westward shift, no longer maintaining distinct independent heat centers. Key historical events in Central Asia during this era included the Arab conquests and the unrest of the late Tang and Five Dynasties, which replaced eastern influences with those of the Arab Empire in the west [37]. In contrast, the Kernel Density heatmap from the 12th to the 15th century depicts a dual-center pattern with Europe and East Asia emerging as focal points on the Eurasian continent's eastern and western sides. These centers operate largely independently, with Europe exhibiting a higher clustering level compared to East Asia. This era marked the rise of the Italian Renaissance and foreshadowed the Age of Discovery, originating from Western Europe. The economic prosperity of the Renaissance was fueled by the ascent of Italian city-states in the late 12th century, potentially planting the seeds of modern capitalist systems [38]. Conversely, the eastern and central regions of the Eurasian continent were besieged by conflicts linked to the Crusades, Mongol conquests, and the Timurid era wars [39]. Economic trade flourished primarily in East Asia and Southeast Asia, bolstering the prosperity of the East Asian cultural sphere (Confucian cultural circle). The disruption of the Silk Road trade severed connections running along the 40th parallel to Central Asia, hindering east–west interactions at the Tibetan Plateau along the 30th parallel. Trade from the Western regions to the expansive Indian trading zone had to navigate the low latitude waters of Southeast Asia to engage with East Asia, depicting a distinct trading pattern from earlier epochs [40].

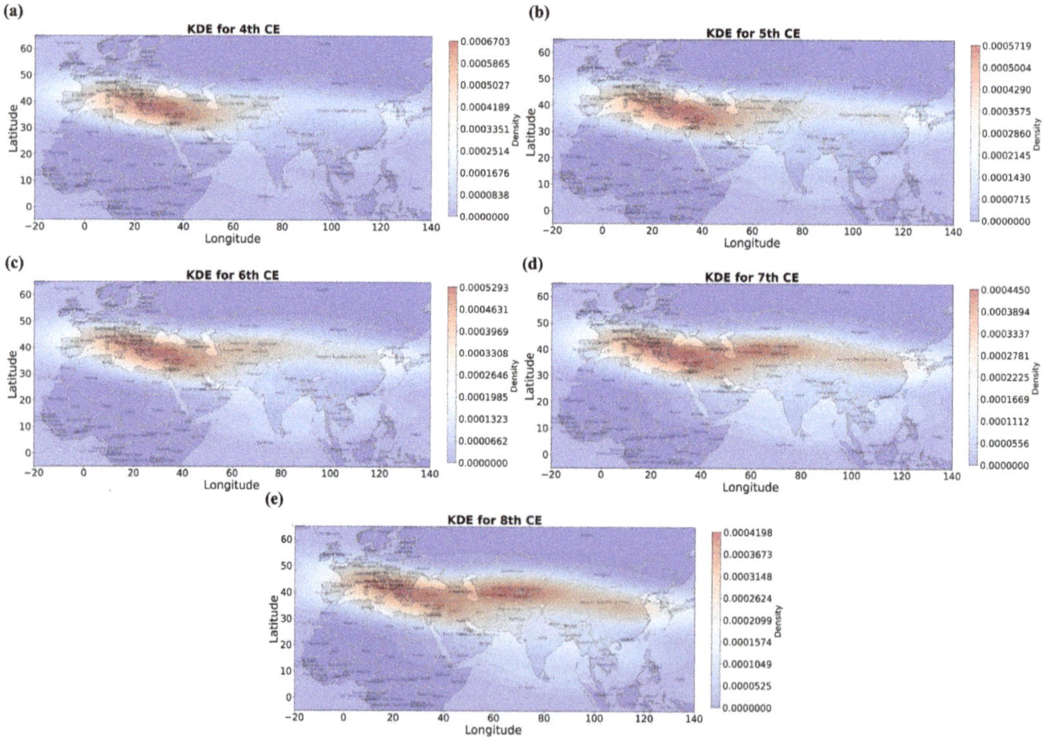

Figure 9. This is a figure illustrating the heatmap of Kernel Density Estimation (KDE) results for heritage sites along the Silk Road network (SRN) from 4th to 8th century. The figure contains five panels, each representing a different time period: (**a**) KDE for 4th century; (**b**) KDE for 5th century; (**c**) KDE for 6th century; (**d**) KDE for 7th century; (**e**) KDE for 8th century.

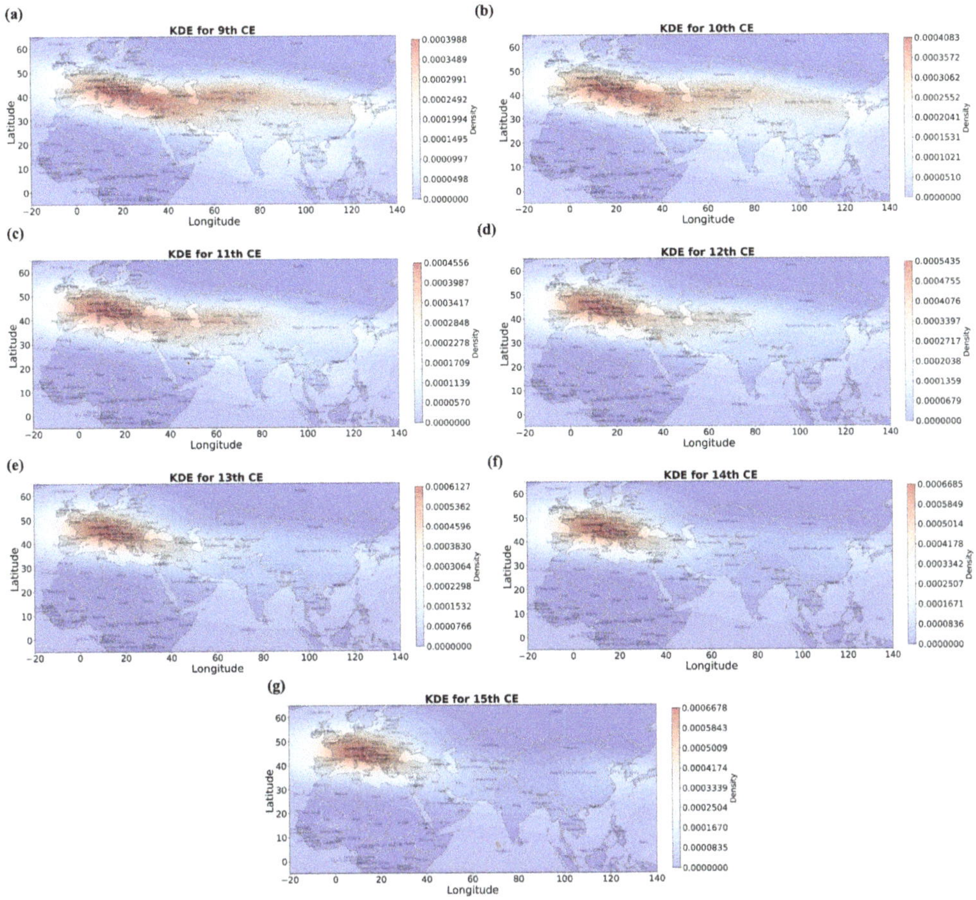

Figure 10. This is a figure illustrating the heatmap of Kernel Density Estimation (KDE) results for heritage sites along the Silk Road network (SRN) from the 9th to 15th Century. The figure contains seven panels, each representing a different time period: (**a**) KDE for 9th century; (**b**) KDE for 10th century; (**c**) KDE for 11th century; (**d**) KDE for 12th century; (**e**) KDE for 13th century; (**f**) KDE for 14th century; (**g**) KDE for 15th century.

With the discovery of the Americas in 1492, the conventional SRN eventually evolved from a closed system to a global exchange network. Europe assumed a central role in this emerging global structure. It is anticipated that extending the research methodology of this study to encompass investigations of the Americas and Oceania post-1492 would require evaluating heritage sites in these regions, contributing to conclusions that reinforce Europe's pivotal position in global exchanges [41].

3.3.2. Scatter Plot Visualization and Interpretation

This section uses scatter plots to visualize changes in heritage sites, providing a clear representation of individual data points and their exact locations on the map. This method is typically used to study the relationships between variables or to understand the spatial distribution of individual observations. In a scatter plot, each data point is represented by a dot on the graph. The position of the dot corresponds to the exact coordinates of the observation. In this study, the red dots represent denser locations, while the blue dots indicate more dispersed area. This section provides a more detailed description of the SRN's heritage site compared to previous section.

The density analysis map of World Heritage Sites along the SRN from the 2nd century B.C.E. to the 3rd century (shown in Figure 11) reveals a significant emergence of new heritage sites in the eastern Mediterranean region, including the Greek city-states of Asia Minor along the Aegean Coast and the Levant region on the eastern Mediterranean coast, compared to other regions. Additionally, there is a notable increase in the density of heritage sites eastward from the Central Asian Transoxiana region between the 2nd century B.C.E and the 3rd century, marking the eastern path of the ancient Silk Road extending to East Asia. The connections among high-density sites exhibit an east–west alignment, stretching from the Western Mediterranean to East Asia [42]. The 3rd to 4th centuries mark a transitional period [43]. During this time, the density of heritage sites in the Western Mediterranean declined, while it continued to grow in the Aegean Sea, the Levant, Iran, and Central Asia [44]. Consequently, the fluctuations in the number of heritage sites from the 4th to the 8th centuries indicate a new trend: a reduction in heritage sites in Western Europe, continuous growth in the eastern Mediterranean, and the emergence of new areas of heritage site growth in Central Asia and Western China.

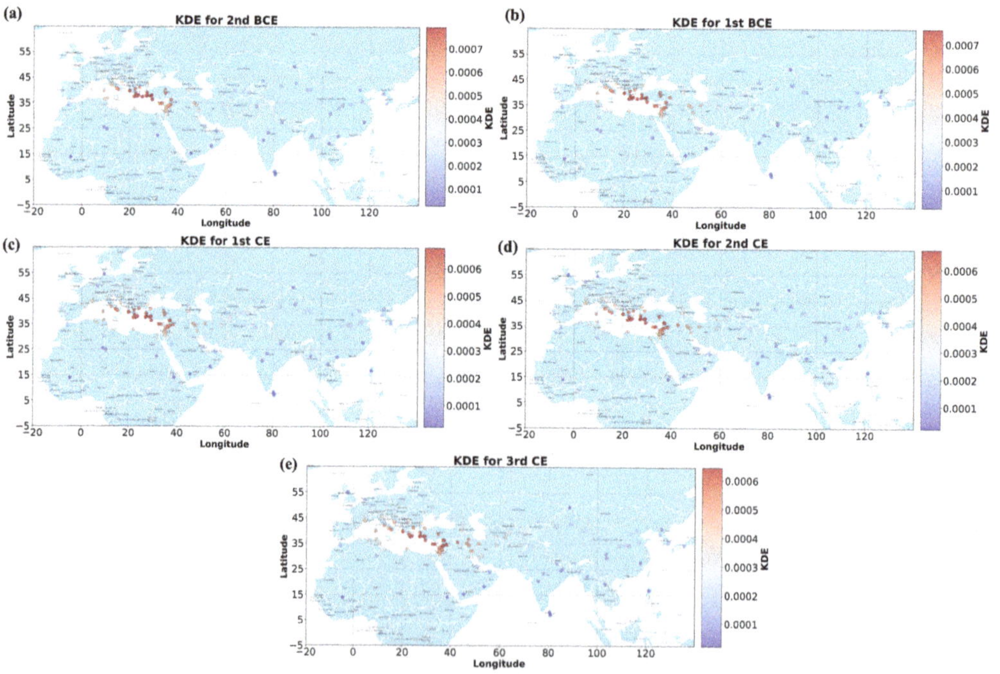

Figure 11. This is a figure illustrating the scatter plot of Kernel Density Estimation (KDE) results for heritage sites along the Silk Road network (SRN) from 2nd century B.C.E. to 3rd century. The figure contains seven panels, each representing a different time period: (**a**) KDE for 2nd century B.C.E.; (**b**) KDE for 1st century B.C.E.; (**c**) KDE for 1st century; (**d**) KDE for 2nd century; (**e**) KDE for 3rd century.

From the 4th to the 8th centuries, the analysis of the increase and decrease in the number of World Heritage Sites within the Silk Road network presents an intriguing scenario (shown in Figure 12). This period reveals noticeable transformations: a substantial decline in heritage site number is observed in the eastern Mediterranean, particularly around the Aegean Sea, North Africa, the Levant, and the Mesopotamian region. In contrast, there is a notable increase in heritage site number in the Central Asian Transoxiana region,

East Asia, and Southeast Asia. The reasons for this change may lie in the fact that during this historical period, the fragmentation and fall of the Roman Empire led to prolonged periods of conflict and disunity in the Mediterranean region [45]. The collapse of the Western Roman Empire in 476 C.E. required the establishment of the Carolingian Empire by the Franks (771–814 C.E.) for stability to return to Western Europe. In the eastern Mediterranean, the Byzantine Empire engaged in protracted conflicts with the Sassanian Persians and the Arab Empire in the Levant and North Africa [46]. Climatically, the primary regions of the Silk Road network encountered cooling events during the 3rd to 6th centuries. A distinctive trend in climate change from lower to higher annual average temperature with notable fluctuations had a notable impact on North Africa, resulting in a continual decrease in heritage sites from the 4th to the 8th centuries [47,48]. Consequently, there was a reduction in the number of Roman heritage sites in the western region. Despite the upheavals in East Asia during the Six Dynasties, factors such as the introduction of Buddhism, the flourishing Sui and Tang dynasties, and the growth of thriving Sogdian merchant communities significantly bolstered trade and cultural exchange in the area [49]. The connections among high-density points formed a spindle-shaped pattern extending from northeast to southwest, with primary concentrations in the eastern Mediterranean and Central Asian Transoxiana region.

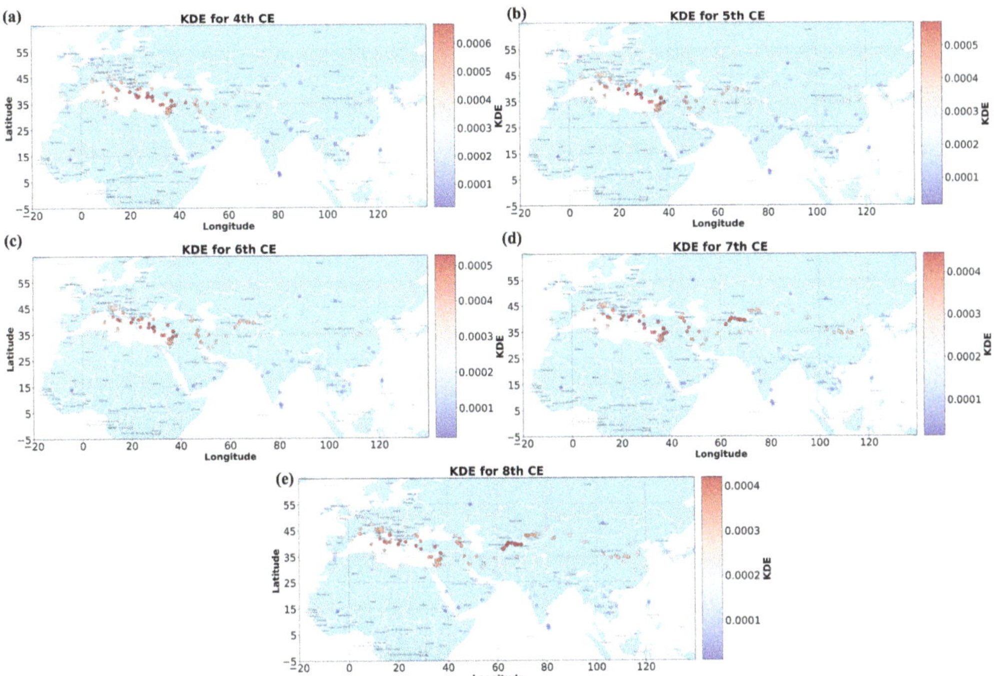

Figure 12. This is a figure illustrating the scatter plot of Kernel Density Estimation (KDE) results for heritage sites along the Silk Road network (SRN) from the 4th to 8th century. The figure contains five panels, each representing a different time period: (**a**) KDE for 4th century; (**b**) KDE for 5th century; (**c**) KDE for 6th century; (**d**) KDE for 7th century; (**e**) KDE for 8th century.

The 8th to 9th centuries signify yet another transitional period [50]. During this time, the increase in heritage sites in Central Asia and Western China stagnated, while the number of sites in Eastern China further declined. As a result, the SRN centered around the eastern Mediterranean, Iran, and Central Asia experienced an interruption in the trends of increasing and decreasing heritage sites observed from the 4th to the 8th centuries in various regions of the Silk Road. Meanwhile, the number of heritage sites in Western Europe, including present-day France, Germany, and Northern Italy, significantly increased. Consequently, the Silk Road underwent a historical decline from prosperity between the 9th and 15th centuries, a significant transformation that can be further divided into two stages: the 9th to 11th centuries and the 12th to 15th centuries.

From the 9th to the 15th century, there was a steady rise in the number of heritage sites rooted in medieval Europe (shown in Figure 13). The pace of growth accelerated notably, surpassing all other regions by the 12th century. Compared to Europe, the number of heritage sites in the Levant, Mesopotamia, Central Asia, and Western China did increase during this era, but the density gap widened significantly. The interaction patterns of the Silk Road network persisted between the 9th and 11th centuries, characterized by fluctuations. While the Silk Road's path remains visible on maps, the eastern segment from Central Asia through Xinjiang to East Asia progressively diminishes between the 12th and 15th centuries until its eventual disappearance [7].

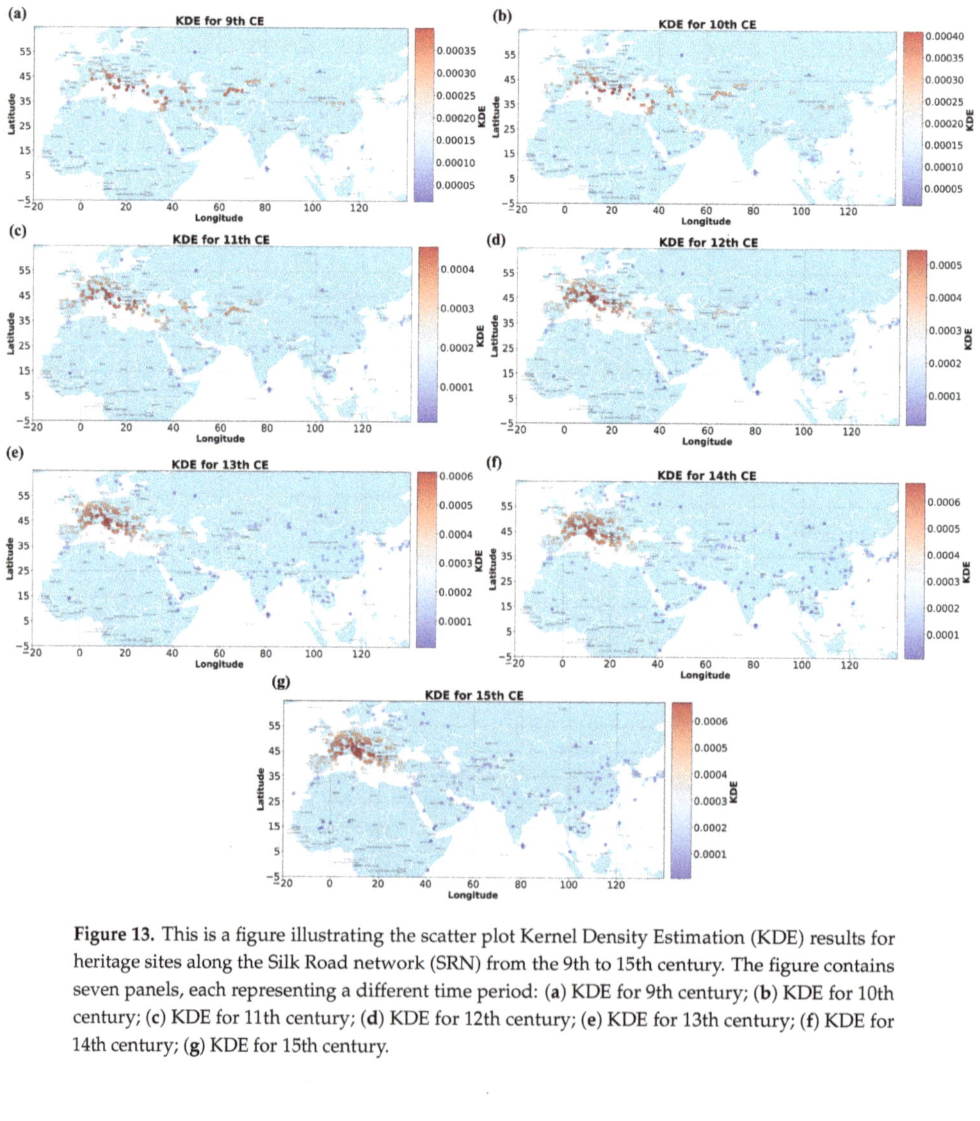

Figure 13. This is a figure illustrating the scatter plot Kernel Density Estimation (KDE) results for heritage sites along the Silk Road network (SRN) from the 9th to 15th century. The figure contains seven panels, each representing a different time period: (**a**) KDE for 9th century; (**b**) KDE for 10th century; (**c**) KDE for 11th century; (**d**) KDE for 12th century; (**e**) KDE for 13th century; (**f**) KDE for 14th century; (**g**) KDE for 15th century.

4. Discussion

4.1. Analysis of Assumptions in the SRN Study

The study on the SRN comprehensively addresses the hypotheses that frame its investigation into the historical development and dynamics of the SRN. Firstly, the SRN is confirmed to have functioned as an enclosed and stable system from its formation after General Zhang Qian's visit to Central Asia in the 2nd century B.C.E. until the end of the 15th century, when the advent of Christopher Columbus's navigation to the Americas in 1492 transformed it into the GRN. This transition integrated a significant portion of the global population and resources, marking the evolution from a closed to an open system. Secondly, the study utilizes heritage sites listed by UNESCO as precise indicators of the SRN's prosperity and historical significance. By creating a geographic information database of these SRN sites, the study reconstructs the SRN routes with a higher degree of accuracy than documentary sources. This approach assumes that heritage sites, which once existed and flourished, effectively reflect the prosperity of the SRN's interaction system. Lastly, the network density between these heritage sites is used to mirror the network density of the SRN itself. By employing spatial analysis, the study illustrates the distribution and evolution of heritage sites, revealing varying levels of route usage and significant historical interactions. This method highlights how different regions within the SRN experienced distinct impacts during periods of climate change, reinforcing the study's assumptions. Overall, the research methodically validates the hypotheses that the SRN was an enclosed and stable system, that heritage sites reflect its prosperity, and that their network density accurately represents the SRN's network density.

4.2. Analysis of Relevant Elements on SRN Heritage Sites

By comparing the evolution of the Silk Road network (SRN) across three major historical periods following its establishment in the 2nd century B.C.E., this study analyzes the intrinsic factors driving SRN changes.

4.2.1. Climate Change on the SRN

Based on the findings, significant spatial changes in many heritage sites are revealed. The heritage sites mentioned in this study can also be understood as traces of past human habitation, such as settlements. Climate change is a major natural factor influencing the evolution of the SRN. The alternation of warm and cold periods, driven by global climate fluctuations, has had a significant impact on SRN regions, particularly in the mid- to low-latitude areas of the Eurasian continent and the Mediterranean coast. Research indicates that climate fluctuations, particularly in precipitation and temperature, significantly influenced the rise and fall of civilizations in the arid regions of Central Asia. Warm and humid climates contributed to the rise of empires, while deteriorating climates led to the decline of civilizations and the abandonment of cities [36]. Moreover, fluctuations in precipitation levels also affected the rise and fall of trade routes and urban centers along the Silk Road in Central Asia. Wet periods promoted urban expansion, while dry periods hindered the development of these networks [51]. Further research reveals that long-term climate changes, such as mega-droughts, delayed or altered patterns of human settlement and cultural exchange along these routes [39]. However, the effects of climate on human activities may display a temporal lag. Furthermore, the effects of short-term climate changes on SRN alterations may not be immediately apparent. Therefore, the timing of climate change does not necessarily align with changes in the SRN. According to the three historical periods delineated in this study, the following general trends emerge [52]:

Classical Period Warming (2nd century B.C.E to 3rd century): During this period, the main SRN regions experienced higher average temperatures compared to the modern climate (represented by 1950). Until the 3rd century, the overall climate trend gradually shifted from warm to cold.

Post-Classical and Early Medieval Climate Fluctuation (4th to 10th century): After experiencing cooling during the 3rd to 4th centuries, the main SRN regions had average temperatures lower than those of modern climate. During these periods, the general trend was a temperature increase from a lower average, accompanied by significant fluctuations. This trend continued until a sudden rise in the latter half of the 8th century, which exceeded modern climate levels and persisted until the 10th century.

Late Medieval Cooling (11th to 15th century): This period was characterized by several major temperature drops (12th, 14th, and 15th centuries), leading to a slow cooling process. Temperatures declined from a peak around 1000 C.E. to a low around 1300 C.E., followed by a gradual increase, with more dramatic fluctuations compared to the first period.

This analysis highlights how climate change significantly influenced the development and shifts in the SRN, with different regions experiencing varied impacts during cooling and fluctuation periods.

Analysis using Kernel Density heatmaps reveals that during the warm period of the Classical Era, the SRN exhibited dominance in the eastern Mediterranean dominance and extended both east and west. This period was characterized by a high density of heritage sites centered in the eastern Mediterranean. During the Late Medieval Cooling, the SRN exhibited western dominance, with a high density of heritage sites in Europe. Conversely, during the climate fluctuation periods, the SRN showed a "dual-core" characteristic, with significant centers in the Near East and Transoxiana (Central Asia).

Considering the movement of the network's central points, it is evident that during the first and third periods, the central points shifted westward, while during the second period, they shifted eastward. This indicates the substantial impact of climate change on different civilization regions within the Eurasian continent.

If the historical periods of the Eurasian continent are classified into the warm period, the cooling period, and the climate fluctuation period, the former corresponds to continuous temperature changes, while the latter corresponds to the fluctuation period. These periods impacted the main regions of the SRN, namely the eastern, central, and western parts of the mid- to low-latitude Eurasian continent and the Mediterranean coast. Notably, during the warm periods, both the eastern and western parts of the Eurasian continent experienced sustained development. However, Western Europe, being closer to the eastern Mediterranean, progressed at a faster rate, and the central points were westward [53]. During the prolonged cooling period, major civilization regions primarily faced significant challenges due to persistent cooling trends. In the western part of the Eurasian continent, characterized by maritime and Mediterranean climates, temperature drops could lead to reduced agricultural production, a shift towards more cold-resistant crops, and the exploitation of marine resources like fish [54]. In the temperate monsoon climate zone of the eastern Eurasian continent, the prolonged cooling period could cause severe cold, threatening agriculture and population, and the central points were also westward. This examination underscores how warm, cooling, and fluctuation periods had distinct and significant impacts on the historical development of the SRN and its regions.

Comparatively, the continental climate of central Eurasia was more severely affected by cooling periods, whether in the context of long-term or short-term cooling, potentially leading to interruptions in agricultural production or shifts to alternative livelihoods [48]. However, the inhabitants of this region were likely better adapted to cold climates than those in other climatic zones. During the cooling phases, they could change their livelihoods or migrate to cope with climate changes. Following the significant temperature drop at the end of the Classical Period, nomadic groups from the inland continental climate zones moved south, threatening the agricultural civilizations of East Asia and the Mediterranean world.

The Near Eastern civilization regions at the junction of Eurasia and Africa, and the Indian subcontinent, fall under subtropical desert climates and tropical monsoon climates, respectively. Cooling periods facilitated cultural development in these regions. Consequently, during the cooling period of the Classical Era, these areas were interconnected via the maritime Silk Road, becoming significant regions within the SRN.

During climate fluctuation periods, the greatest challenge faced by major civilizations was sudden cooling. There are records of abrupt cooling in the post-Classical and early Medieval periods, often linked to significant historical events. The severe impact of climate fluctuations during these periods frequently led to cross-civilization interactions. Compared to Western Europe and the Mediterranean, East Asia and Central Asia may have exhibited a greater adaptive capacity to sudden climate fluctuations. For example, during the severe cooling of the 3rd to 4th centuries, the Western Roman Empire completely collapsed, leading to significant decline in Western Europe and North Africa [55]. These areas gradually recovered with the arrival of another prolonged warm period in the 8th century. In contrast, while East Asia also experienced political crises due to the cooling in the 3rd to 4th centuries, the region's greater longitudinal extent, crossing multiple climate zones, allowed for the preservation of governance and civilization in the southern regions, enabling rapid recovery in the next warm period [56].

During climate fluctuation periods, the short durations of each warming and cooling phase meant that East Asian states, benefiting from a larger north–south expanse, were not uniformly affected by climate changes across all zones [57]. The variability in climate impact allowed some regions within East Asia to adapt better to changes, leading to a more resilient response to these fluctuations [58]. Consequently, they exhibited greater adaptability to climatic variations than Western Europe and the Mediterranean. If the period surrounding the climate fluctuation period (the second period) was characterized by a gradual rise in average temperatures, the most significant difference between the two 800-year-apart continuous temperature-changing periods is the substantial climate fluctuations in the latter. The combined effects of warming, prolonged cooling, and significant climate fluctuations could lead to paradigm shifts in historical development. This period facilitated the transformation of the SRN into the GRN, establishing Europe's central role within the SRN.

During the two continuous temperature changing periods, especially the late medieval period with significant climate fluctuations, Europe emerged as the SRN hub. This reflects the resilience of the maritime and Mediterranean climates in Western Eurasia to climate change. This resilience is due to several factors: First, Europe's proximity to the North Atlantic Drift brings a stable, warm, and humid climate, making it milder than other regions at similar latitudes, thus maintaining stability during long lasting periods [59]. Second, the main characteristics of a maritime climate are warm winters and cool summers. The ocean's high heat capacity moderate's temperature changes, allowing Europe to remain stable during sudden cooling [60]. Lastly, the Mediterranean climate features wet winters and dry summers, providing the region with abundant agricultural and natural resources. These resources can be stored, helping society and the economy better withstand adverse climate impacts.

4.2.2. Historical Events and Heritage Site Dynamics on the SRN

Human activities had a profound impact on the changes in the SRN and heritage site locations [61]. Key driving factors included trade, the spread of religion, war, and political changes [62]. This study organizes these factors into a table and uses visualizations to display the timelines of different events (as seen in Figure 14), aiming to connect these with the developments of SRN heritage sites, particularly with the findings discussed in the previous section. Special attention is given to political events: the rise and fall of various civilizations and empires continually altered the main routes and nodes of the Silk Road. For example, the expansion of the Arabian Empire and the prosperity of the Tang Dynasty facilitated exchanges between Central Asia and China, while the unification under the Mongol Empire further strengthened East–West links [63]. These activities caused the rise and fall of cities and sites along the route, exemplified by the prosperity of Samarkand and Balkh during different periods [64].

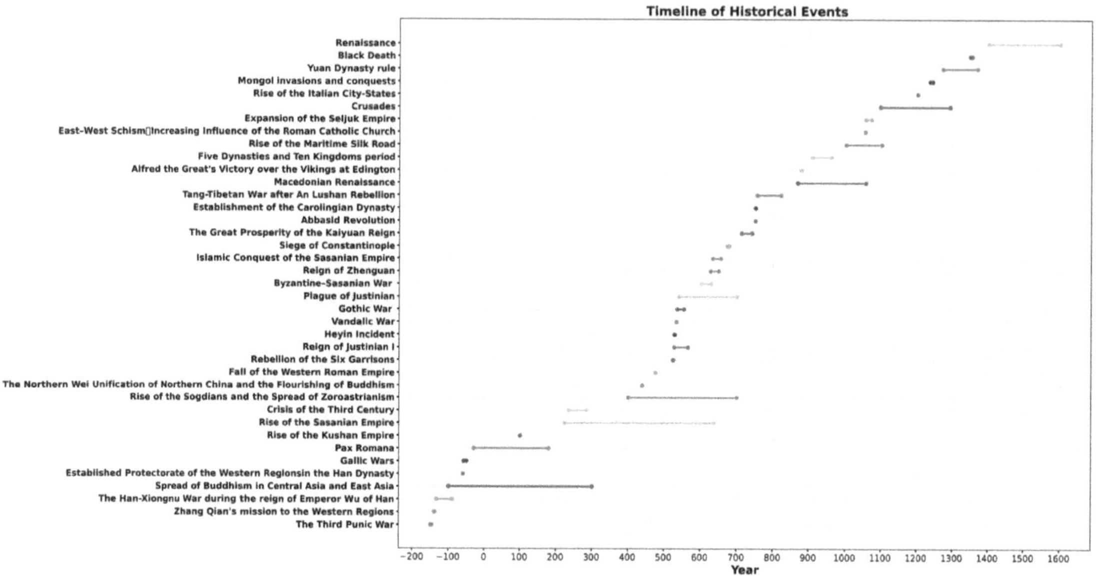

Figure 14. Timeline of historical events along the SRN.

From the 2nd century B.C.E. to the 3rd century, key historical events significantly influenced the establishment and consolidation of the SRN. Zhang Qian's mission to the Western Regions (139 B.C.E.) marked Han China's direct involvement in the Silk Road trade, promoting East–West exchange of goods and culture [65]. The Han Dynasty's establishment of the Protectorate of the Western Regions (60 B.C.E.) further secured and facilitated Central Asian trade routes [65].

The Roman Republic's expansion, through the Third Punic War (149–146 B.C.E.) and the conquest of Gaul (58–50 B.C.E.), established Roman dominance in the western Mediterranean [66,67]. This period also witnessed the spread of Buddhism into Central Asia and China, enhancing cultural exchanges along the Silk Road.

Heritage sites from this period are concentrated in the eastern Mediterranean, Central Asia, and China, reflecting vibrant trade and cultural interactions facilitated by these significant events.

From the 4th to the 8th century, SRN site density fluctuated due to climate changes and political upheavals. The decline of the Roman Empire and the rise of the Byzantine Empire marked significant shifts in the SRN's western regions. The Justinian Plague (541 C.E.) and prolonged Byzantine–Sassanian conflicts disrupted stability, reducing heritage site density in the eastern Mediterranean [68].

In contrast, the stability of the Sassanian Empire in Iran and the flourishing Sogdian trade network in Central Asia contributed to the persistence and growth of heritage sites in these regions [69,70]. The spread of Buddhism continued to influence the establishment of religious sites along the SRN [71].

From the 9th to the 15th century, significant transformations in the SRN occurred due to the rise and fall of various empires and the impact of pandemics. Arab conquests and the establishment of trade routes under Islamic rule enhanced connectivity between the Mediterranean, Central Asia, and China [72]. The reunification of China by the Tang Dynasty and the establishment of the Silk Road as a major trade route promoted the growth of heritage sites in East Asia [73]. Interactions between the Byzantine Empire, Islamic Caliphates, and Tang Dynasty facilitated the exchange of goods, ideas, and culture, resulting in a high density of heritage sites in the Near East, Central Asia, and China.

From the 11th century onwards, significant transformations occurred in the SRN. The rise of the Mongol Empire under Genghis Khan and his successors created a vast, unified territory, enhancing the safety and efficiency of the Silk Road trade routes [74]. The Pax Mongolica (13th to 14th Century) increased commercial and cultural exchanges across Eurasia, reflected in the high density of heritage sites in Central Asia and China [75]. The decline of the Mongol Empire and the onset of the Black Death (1346–1353) caused trade disruptions and shifted the SRN center towards Western Europe [76]. The Renaissance in Italy and the Age of Discovery marked the decline of the overland Silk Road and the rise of maritime trade routes [77].

The evolution of the SRN's sites is closely linked to significant human activities and specific historical events. From the expansion of empires and the spread of religions to climatic changes and pandemics, these factors shaped the spatial patterns and density of heritage sites along the SRN. Understanding these relationships provides valuable insights into the long-term impact of the Silk Road on cultural and economic exchanges between the East and the West [7].

4.3. Limitations

While this study provides a comprehensive analysis of the Silk Road network (SRN) using UNESCO World Heritage sites to reconstruct SRN routes, it has notable limitations. One significant limitation is the uneven distribution of heritage sites across different regions. Europe has a disproportionately high number of heritage sites compared to regions like Central Asia, the Near East, and East Asia. This imbalance could introduce bias, leading to an overrepresentation of European sites and potentially skewing the interpretation of the SRN's historical significance and network density. Consequently, the heritage site density in Europe might not accurately reflect the actual historical network density of the SRN across all regions.

Despite valuable insights into the clustering and dispersion patterns of sites from the 2nd century B.C.E. to the 15th century, several limitations persist. The sensitivity of the DBSCAN algorithm to parameter settings can yield different clustering outcomes based on the chosen parameters. Interpreting silhouette scores is challenging, as low scores may reflect the intrinsic nature of spatial distributions rather than poor clustering quality. Furthermore, variations in data collection methods and preservation conditions may introduce biases, and the complexity of environmental and social factors is difficult to fully capture, leading to potential interpretative biases.

Future research might consider incorporating additional data sources or applying normalization techniques to address discrepancies, ensuring a more balanced and representative analysis of the SRN's historical development and interactions. Additionally, future studies should integrate finer temporal and spatial data, ensure data quality and consistency, and employ multiple methods to validate results. This approach will enhance the reliability and accuracy of clustering analysis.

4.4. Recommendations for Enhanced Conservation and Management

This study analyzes SRN heritage distribution across centuries, revealing changes in clustering patterns over time. Using multiple methods, the research mapped the spatiotemporal evolution of settlements and identified areas of significant historical activity. The discussion links these findings to relevant historical events, hypothesizing reasons behind the observed changes in settlement patterns. These findings highlight the significant impact of historical trade routes on settlement development and land use patterns. Analysis of site clustering reveals that these clusters are not merely geographic concentrations but represent key nodes in the historical Silk Road network. The connections between these nodes form a crucial network structure for trade and cultural exchange. Based on the identified settlement clusters, recommendations are made to preserve and enhance the connectivity of cultural landscape heritage. The proposed conservation strategies focus not only on preserving individual sites but also on maintaining the integrity of the entire SRN [78]. Here are three recommendations:

1. Prioritize Conservation Based on Spatial Clustering: Prioritizing conservation efforts using spatial clustering is vital for managing heritage landscapes along the SRN. This method preserves not only the physical integrity of sites but also the continuity of the heritage landscape and its historical narrative. Spatial clustering analysis allows for a transition from protecting isolated sites to adopting a more comprehensive strategy that safeguards entire clusters of heritage sites. By concentrating on clusters of significant cultural and economic activity, conservation efforts can more effectively preserve historically interconnected sites, thus maintaining the network's overall integrity.
2. Enhance and Protect Regional Connectivity: Regional connectivity is crucial to the formation and sustainability of the Silk Road network. Preserving connections between historical sites is not only essential for individual site protection but also for maintaining the coherence of the entire network. This helps in comprehensively understanding the cultural and historical significance of the heritage. To enhance protection, establishing buffer zones around high-density clusters may be necessary. These zones act as safeguards, preventing modern developments from disrupting the historical context and spatial relationships critical to the evolution of these heritage landscapes.
3. Foster Transnational Collaboration: Beyond these protective measures, strategies to maintain and strengthen linkages between heritage sites are crucial. Notably, transnational linkages, such as the bid for the Routes Network of the Chang'an–Tianshan Corridor, highlight the importance of multinational cultural heritage and cooperation [79]. Preserving historic routes and preventing modern infrastructure from disrupting these connections is vital to maintaining the Silk Road network's historical and cultural integrity. Such strategies are integral to a holistic approach to heritage conservation, ensuring that both the tangible and intangible aspects of the Silk Road's cultural heritage are preserved for future generations.

5. Conclusions

This study provides a detailed analysis of the Silk Road network (SRN) by utilizing data from the UNESCO's World Heritage Sites to reconstruct its historical routes and understand its evolution. The research demonstrates the SRN's function as a closed and stable system from its establishment in the 2nd century BCE until the 15th century when it transformed into a global network. The use of heritage sites offers a precise reflection of the SRN's prosperity and historical significance, and the network density between these sites effectively mirrors the network density of the SRN itself. The understanding of the SRN's historical development and its role in facilitating cross-cultural exchanges is significantly enhanced by these findings.

The study utilizes spatial and statistical analyses, including Kernel Density heatmaps and spatial clustering, to draw its conclusions. The study reveals the spatial distribution of site clusters, offering an in-depth understanding of their key role in the Silk Road's cultural network. The main findings validate previous assumptions and identify key factors driving changes in SRN heritage sites across different centuries, particularly climate change and human activities. It identifies trends in the evolution of the SRN over time, along with its changing spatial characteristics. These changes are characterized by a core area in the eastern Mediterranean during the Classical Period, a shift to a dual core in the eastern Mediterranean and Central Asia during the medieval period, and a subsequent transition to a European core in the late medieval period.

Preserving this network is crucial for maintaining its historical value; hence, the proposed conservation strategy emphasizes this aspect. Specifically, conservation efforts should focus on clusters of heritage sites with significant cultural and economic activity to preserve historically interconnected sites and maintain the integrity of the entire Silk Road network. Additionally, we should preserve and strengthen connections between heritage sites by implementing buffer zones around high-density clusters, safeguarding the network's coherence, and preventing modern developments from disrupting historical routes and spatial relationships. Finally, we should encourage and support transnational linkages, such as multinational cultural heritage initiatives, to ensure the continued protection of the Silk Road's historical and cultural integrity through cooperative efforts.

Overall, the practical value lies in providing a more accurate representation of the SRN compared to traditional documentary sources, thanks to the precise geographic and historical data of the heritage sites. The innovative insights include demonstrating how different regions within the SRN experienced distinct impacts during periods of climate change, and how the network density between heritage sites reflects the SRN's overall network density. This research addresses the practical problem of reconstructing historical trade routes and has potential applications in historical geography, archeology, and cultural heritage preservation.

Author Contributions: Conceptualization, X.W.; methodology, Y.S.; software, Y.S.; validation, Y.S., J.L., J.H. and X.W.; formal analysis, J.L. and J.H.; investigation, J.H.; resources, J.H.; data curation, J.L. and J.H.; writing—original draft preparation, Y.S. and J.H.; writing—review and editing, X.W. and J.L.; visualization, Y.S.; supervision, X.W.; project administration, X.W.; funding acquisition, X.W. All authors have read and agreed to the published version of the manuscript.

Funding: This research was funded by the 2023 Shaanxi Provincial Cultural Relics Protection and Utilization Major Issues (2023HZ1519).

Data Availability Statement: The original contributions presented in the study are included in the article; further inquiries can be directed to the corresponding author.

Acknowledgments: We would like to express our sincere gratitude to all those who have contributed to this research. Special thanks to the Shaanxi Provincial Government for their financial support. We also extend our heartfelt appreciation to every friend and professors who has assisted us. Lastly, we thank our families and friends for their unwavering support and encouragement during this study.

Conflicts of Interest: The authors declare no conflicts of interest.

Appendix A

This section illustrates the changes in the scatter plot for different years in detail.

- 2nd Century B.C.E. to 3rd Century

From the 2nd century B.C.E. to the 1st century B.C.E., a notable increase in World Heritage Sites is evident at both the eastern and western ends, particularly in the western regions of China and along the western Mediterranean coast. This trend reflects the eastward expansion of the ha network into China's heartland and westward to the coastal Mediterranean regions. Such transformations likely stem from significant historical events during that era, for instance, Zhang Qian's mission to the Western Regions in 139 B.C.E. [80], and the Han Dynasty's establishment of the Protectorate of the Western Regions in 60 B.C.E. solidified control [65], enhanced security, and promoted trade and cultural interactions. These efforts resulted in the development of new trade routes and urban centers.

Concurrently, the Roman Republic asserted its influence in the Mediterranean region from the 2nd century B.C.E. to the 1st century B.C.E. The decisive victory of Rome over Carthage in the Third Punic War (149–146 B.C.E.) led to Roman control over areas that include present-day Algeria, Tunisia, and the Iberian Peninsula. Subsequently, from 58 to 50 B.C.E., Gaul was conquered by Rome, the latter establishing dominance over the entire western Mediterranean region [59,60]. This period of Roman governance in the western Mediterranean not only enhanced security and facilitated urban growth but also integrated these regions into the trade network later known as the Silk Road.

During the period from the 1st century B.C.E to the 2nd century, there is a noticeable increase in World Heritage Sites, with the eastern Mediterranean remaining a central focus, expanding towards both the east and the west. The Silk Road network expanded eastward into present-day Afghanistan, Pakistan, and northern India, leading to a sustained rise in the concentration of heritage sites in China and Central Asia. This growth can be attributed to the Han Dynasty's maintenance of East–West trade routes following internal upheavals, especially after the establishment of the Eastern Han Dynasty, which ensured control over the Western Regions [65]. Furthermore, the founding of the Kushan Empire in the early 1st century resolved tumult in Central Asia, especially in the Transoxiana region [81]. The number of heritage sites in the western Mediterranean region significantly increased, notably in southern France and the Iberian Peninsula. This expansion may be associated with the establishment of the Roman Empire in 27 CE, heralding a period of "Pax Romana" across the Mediterranean region [68]. This era of intercontinental tranquility witnessed substantial economic and cultural advancements, with unity and peace supporting the safeguarding and upkeep of trade routes.

During the 2nd to 3rd centuries, the eastern Mediterranean region, encompassing the Greek city-states of Asia Minor along the Aegean coast and the Levant region on the eastern Mediterranean coast, remained pivotal in the Silk Road network. A notable shift occurred as heritage sites expanded eastward towards East Asia, leading to an increase in heritage density along the eastern Silk Road, spanning the Central Asian Transoxiana region, India, and China. Simultaneously, an increase in heritage sites is noted in central and eastern Europe. The primary drive behind the eastward expansion of heritage sites can be linked to the rapid spread of Hinduism and Buddhism across the Indian subcontinent, extending to East Asia [71]. Buddhism initially penetrated Central Asia via the Silk Road in the 1st century B.C.E. and thrived in East Asia from the 2nd to 3rd centuries, most notably in China. Influential Buddhist missionaries during the late Eastern Han Dynasty, including An Shigao and Zhu Ḍaosheng, emerged as key historical figures [82]. The growth of heritage sites in central and eastern Europe may be linked to the political challenges faced by the Roman Empire in the late 2nd and 3rd centuries, prompting Emperors Septimius Severus and Marcus Aurelius to bolster military activities in the Danube region and establish new military outposts [83].

- 4th to 8th Century

The analysis of data from the 4th to 5th centuries provides insights into the changes in heritage sites along the Silk Road network. While the overall number of heritage sites in the Mediterranean region shows minimal alterations, a significant decline in sites in regions like Greece, Libya, and the Iberian Peninsula can be seen from the map. In contrast, a notable increase in core site density in northern and western China, Iran, and especially in the Central Asian Transoxiana region is well attested. This indicates a considerable enhancement in the significance of the eastern part of the Silk Road network during this era. The transformations in heritage sites in the Mediterranean region are likely due to the fragmentation and collapse of the Western Roman Empire during the 4th and 5th centuries, leading to extended periods of political and economic instability [68]. Concurrently, the European domains of the Byzantine Empire experienced invasions by nomadic groups, potentially resulting in the destruction or desertion of certain heritage sites. Nonetheless, the Byzantine Empire maintained a relatively serene presence in Asia Minor and the Levant region, preserving high-density heritage sites and playing an essential role in the Silk Road network. In China, the uptick in heritage sites in the northern and western regions can be associated with the consolidation of northern China by the Northern Wei dynasty in the early 5th century, facilitating stability [84]. Regarding religion, the dissemination of Buddhist art along the Silk Road reached new heights, exemplified by locations like the Yungang Grottoes established in northern and western China from the 4th to the 5th century. The surge in heritage sites in Iran and Central Asia, particularly in the Transoxiana region, may be attributed to the ascendancy of the Sasanian Empire (224–651 C.E.) and the emergence of the Sogdians (4th to 7th centuries) [69,70]. In contrast to the tumult experienced by the Roman Empire and China during nomadic incursions, Central Asia witnessed the emergence of a series of political entities with nomadic features, such as the Parthians, Kushans, and Hephthalites, blending with classical Persian and Central Asian civilizations to evolve amidst the influx of nomadic groups [71].

During the 5th to 6th centuries, the growth trajectory of significant heritage sites was predominantly focused along the eastern Mediterranean coast, Iran, and the Central Asian Transoxiana region, with emerging growth centers in southern India and Southeast Asia. While minor fluctuations in heritage site density were observed in Western Europe, North Africa, and the Yellow River Basin in China, the magnitude of these changes was not substantial. The Silk Road network gradually demonstrated a pattern centered with the eastern Mediterranean and Iran–Central Asian regions. The uptick in heritage sites along the eastern Mediterranean coast can be associated with the reign of Byzantine Emperor Justinian I from 527 to 565 C.E., which triggered economic and political revitalization [67]. The expansion of Byzantine territories led to the reintegration of some coastal urban centers, fostering a phase of relative calm in the eastern Mediterranean and facilitating the founding of new heritage sites and the restoration of existing ones. In the Iranian plateau, the rise in heritage site density may be linked to the political stability under the Sassanian Empire, characterized by centralized governance and efficient administrative practices that strengthened fiscal stability [71]. Religious advancements during this era positioned Iran as the nucleus of Zoroastrianism, with the official establishment of Zoroastrianism as the state religion under the Sassanian Empire, fostering its prosperity in Iran and the Sogdian-administered zones of Central Asia. The emergence of new religious establishments and artistic productions aligned with the Sassanian Empire's focus on commerce, playing a pivotal role in Silk Road trade and stimulating the ascension of commercial hubs along the Silk Road. In North Africa, heritage sites continued to decline due to the ravages of warfare and climatic shifts. Western Europe encountered political instability during the Vandals' War in 533 C.E. and the Gothic Wars spanning from 535 to 554 C.E. during the reign of Justinian I [68]. The Yellow River Basin in China grappled with tumult and fragmentation at the end of the Northern Wei dynasty in the early 6th century, resulting in social stagnation [24,85].

During the 6th to 7th centuries, remarkable shifts in heritage sites along the Silk Road network indicate a gradual eastward movement. The quantity of heritage sites in the western Mediterranean remains stable with no significant alterations, whereas the density of heritage sites in the eastern Mediterranean diminishes, showing a continual decline in North Africa, particularly present-day Tunisia and Algeria. Concurrently, the density of heritage sites in the eastern sections of the Silk Road network, encompassing Iran, Central Asia, East Asia, and Southeast Asia, undergoes a substantial increase. Throughout this period, the trend of the Silk Road network's central regions shifting eastward remains prominent, leading to the emergence of a new localized central hub in the Yellow River Basin of China. The decline in heritage site density in the eastern Mediterranean region can be traced back to the 6th and 7th centuries, characterized by the initial decline of the Byzantine Empire. The outbreak of the "Justinian Plague" in 541 C.E. affected the Byzantine Empire for over 70 years, followed by more than two decades of Byzantine–Sassanian Wars (602–628 C.E.) [86]. Subsequently, in the mid-7th century, the rise of the Arab Caliphate initiated prolonged conflicts with the Byzantine Empire, culminating in the siege of Constantinople in 678 C.E., resulting in renewed turmoil in the eastern Mediterranean region [71,86]. The increase in the number of heritage sites in Iran may be attributed to the rapid spread of Islam following the Arab conquest of the Sassanian Empire in 638 C.E., leading to the introduction of new religious structures and enriching the cultural landscape of the region. The rise in heritage site density in the Central Asian Transoxiana region can be credited to the political stability under the governance of the Western Turkic Khaganate, with the Sogdians forming alliances with the Turks to safeguard commercial trade along the Silk Road, fostering economic prosperity in Central Asia. The gradual establishment of a new localized central hub in East Asia can be linked back to the reunification of China in the 6th and 7th centuries. The unification of China under the Sui Dynasty in 589 C.E. and the establishment of the Tang Dynasty in 618 C.E. ushered in a prosperous era, especially during the "Reign of Zhenguan" from 627 to 649 C.E. This period is regarded as the heyday in Chinese culture and Buddhist art, marked by political stability, economic affluence, and cultural openness that facilitated the construction of numerous temples, imperial mausoleums, and palaces, consolidating the eastern section of the Silk Road network and establishing a new localized central hub within China [73,85].

During the 7th and 8th centuries, changes in heritage sites along the Silk Road network became apparent. There was a notable increase in the number of heritage sites in the western Mediterranean and Western Europe, particularly in regions such as France and Italy. In contrast, the density of heritage sites in the eastern Mediterranean did not show significant growth during the 8th century, with some areas experiencing a decline along the Aegean coast, the Levant, and Egypt. While the number of heritage sites in North Africa stabilized, there was a slight decline in heritage site density in Western Asia and Iran. Concurrently, the density of heritage sites in the Central Asian Transoxiana region near western China continued to rise, alongside with an overall increase in heritage site density in South Asia, East Asia, and Southeast Asia. The alterations in heritage site distribution along the Silk Road illustrate a shift of the central focus from the eastern Mediterranean towards Italy and France, while the core region of the Silk Road network expanded further eastward. The Yellow River Basin in China, serving as a core area, initiated the extension of the Silk Road towards Southeast Asia, Korea, and Japan, marking the early development of the maritime Silk Road in this era. The upsurge in heritage sites in the western Mediterranean and Western Europe can be linked to the expansion of the Frankish Kingdom. During the 7th and 8th centuries, this kingdom underwent significant growth, eventually unifying Western Europe following the establishment of the Carolingian dynasty in 751 C.E., ushering in stability to a previously turbulent region [87]. The decline in heritage site density in the eastern Mediterranean region may be attributed to political instability, conflicts, and the impact of epidemics on the Byzantine Empire. Ongoing confrontations with the Arab Caliphate and the Eastern Epidemic of 746 C.E. further destabilized the empire. The decrease in heritage site density in Western Asia and Iran may be linked to the destruction of heritage sites due to the Arab conquest of the Sassanian Empire in the mid-7th century and the Abbasid Revolution in the 8th century, leading to prolonged conflicts [72]. The increase in heritage site density in the Central Asian Transoxiana region adjacent to western China and within the Chinese territory may be a result of the "Golden Age of the Tang Dynasty" in the early 8th century and the expansion of the maritime Silk Road towards Southeast Asia and Japan [73,85,88].

- 9th to 15th Century

During the 9th and 10th centuries, heritage site density significantly increased in the Italian Peninsula and the British Isles, with a slight upsurge along the Aegean coast and in northern Iran. In contrast, heritage density declined in Central Asia (especially territories of Kazakhstan and Uzbekistan) and the northern Yellow River Basin area in China. This shift indicates a movement of the central hub of the Silk Road network westward, contributing to the gradual decline of the eastern focal points in the Yellow River Basin of China and the Central Asian Transoxiana region. The substantial rise in heritage density in the Italian Peninsula may be attributed to the foundation of the Holy Roman Empire. Similarly, the growth in heritage density in the British Isles could be connected to King Alfred's actions that compelled the Vikings to retreat in the late 9th to 10th centuries, fostering a relatively stable political environment, enhancing Christian influence, and facilitating the establishment or expansion of notable institutions such as the Canterbury Cathedral [88,89]. The escalation of heritage site density along the Aegean coast may be linked to the Byzantine Empire's revival under the Macedonian dynasty, symbolizing a period of resurgence as the empire reclaimed control over the eastern Mediterranean, regained territories, and maintained stability [68]. The rise in heritage density in northern Iran correlates with the emergence and consolidation of kingdoms in specific regions of Georgia and Armenia, exemplified by the Bagratuni Kingdom of Armenia, resulting in significant cultural and architectural advancements in the area [88,89]. The decline in heritage site density in Central Asia (including Kazakhstan and Uzbekistan) may be ascribed to the upheaval in China during the late Tang and Five Dynasties period, reduced trade activities, and the closure of trade routes by nomadic regimes between the Central Plains and the Western Regions. This led to diminished trade flow along the eastern part of the Silk Road and a decline in the significance of cities in Central Asia, such as Samarkand and Bukhara [71,73,85].

During the 10th and 11th centuries, the Silk Road network underwent significant development: the center shifted westward, leading to the emergence of independent hubs in East Asia that were originally part of a connected network. Italy and Greece experienced a notable increase in heritage site density, while Western Europe, particularly in France and Germany, saw significant growth in density. Changes in heritage sites in western China, Central Asia, and Iran were less prominent, contrasting with the concentration of heritage sites in East Asia along the eastern coast, the middle-lower reaches of the Yangtze River, Japan, and Southeast Asia. The number of heritage sites along the eastern routes of the Silk Road network in Central Asia and western China remained relatively stable, indicating a declining trend compared to the rapid growth in other regions. The maritime Silk Road saw swift progress, resulting in a southward shift in the Silk Road network and the emergence of independent centers in East Asia [88,89]. The upsurge in heritage density in Western Europe and the eastern Mediterranean coast can be attributed to the urban development in Italy, Germany, and France, aided by the influence of the Roman Catholic Church, leading to the construction of momentous religious edifices like St. Mark's Basilica in Venice [90]. In the early 11th century, the Byzantine Empire enjoyed a revival, exemplified by Emperor Basil II's expansion into southern Italy and northern Syria [68]. The increase in heritage sites in Central Asia (specifically present-day Uzbekistan and Turkmenistan) may be linked to the westward expansion of the Seljuk Empire during the 11th century [71]. The rise in heritage density along the eastern coast of China, Japan, and Southeast Asia could be associated with the thriving commercial economy of the period, leading to increased maritime trade and interactions with neighboring regions [73].

During the 11th to 12th centuries, the focus of the Silk Road network shifted towards Western Europe, while independent hubs in East Asia began to emerge. This era witnessed significant growth in northern and central Italy, northern France, and England, with only partial increases in heritage sites in Central Asia. Conversely, there was a decrease in heritage numbers in the Middle East and North Africa, while East Asia saw little significant change in heritage site density. The European side displayed extensive clustering and growth rates that surpassed those of other regions. These transformations can be attributed to various factors: the expansion in Europe was fueled by economic and urban population growth in the 11th to 12th centuries, notably flourishing Italian cities in the 12th century [90]. The decline in heritage in the Middle East and North Africa could be attributed to the substantial destruction caused by the Crusades beginning in 1096, impacting societies of the Islamic cultural zone along the eastern Mediterranean coast and the Byzantine Empire [91]. The rise of coastal city-states in Italy and the consolidation of papal authority in Europe led to the demise of existing urban centers and settlements. In East Asia, internal conflicts within the Song Dynasty disrupted connections with the Western Regions, while frequent conflicts between the Seljuk Empire and the Kara Khitay Khanate in Central Asia contributed to a period of stagnation in development [71].

During the 12th to 13th centuries, the focal point of the Silk Road network continued its westward shift, resulting in a gradual decline in activities along the eastern segment of the Silk Road. Growth was prominently centered in Western Europe, notably in France, the central–western regions of Germany, and northern and central Italy, accompanied by a slight rise in heritage density in the Byzantine territory. In contrast, heritage density in Central Asia and East Asia remained relatively stable, with a minor increase observed in East Asia where the establishment of autonomous centers progressed at a slow pace. The influence of the Mongol expansion towards the West on heritage sites in Central and East Asia during the 13th century did not have a significant impact [74]. The total number of heritage sites in Central Asia remained static, particularly in China under Mongol governance, where a modest increase in heritage sites occurred during a period of relative tranquility [75]. The Mongol conquests resulted in a net reduction rather than an annexation of territories, slowing the rate of development in these regions. Simultaneously, Europe stood at the threshold of the flourishing Italian Renaissance and the Age of Discovery. The economic prosperity of the Renaissance was rooted in the prosperity of Italian city-states in the late Middle Ages, a trend originating in the 12th century, believed to mark the initial emergence of modern global trade systems [77]. Furthermore, temporary prosperity was brought by the new settlements established during the Crusades in Anatolia and the Levant region [91].

During the 13th to 14th centuries, minor fluctuations in overall heritage density occurred across the Silk Road network, with slight increases in heritage density noted in northern Italy, southern Germany, and the eastern coastal regions of China, while heritage numbers in North Africa saw a slight decline. The deceleration of heritage site growth in Europe during this period can be attributed to the outbreak of the Black Death from 1346 to 1353 [76]. This devastating bubonic plague originated in Europe and ravaged the continent for eight years, resulting in the deaths of almost half the population. Under the governance of the Yuan Dynasty, in contrast, China enjoyed a relatively stable era during the 13th to 14th centuries, experiencing prosperity in trade, particularly in maritime Silk Road commerce. Port cities like Quanzhou (Zayton) thrived during this prosperous period [75].

During the 14th to 15th centuries, Europe became the central hub of the Silk Road network, resulting in the fading depiction of the original overland Silk Road on maps. The rise in heritage density in this period was predominantly focused on central Europe, notably in the German territories, northern and central Italy, and the northern Iberian Peninsula, accompanied by the emergence of new heritage sites in Eastern Europe. While there was a minor uptick in heritage density observed in areas like the eastern coastal regions of China in East Asia, the growth rate was modest. Throughout this period, the high-density heritage sites formed an elongated pattern stretching from the northwest to the southeast [88,89].

Appendix B. A Summary Presentation of the Major Cities in Figure 3

City	Latitude	Longitude
Karakorum	47.1922	102.8193
Nara	34.5666	135.7666
Quanzhou	24.9139	118.5858
Xi'an	34.2666	108.9333
Dunhuang	40.0999	94.6499
Urumqi	43.8009	87.6004
Madras (Chennai)	13.0878	80.2784
Fatehpur-Sikri	27.0937	77.66
Kashgar	39.4707	75.9895
Almaty	43.2566	76.9286
Samarkand	39.6542	66.9597
Balkh	36.7564	66.8972
Shahrisabz	39.0537	66.8201
Bamiyan	34.8216	67.8273
Herat	34.3481	62.1996
Bam	29.106	58.357
Yazd	31.8972	54.3675
Isfahan	32.6524	51.6746
Baku	40.3776	49.892
Muskat	23.6138	58.5922
Aleppo	36.2166	37.1666
Jeddah	21.5423	39.1979
Zanzibar	−6.1333	39.3167
Alexandria	31.2156	29.9553
Ephesus	37.9372	27.3378
Bursa	40.1955	29.0601
Venice	45.4371	12.3326
Valencia	39.4698	−0.3374

References

1. Mertens, M. Did Richthofen Really Coin "the Silk Road". *Silk Road* **2019**, *17*, 1–9.
2. Herrmann, A. *Die Alten Seidenstrassen Zwischen China Und Syrien: Beiträge Zur Alten Geographie Asiens;* Weidmann: Vienna, Austria, 1910; Volume 1.
3. Hedin, S. *The Silk Road: Ten Thousand Miles through Central Asia;* Bloomsbury: New York, NY, USA, 2009; ISBN 978-1-84511-898-3.
4. Boulnois, L. Loveday, H., Translator; *Silk Road: Monks, Warriors & Merchants on The Silk Road;* Odyssey Books & Guides: Hongkong, China, 2005.
5. Wood, F. *The Silk Road: Two Thousand Years in the Heart of Asia;* Univ of California Press: Berkeley, CA, USA, 2002.
6. Waugh, D.C. Richthofen's 'Silk Roads': Toward the Archaeology of a Concept. *Silk Road* **2007**, *5*, 1–10.
7. Hansen, V. *The Silk Road. A New Histor;* Oxford University Press: New York, NY, USA, 2012.
8. Chin, T. The Invention of the Silk Road, 1877. *Crit. Inq.* **2013**, *40*, 194–219. [CrossRef]
9. Hoshmand, A.R. Eurasian Connection via the Silk Road: The Spread of Islam. In *Silk Road to Belt Road;* Islam, M.N., Ed.; Springer Nature: Singapore, 2019; pp. 95–104. ISBN 9789811329975.
10. Khan, K.A.; Doszhanova, A.I. *Religion and the Great Silk Road: Means of Communication, Contours of Interaction, Trade;* Academic Council: Cambridge, MA, USA, 2022; pp. 231–238. [CrossRef]
11. Hansen, V. The Classical Silk Road: Trade and Connectivity across Central Asia, 100 BCE–1200 CE. In *Oxford Research Encyclopedia of Asian History;* Oxford University Press: Oxford, UK, 2021; ISBN 978-0-19-027772-7.
12. Høisæter, T.L. Polities and Nomads: The Emergence of the Silk Road Exchange in the Tarim Basin Region during Late Prehistory (2000–400 BCE). *Bull. Sch. Orient. Afr. Stud.* **2017**, *80*, 339–363. [CrossRef]
13. Andrea, A.J. The Silk Road in World History: A Review Essay. *Asian Rev. World Hist.* **2014**, *2*, 105–127. [CrossRef]
14. Xin, N.; Liao, J. History of the Silk Road. *Chronos* **2018**, *36*, 167–195. [CrossRef]
15. Eğrï, S. The Silk Road Impressions in Hitây Sefâretnâmesi and Kanunnâme-i China u Hitây. *J. Turk. Stud.* **2012**, *7*, 411–422. [CrossRef]
16. Whitfield, S. *Life along the Silk Road;* University of California Press: Berkeley, CA, USA, 1999.
17. Tosheva, D. Qarakhanid Roads to China: A History of Sino-Turkic Relations: By Dilnoza Duturaeva, Leiden, Brill, 2022, 278 pp, US$119.00 (Hbk), ISBN: 978-90-04-50852-1. *Cent. Asian Surv.* **2023**, *42*, 211–213. [CrossRef]
18. UNESCO. World Heritage Global Strategy. Available online: https://whc.unesco.org/en/globalstrategy/ (accessed on 28 July 2024).
19. Williams, T. *Final Technical Report on the Results of the UNESCO/Japanese Funds-in-Trust Project: Support for Silk Roads World Heritage Sites in Central Asia (Phase II), April 2015—November 2019;* Williams, T., Lin Chih-Hung, R., Jorayev, G., Eds.; UCL Institute of Archaeology: London, UK, 2020.
20. Vileikis, O. Monitoring Serial Transnational World Heritage—The Central Asian Silk Roads Experience. *Hist. Environ. Policy Pract.* **2016**, *7*, 260–273. [CrossRef]
21. Williams, T. *Final Technical Report on the Results of the UNESCO/Korean Funds-in-Trust Project: Support for the Preparation for the World Heritage Serial Nomination of the Silk Roads in South Asia, 2013–2016;* UCL Institute of Archaeology: London, UK, 2016.
22. Zhan, G. Nomination of the Silk Road to the World Heritage List. *Hist. Environ.* **2014**, *26*, 88–100.
23. Vileikis, O.; Dumont, B.; Serruys, E.; Van Balen, K.; Tigny, V.; De Maeyer, P. Connecting world heritage nominations and monitoring with the support of the silk roads cultural heritage resource information system. *ISPRS Ann. Photogramm. Remote Sens. Spatial Inf. Sci.* **2013**, *II-5/W1*, 319–324. [CrossRef]
24. Sima, Q. *Records of the Grand Historian of China;* Columbia University Press: New York, NY, USA, 1971; Volume 2, ISBN 978-0-231-08167-2.
25. Peck, D.T. The Empirical Reconstruction of Columbus' Navigational Log and Track of His 1492–1493 Discovery Voyage. *J. Navig.* **2011**, *64*, 193–205. [CrossRef]
26. Byars, J. The Venetian Discovery of America: Geographic Imagination and Print Culture in the Age of Encounters. *J. Early Mod. Hist.* **2020**, *24*, 571–572. [CrossRef]
27. Wilson, G. Multifunctional "quality" and Rural Community Resilience. *Trans. Inst. Br. Geogr.* **2010**, *35*, 364–381. [CrossRef]
28. Thacher, D.; Milne, S.; Park, R. Applying GIS and Statistical Analysis to Assess the Correlation of Human Behaviour and Ephemeral Architectural Features among Palaeo-Eskimo Sites on Southern Baffin Island, Nunavut. *J. Archaeol. Sci. Rep.* **2017**, *14*, 21–30. [CrossRef]
29. Lang, G.; Marcon, E. Testing Randomness of Spatial Point Patterns with the Ripley Statistic. *Esaim Probab. Stat.* **2010**, *17*, 767–788. [CrossRef]
30. Haase, P. Spatial Pattern Analysis in Ecology Based on Ripley's K-Function: Introduction and Methods of Edge Correction. *J. Veg. Sci.* **1995**, *6*, 575–582. [CrossRef]
31. Díaz-Rodríguez, M.; Fábregas-Valcarce, R. Evaluating the Effectiveness of Three Spatial Cluster Analysis Methods on Palaeolithic Site Location Distributions in Galicia, NW Iberian Peninsula. *J. Archaeol. Sci. Rep.* **2022**, *41*, 103323. [CrossRef]
32. UNESCO. Silkroad Interactive Map. Available online: https://en.unesco.org/silkroad/silkroad-interactive-map (accessed on 26 July 2024).
33. Spengler, R.N.; Maksudov, F.; Bullion, E.; Merkle, A.; Hermes, T.; Frachetti, M. Arboreal Crops on the Medieval Silk Road: Archaeobotanical Studies at Tashbulak. *PLoS ONE* **2018**, *13*, e0201409. [CrossRef]

34. Daryaee, T. The Persian Gulf in Late Antiquity. *J. World Hist.* **2003**, *14*, 1–16. [CrossRef]
35. Golden, P. *Central Asia in World History*; Oxford University Press: Oxford, UK, 2011. [CrossRef]
36. Chen, F.; Chen, J.; Dong, G.; Huang, W.; Hou, J.; Wang, T. Climate Change and Silk Road Civilization in the Arid Central Asia. 2020. Available online: https://meetingorganizer.copernicus.org/EGU2020/EGU2020-7335.html (accessed on 28 July 2024).
37. Hao, Z.; Zheng, J.; Yu, Y.; Xiong, D.; Liu, Y.; Ge, Q. Climatic Changes during the Past Two Millennia along the Ancient Silk Road. *Prog. Phys. Geogr. Earth Environ.* **2020**, *44*, 605–623. [CrossRef]
38. Babajanyan, A.; Franklin, K. Everyday Life on the Medieval Silk Road: VDSRS Excavations at Arpa, Armenia. *Aram. Armen. J. Near East. Stud.* **2018**, *12*, 155–184. [CrossRef]
39. Tan, L.; Dong, G.; An, Z.; Lawrence Edwards, R.; Li, H.; Li, D.; Spengler, R.; Cai, Y.; Cheng, H.; Lan, J.; et al. Megadrought and Cultural Exchange along the Proto-Silk Road. *Sci. Bull.* **2021**, *66*, 603–611. [CrossRef]
40. Yao, Y.-F.; Wang, X.; Guo, W.; Jiang, H.-E.; Li, M.; Ferguson, D.K.; Wang, Y.-F.; Li, C.-S. Archaeobotanical Evidence Reveals the Human–Environment Interactions during the 9th–13th Centuries at Turpan, Xinjiang on the Ancient Silk Road. *Vege. Hist. Archaeobot.* **2020**, *29*, 539–552. [CrossRef]
41. Nunn, N.; Qian, N. The Columbian Exchange: A History of Disease, Food, and Ideas. *J. Econ. Perspect.* **2010**, *24*, 163–188. [CrossRef]
42. Knutson, S.A. Archaeology and the Silk Road Model. *World Archaeol.* **2020**, *52*, 619–638. [CrossRef]
43. Meng, X.; Wu, J.; Zhu, Q.; Hu, S.; Zhu, L. Analytical Method to Determine the Environmental Impact on the Stability Safety of a Bulk Carrier Navigating in the Maritime Silk Road. *J. Tsinghua Univ. (Sci. Technol.)* **2024**, *64*, 1060–1069. [CrossRef]
44. Gu, H. *Frustrated Relations?: China with Central and Eastern Europe*; Contributions to International Relations; Springer Nature: Singapore, 2023; ISBN 978-981-9937-29-5.
45. Brunn, S.D. *Geography of Time, Place, Movement and Networks, Volume 3: Mapping Time Journeys in Music, Art and Spirituality*; Springer Nature: Singapore, 2024; ISBN 978-3-031-58033-8.
46. Agnew, N. *Conservation of Ancient Sites on the Silk Road: Proceedings of an International Conference on the Conservation of Grotto Sites*; Getty Conservation Institute: Los Angeles, CA, USA, 1997.
47. Yao, Y.; Wang, X.; Lu, L.; Liu, C.; Wu, Q.; Ren, H.; Yang, S.; Sun, R.; Luo, L.; Wu, K. Proportionated Distributions in Spatiotemporal Structure of the World Cultural Heritage Sites: Analysis and Countermeasures. *Sustainability* **2021**, *13*, 2148. [CrossRef]
48. Che, P.; Lan, J. Climate Change along the Silk Road and Its Influence on Scythian Cultural Expansion and Rise of the Mongol Empire. *Sustainability* **2021**, *13*, 2530. [CrossRef]
49. Skaff, J.K. The Sogdian Trade Diaspora in East Turkestan during the Seventh and Eighth Centuries. *J. Econ. Soc. Hist. Orient* **2003**, *46*, 475–524. [CrossRef]
50. Peng, Q.; Moradi, H. Studies on the History and Culture along the Continental Silk Road. *Eurasian Geogr. Econ.* **2024**, *203*, 1–3. [CrossRef]
51. Hill, D.J. Climate Change and the Rise of the Central Asian Silk Roads. In *Socio-Environmental Dynamics along the Historical Silk Road*; Yang, L., Bork, H.R., Fang, X., Mischke, S., Eds.; Springer: Cham, Switzerland, 2019; pp. 247–259. [CrossRef]
52. Ge, Q.; Hua, Z.; Zheng, J.; Fang, X.; Xiao, L.; Liu, J.; Yang, B. Forcing and impacts of warm periods in the past 2000 years. *Chin. Sci. Bull.* **2015**, *60*, 1728–1735. [CrossRef]
53. Giannakopoulos, C.; Sager, P.; Bindi, M.; Moriondo, M.; Kostopoulou, E.; Goodess, C. Climatic Changes and Associated Impacts in the Mediterranean Resulting from a 2 °C Global Warming. *Glob. Planet. Chang.* **2009**, *68*, 209–224. [CrossRef]
54. Maracchi, G.; Sirotenko, O.; Bindi, M. Impacts of Present and Future Climate Variability on Agriculture and Forestry in the Temperate Regions: Europe. *Clim. Chang.* **2005**, *70*, 117–135. [CrossRef]
55. Erdkamp, P. War, Food, Climate Change, and the Decline of the Roman Empire. *J. Late Antiq.* **2019**, *12*, 422–465. [CrossRef]
56. Zhang, H.; Werner, J.; García-Bustamante, E.; González-Rouco, F.; Wagner, S.; Zorita, E.; Fraedrich, K.; Jungclaus, J.; Ljungqvist, F.; Zhu, X.; et al. East Asian Warm Season Temperature Variations over the Past Two Millennia. *Sci. Rep.* **2018**, *8*, 7702. [CrossRef]
57. Wang, L.; Chen, W.; Huang, G.; Zeng, G. Changes of the Transitional Climate Zone in East Asia: Past and Future. *Clim. Dyn.* **2017**, *49*, 1463–1477. [CrossRef]
58. He, S.; Gao, Y.; Li, F.; Wang, H.; He, Y. Impact of Arctic Oscillation on the East Asian Climate: A Review. *Earth-Sci. Rev.* **2017**, *164*, 48–62. [CrossRef]
59. Trouet, V.; Esper, J.; Graham, N.; Baker, A.; Scourse, J.; Frank, D. Persistent Positive North Atlantic Oscillation Mode Dominated the Medieval Climate Anomaly. *Science* **2009**, *324*, 78–80. [CrossRef]
60. Gronenborn, D.; Strien, H.-C.; Dietrich, S.; Sirocko, F. 'Adaptive Cycles' and Climate Fluctuations: A Case Study from Linear Pottery Culture in Western Central Europe. *J. Archaeol. Sci.* **2014**, *51*, 73–83. [CrossRef]
61. Zhang, S.; Liu, H.; Li, G.; Zhang, Z.; Chen, X.; Shi, Z.; Zhou, A.; Dong, G. Warfare Impact Overtakes Climate-Controlled Fires in the Eastern Silk Roads since 2000 B.P. *Proc. Natl. Acad. Sci. USA* **2023**, *2*, pgad408. [CrossRef]
62. Yu, J.; Safarov, B.; Yi, L.; Buzrukova, M.; Janzakov, B. The Adaptive Evolution of Cultural Ecosystems along the Silk Road and Cultural Tourism Heritage: A Case Study of 22 Cultural Sites on the Chinese Section of the Silk Road World Heritage. *Sustainability* **2023**, *15*, 2465. [CrossRef]
63. Zandi, S. Revival of the Silk Road Using the Applications of AR/VR and Its Role on Cultural Tourism. *arXiv* **2023**, arXiv:bs/2304.10545. [CrossRef]

64. Allaberganov, A.; Catterall, P. Using Social Exchange Theory to Examine Residents' Responses to Heritage Tourism: Case Studies of Samarqand and Bukhara in Uzbekistan. *J. Herit. Tour.* **2023**, *18*, 846–863. [CrossRef]
65. Ban, G. *The Book of Han-Biography of Zhang Qian*; Zhonghua Book Company: Beijing, China, 2012.
66. Scullard, H.H. *A History of the Roman World: 753 to 146 B*, 4th ed.; Routledge Classics; Routledge: Abingdon, UK; Oxon, UK; New York, NY, USA, 2013; ISBN 978-0-415-52227-4.
67. Caesar, J.; Handford, S.A.; Stanley, A.; Gardner, J.F. *The Conquest of Gaul*; Penguin Books: London, UK; New York, NY, USA, 1982; ISBN 978-0-14-044433-9.
68. Gibbon, E.; Milman, H.H. *The Decline and Fall of the Roman Empire*; Modern Library: New York, NY, USA, 2003; Volume 3.
69. Fisher, W.B. (Ed.) *The Cambridge History of Iran: Volume 1: The Land of Iran*; The Cambridge History of Iran; Cambridge University Press: Cambridge, UK, 1968; Volume 1, ISBN 978-0-521-06935-9.
70. Boyle, J.A. (Ed.) *The Cambridge History of Iran: Volume 5: The Saljuq and Mongol Periods*; The Cambridge History of Iran; Cambridge University Press: Cambridge, UK, 1968; Volume 5, ISBN 978-0-521-06936-6.
71. UNESCO. *History of Civilizations of Central Asia*; UNESCO Publications: Paris, France, 1992.
72. Jackson, P.; Lockhart, L. (Eds.) *The Cambridge History of Iran: Volume 6: The Timurid and Safavid Periods*; The Cambridge History of Iran; Cambridge University Press: Cambridge, UK, 1986; Volume 6, ISBN 978-0-521-20094-3.
73. Ouyang, X. *New Book of Tang*; Song, Q., Ed.; Zhonghua Book Company: Beijing, China, 1997.
74. Saunders, J.J. *The History of the Mongol Conquests*; University of Pennsylvania Press: Philadelphia, PA, USA, 2001.
75. Song, L. *History of the Yuan Dynasty*; National Library Press: Beijing, China, 2014.
76. Frith, J. The History of Plague-Part 1: The Three Great Pandemics. *J. Mil. Veterans Health* **2012**, *20*, 11–16.
77. Burckhardt, J. *The Civilization of the Renaissance in Italy*; Penguin classics; Penguin Books: London, UK ; New York, NY, USA, 1928; ISBN 978-0-14-044534-3.
78. Li, L.; Shao, M.; Wang, S.; Li, Z. Preservation of Earthen Heritage Sites on the Silk Road, Northwest China from the Impact of the Environment. *Environ. Earth Sci.* **2011**, *64*, 1625–1639. [CrossRef]
79. Sciorati, G. 'Constructing' Heritage Diplomacy in Central Asia: China's Sinocentric Historicisation of Transnational World Heritage Sites. *Int. J. Cult. Policy* **2022**, *29*, 94–112. [CrossRef]
80. Si Ma, Q. *Shi Ji*; Zhong Hua Shu Ju: Beiing, China, 2004; ISBN 978-7-101-00304-8.
81. Beckwith, C.I. Empires of the Silk Road: A History of Central Eurasia from the Bronze Age to the Present. In *Empires of the Silk Road*; Princeton University Press: Princeton, NJ, USA, 2009.
82. Fan, Y. *Book of the Later Han*; Zhonghua Book Company: Beijing, China, 2012.
83. Brown, P. *The World of Late Antiquity: AD 150-750*; Thames and Hudson: London, UK, 2024.
84. Wei, S. *Book of Wei*; Zhonghua Book Company: Beijing, China, 2017.
85. Liu, X. *Old Book of Tang*; Zhonghua Book Company: Beijing, China, 1997.
86. Dignas, B.; Winter, E. *Rome and Persia in Late Antiquity: Neighbours and Rivals*; Cambridge University Press: Cambridge, UK, 2007.
87. Davis, J.R. *Charlemagne's Practice of Empire*; Cambridge University Press: Cambridge, UK, 2015.
88. Bentley, J.H.; Ziegler, H.F.; Streets-Salter, H. *Traditions & Encounters: A Brief Global History*; McGraw-Hill Education: New York, NY, USA, 2008.
89. Stavrianos, L.S. *A Global History: From Prehistory to the 21st Century*; Prentice Hall: Upper Saddle River, NJ, USA, 1999.
90. Bryce, J. *The Holy Roman Empire*; Macmillan and Co.: London, UK; New York, NY, USA, 1889; Available online: https://archive.org/details/cu31924019888258/page/n7/mode/2up (accessed on 29 July 2024).
91. Riley Smith, J. *The Oxford History of the Crusades*; Oxford University Press: New York, NY, USA, 2002.

land

MDPI

Article

Land Finance, Local Government Debt and Economic Green Transformation

Yinglan Zhao [1], Song Peng [1], Qian Zhang [1], Yao Wang [1], Chi Gong [2,*] and Xiaoye Lu [2,*]

[1] School of Economics, Sichuan University, Chengdu 610000, China; ylzhao@scu.edu.cn (Y.Z.); 2022221010089@stu.scu.edu.cn (S.P.); zhangqian6@stu.scu.edu.cn (Q.Z.); wangyao51@stu.scu.edu.cn (Y.W.)
[2] Department of Global Business, Kyonggi University, Suwon 16227, Republic of Korea
* Correspondence: gongchi@kyonggi.ac.kr (C.G.); 85164885@kyonggi.ac.kr (X.L.)

Abstract: As economic development continues to advance globally, countries are increasingly focused on the green transformation of their economies. This paper employs a data envelopment analysis (DEA) model and entropy weighting methodology to construct and assess an indicator system for economic green transformation, taking into account environmental pollution. The analysis is based on panel data from 215 prefecture-level cities in China between 2015 and 2019. The two-way fixed effects model and moderating effect model are employed to investigate the influence of land finance on economic green transformation and to ascertain the moderating effect mechanism of local government debt. The study's conclusions are as follows: (1) Land finance impedes the transition to a green economy. (2) Local government debt is a major factor that restricts the influence of land finance on the transition to a green economy; as local government debt levels rise, land finance's role in impeding the transition to a green economy rises, and a positive moderating effect occurs. (3) Depending on the urban geographic location, the size of the economy and other factors, the impact of local government debt and land finance on the economic green transition varies. This empirical study demonstrates that the more local government debt there is in an eastern region or city with a bigger economic output scale, the more of an inhibitory influence land finance has on the economic transition to a green economy. In light of this, the paper suggests that the land finance policy be updated at the appropriate time, that the land market be regulated, that the government's reliance on land finance be gradually decreased, that the nature and amount of public debt be actively optimised, that the industrial infrastructure be enhanced to facilitate the transition towards a more environmentally sustainable economy, and that other suggestions be made.

Keywords: land finance; local government debt; green economy transition; moderating effect

1. Introduction

From the inception of the United Nations Framework Convention on Climate Change to the ratification of the Kyoto Protocol and the Copenhagen Climate Conference, environmental concerns, particularly those related to global warming caused by carbon emissions and particulate matter emissions, have garnered increasing global attention. In 2009, the United Nations Environment Programme published the Global Green New Deal Policy Brief, which advocated a transition to a green economy within the global economy. In 2011, Resolution 236 of the 64th United Nations General Assembly proposed that the concept of a "green economy in the context of sustainable development and poverty eradication" be included as one of the themes of the 2012 United Nations Conference on Sustainable Development. In 2012, on the 20th anniversary of the Rio Conference on Environment and Development, the "Rio+20" United Nations Conference on Sustainable Development proposed a new concept of green economy, which can be understood as a call for a paradigm shift in the economy. It can be observed that the transition to a green economy is becoming increasingly serious for governments. Since the "reform and opening-up" of China, i.e., its economic reform, the country has experienced rapid economic growth and has become the world's second largest economy. However, it faces some challenges. Certain industries with high energy consumption, high emissions and high pollution have a significant impact on the ecological environment and also produce certain negative externalities on a global scale. In September 2020, China announced to the world at the United Nations General Assembly that it would achieve the goal of "carbon peaking" by 2030 and "carbon neutrality" by 2060. Local governments in China have promoted economic growth and social development through land finance and massive borrowing, which has also affected climate change and environmental pollution. The Chinese government is actively exploring the adoption of green economic policies to improve environmental issues. In this context, it is crucial to examine the interrelationships among land finance, local government debt and the green transformation of the economy.

Land finance represents a significant source of revenue for local governments, deriving from the transfer of land use rights. In the short term, land finance can facilitate the green transformation of the local economy, stimulate economic growth, enhance production efficiency and optimise the urban environment. However, as the reliance on land finance increases, its long-term negative effects begin to emerge, including the sacrifice of environmental quality, the distortion of the allocation of land resources, the hindrance of the rationalisation of the industrial structure and the crowding out of social resources, thereby inhibiting the green transformation of the regional economy. In addition, the specific areas in which local governments invest the funds raised through debt, as well as the methods of repayment guarantees, will all have an impact on the manner in which land finance affects the green transformation of the regional economy. The expansion of local government debt has resulted in accelerated urbanisation, a bias towards non-environmental infrastructure investment, and increased environmental pollution. It has also promoted the reliance on land finance, hindered the upgrading of the industrial structure and squeezed social resources, affecting the green innovation of enterprises and inhibiting the green transformation of the economy. Consequently, a more profound comprehension of the influence of local government debt on the ecological transformation of the economy through land finance can assist in a more comprehensive understanding of the economic policies of local governments in China in the context of the sustainable development goals.

The objective of this paper is to examine the influence of land finance on the green transformation of the economy, as well as the role of local government debt as a highly correlated variable with land finance in the mechanism between land finance and the green transformation of the economy. Empirical research indicates that land finance has impeded the advancement of the green transformation of the economy at this juncture. Furthermore, the use of local government debt as a moderating variable in empirical research indicates that local government debt plays an important role in the relationship between land finance and the green transformation of the economy. This conclusion is also valid in the context of heterogeneity in geographical regions and economic output. In addition to giving local governments experience in formulating land policies and resolving local government debt policies, the research given in this paper enhances our comprehension of the logical connection among land finance, local government debt and economic green transformation. It also provides a theoretical foundation for promoting the region's economic green transformation in order to achieve sustainable development. The primary contribution of this study to the field of research it belongs to can be summed up in the following three aspects.

Research perspective: Firstly, the green transformation of the economy is currently a hot topic in academia, and many papers have studied the factors influencing the green transformation of the economy. Some scholars have begun to study and discuss the impact of land finance on green economic growth or environmental pollution, but there has been very little research on the mechanism of the interaction among land finance, local government debt and the green transformation of the economy. This paper introduces the variable of local government debt into the framework of the impact of land finance on the green transformation of the economy, enriching the research perspective on land use policy and the green transformation of the economy.

Theoretical and empirical analysis: Secondly, this paper constructs a theoretical framework among land finance, local government debt and economic green transformation earlier than the current literature and empirically examines the moderating effect of local government debt on the impact of land finance on economic green transformation, enriching research on land use policy and economic green transformation.

Practice: Finally, this paper provides practical suggestions for government decision makers on how to promote green economic development through the formulation of land finance policies and local government debt policies. It can help the government better reduce its dependence on land finance; effectively resolve local government debt; maintain a balance between local financial sustainability and environmental sustainability; reduce environmental pollution, natural resource depletion and ecosystem damage; and better achieve high-quality economic development.

2. Literature Review

2.1. Land Finance

Since the reform of China's tax system, land finance has greatly promoted local economic growth. There are two main views in academia on the causes of local government land finance [1]. First, local financial pressures. After China's tax system reform, local governments faced the mismatch between financial and administrative powers, which led to pressure on expenditure. They compensate for the financial gap by developing land finance [2]. Second, the incentive for officials to be promoted and competition among governments. China's GDP-oriented and short-term-oriented official assessment mechanism has led to local officials' political incentives to quickly obtain financial revenue by selling land and accelerating land finance [3]. Competition among governments also promotes land finance and drives the continuous expansion of land finance [4].

Local governments can monopolise the primary market for land supply, and they are both operators and managers. Land, as a factor of production, can bring land transfer fees and related taxes, becoming the "second finance". In terms of the economic role of land finance, the impact of land finance on economic development is complex and diverse. It is generally believed that land finance has a short-term effect on economic development, but it is not sustainable in the long term. For example, Mo (2018) used a land finance model to study the long-term and short-term effects of land finance on the Chinese economy, and land policy plays a special role in stimulating the economy [5]. Zheng (2014) empirically demonstrated that land resources have a significant positive impact on economic growth [6]. Su (2024) used Hangzhou as an example and studied the pros and cons of TOD projects implemented by local governments in China. The results of the study show that TOD under land financing leads to high housing prices and the expansion and development of cities. The model of promoting economic growth through land finance has obvious drawbacks and can lead to many hidden risks in the economic operation process. Therefore, reducing the degree of dependence on land finance and transforming the development model are the only effective ways to overcome the obstacles to high-quality economic development in China [7]. Regional coordinated development policies can be an effective tool to reduce the reliance on land finance [8].

2.2. Local Government Debt

With regard to local government debt, a great deal of research and discussion have been conducted by academics, who have developed a number of views and theoretical frameworks on the causes of local government debt and its economic impact. From the perspective of the causes of local government debt, Yang highlight that since the 1994 tax-sharing reform, local governments have been granted greater financial autonomy and diminished accountability, yet their financial expenditure responsibilities have not been reduced, placing pressure on local government finances. This, in conjunction with the intensifying economic competition among local governments, has resulted in a mounting demand for local government financing, which has consequently led to an escalation in local government debt [9]. Chen (2020) also highlight that local governments have incurred hidden deficits through local government financing platforms in order to stimulate the economy, which has also led to an increase in local debt [10].

The economic impact of local government debt includes its impact on economic growth and its impact on systemic risk. There are three different views in academia on whether local government debt can promote economic growth. On the one hand, some scholars argue that local government debt can promote local economic growth by financing public infrastructure investment through debt [11]. Xu (2016) conducted an empirical analysis of panel data on local government debt in cities at the prefecture level and above in China from 2006 to 2013 and found that local government debt has a positive correlation with regional economic development, but this effect is less effective in economically relatively underdeveloped regions [12]. Conversely, while some scholars maintain that local government debt within a certain size can have a beneficial impact on economic growth, others have begun to discuss the adverse effects of excessive indebtedness. Long-term, large-scale debt increase can have a detrimental impact on investment [13], economic growth [14,15] and the stability of the financial system [16,17]. K. Tsui (2011) noted that China's local debt concerns mostly stem from local governments' inclination towards long-term debt-financed infrastructure investment [18]. Moreover, the presence of substantial amounts of local debt hinders efforts to rectify the structural imbalance of the economy. In addition, some scholars believe that there is no non-linear relationship between local government debt and economic growth. When the scale of government debt exceeds a certain threshold, the economic burden it brings will reduce the government budget, reduce the public investment rate and thus have a negative impact on economic growth [19,20].

2.3. *Economic Green Transformation*

The green transformation of the economy is derived from green development. This reflects a gradual shift from static "green development" to dynamic "green transformation". It emphasises the dynamic evolution of the economy from a "black" or "brown" economy to a green economy. In order to assess the economic green transformation, some scholars have proposed the construction of different evaluation index systems in different dimensions for analysis. These systems are constructed by using weighted subjective assignment methods, such as the entropy-OWA operator method and grey correlation improved TOPSIS model [21–23]. Additionally, scholars have developed models such as the SBM model and the Luenberger index to assess the advancement of China's economic green transformation based on the efficiency of this transformation [24,25]. With regard to the dimension of economic green transformation, a number of scholars have sought to measure the level of economic green transformation from different perspectives. These include the green transformation of resource cities, the green transformation of industries, the green transformation of enterprises and the green low-carbon and recycling development of the economy [26–29].

In terms of influencing factors, the green transformation of the economy is affected by the industrial structure, the level of openness to the outside world, the level of finance and innovation factors. The upgrading and transformation of the industrial structure can promote green economic growth [30], and there is significant regional heterogeneity [31]. The main reason is that the rationalisation and upgrading of the industrial structure is conducive to improving labour and capital output efficiency, enhancing the level of resource allocation and thus promoting green economic growth [32]. Foreign direct investment (FDI) is an important indicator of the level of openness to the outside world. Many scholars have found that FDI is closely related to environmental pollution, forming two views: the "pollution refuge" and "pollution halo" hypotheses. The "pollution refuge" hypothesis holds that in order to reduce the cost of pollution discharge, enterprises in developed countries with higher environmental standards often transfer polluting or sunset industries to host countries with lower environmental protection requirements [33]. The "pollution halo" hypothesis holds that foreign-invested enterprises usually have more advanced environmental protection technology, pollution control funds and effective management systems, which are conducive to the introduction of environmentally friendly technologies, the realisation of energy conservation and emission reduction and green economic growth [34]. The impact of finance on the green transformation of the economy is that a high quality level of financial development is conducive to economic transformation and upgrading, quality improvement and efficiency enhancement, and pollution reduction and emission reduction [35]; the impact of innovation factors is mainly achieved through green scale efficiency and green pure technical efficiency [36]. In addition, some scholars have also found that the economic development level, education level, fiscal decentralisation and financial level all have an impact on green economic efficiency.

2.4. Land Finance, Local Government Debt and Economic Green Transformation

The existing literature contains a limited number of studies that analyse the development of land finance, local government debt and the green transition of the economy in a unified framework. Furthermore, the majority of these studies discuss the impacts of land finance and local government debt on the green transition of the economy separately.

Current academic research on land finance and economic green transition is focused on the economic growth effect and environmental effect of land finance. In terms of economic growth, under the framework of the promotion tournament theory, the scale of land finance is closely related to the promotion of officials. In order to obtain promotion, local officials are keen to sell land and win economic performance evaluations, which has a significant negative effect on economic growth [37,38]. The reason why land finance causes environmental pollution is inseparable from China's pursuit of rapid economic growth. Some scholars have studied the environmental effects of land finance from the perspective of carbon emissions. Wang (2020) studied the relationship between land finance and carbon emissions based on fiscal decentralisation and environmental federalism. The empirical results show that the overall impact of land finance on carbon emissions is non-linear and depends on the level of per capita GDP [39]. Xu (2016) also verified the bidirectional causal relationship between the quality of land urbanisation and carbon emissions in China [40]. Further, when land financing becomes the main driving force for urban development, the construction of infrastructure will accelerate the process of urbanisation, which will in turn cause ecological and environmental pollution. Wang (2019) found that there is an environmental Kuznets inverted U-shaped curve between urbanisation and environmental pollution [41].

Additionally, there is a paucity of research on the direct impact of local debt and economic green transformation. Currently, scholars are concentrating their efforts on the influence of local government debt on economic growth and economic development. Nevertheless, as China has initiated the implementation of a green development strategy, some scholars have begun to examine the relationship between local government debt and the development of a green economic transition. Firstly, government debt investment has an investment crowding-out effect [42] and cannot effectively improve the investment environment for environmentally friendly firms [43], thus hindering private investment and the potential for long-term green development of the economy. Secondly, local governments, confronted with the necessity of achieving GDP-based performance appraisals and competing with other regions, will lower environmental pollution enforcement standards and relax restrictions on highly polluting firms in exchange for economic growth [44–47], thereby increasing government revenues to alleviate fiscal pressures.

2.5. Literature Gap

In summary, existing research has discussed land finance, local government debt and economic green transformation, providing a rich theoretical framework and empirical testing methods for further research on the relationship among the three. Most of the existing literature focuses on land finance and economic green transformation, and few scholars have studied the impact of local government debt on economic green transformation. Some scholars have also begun to study and discuss the unilateral impact of land finance on economic growth or environmental pollution, but there are relatively few studies that incorporate land finance, local government debt, and economic green development into a unified framework. Research at the level of the impact mechanism also needs to be supplemented and improved. Although some scholars, such as Hou (2021) [48], have used spatial effect models to study the spatial spillover effects of land finance on green economic growth, the mechanisms and pathways of the effects of land finance and economic green transformation are not very clear. In particular, there is very little research on the effects of land finance and economic green transformation from the perspective of local government debt. Therefore, this paper will incorporate land finance, local government debt and economic green transformation into a unified research framework to more comprehensively explore the relationship among land finance, local debt and economic green development, as well as the impact mechanisms, which is an innovative contribution of certain practical significance.

3. Theoretical Analysis and Research Hypotheses

3.1. The Influence of Land Finance on the Green Transition of the Economy

Land finance is a form of revenue for local governments to meet financial needs and maintain financial expenditure by relying on the transfer of land use rights to obtain land transfer fees. It is undeniable that the economic development model based on land has brought positive economic development to China since the beginning of the new century. Therefore, in the early stages of development, land finance can temporarily promote the green transformation of the local economy in the short term. On the one hand, local governments choose to sell land to ensure that financial expenditure needs are met, which stimulates investment growth and economic expansion; promotes industrial agglomeration and thus forms economies of scale; improves factor production efficiency, including green total factor productivity; and drives the green transformation of the local economy. On the other hand, after local governments sell land, they have more funds to control local pollution. Urban environmental optimisation will promote the healthy development of local green industries and promote the green transformation of the economy.

However, as local governments' dependence on land finance continues to increase, its long-term negative impact is gradually becoming apparent. Because of its unsustainable and predatory nature, land finance will inevitably hinder economic development [49]. The same is true for the green transformation of the urban economy. As local governments' financial pressures continue to increase and the scale of land finance continues to expand, the "original intention" of land sales has gradually "deteriorated", leading to a distortion of land resource mismatches and a shift from promoting to inhibiting the green transformation of the urban economy.

First, land finance has become a financing tool for urban expansion at the expense of environmental quality, inhibiting the green transformation of the regional economy. Land finance has strengthened the power of local governments in the allocation of land resources. The original intention was to increase investment in regional environmental governance and alleviate the pressure on green fiscal expenditure. However, the land finance development model is not sustainable. With the rapid expansion of cities and the expansion of resource-intensive industries, local governments face greater challenges in balancing urban expansion and environmental governance. In addition, the fiscal decentralisation system has weakened local financial power, and the local official assessment system emphasises GDP over environmental performance. Local governments have chosen to invest a large amount of capital in heavy industry in order to pursue rapid economic growth. However, the price is the accelerated depletion of resources, a significant increase in the discharge of three waste types and a continuous decline in environmental quality. The failure of various green transition power indicators inhibits the green transformation of the local economy. In addition, the continuous expansion of cities will also lead to the rapid urbanisation of land. While high-rise buildings are being built, the area of urban green space is being reduced, leading to a decline in the ecological environment conservation and restoration functions and a deterioration of the urban green ecosystem, further hindering the green transformation of the regional economy.

Second, land finance may lead to distortions in the allocation of land resources and hinder the rationalisation of the industrial structure, thereby inhibiting the green transformation of the local economy. The government's over-reliance on land transfer fees may lead to the irrational allocation of land resources, raising the price of residential land while lowering the price of industrial land. On the one hand, the mismatch of land resources may increase the incentive for industrial enterprises to seek rent from land, making the transformation and upgrading process insufficient and blocking the industrial upgrading process. Enterprises attempt to achieve the goal of reducing costs and increasing profits by increasing investment in low-priced industrial land. Since land investment is less complex than industrial restructuring and the business objectives can be achieved more quickly, the existence of this potential path will significantly reduce the subjective willingness of industrial enterprises to carry out technological innovation, the upgrading of production and the optimisation and adjustment of industrial structure [50], hindering the process of green industrial development and thus having a negative impact on the green transformation of the entire regional economy. On the other hand, land resource mismatch will also lead to the distortion of the price signals faced by enterprises. Industrial enterprises with lagging production efficiency can easily obtain land at low cost, while enterprises with higher production efficiency cannot obtain land at ideal prices due to initial land mismatches and poor redistribution channels. The prominent land use conflicts between enterprises and uneven resource allocation among industries hinder the transformation and upgrading of enterprises in other industries and the optimisation of the industry, making industrial agglomeration and economies of scale fail to play a driving role, and ultimately may cause the green transformation of the regional economy to stagnate.

Third, land finance may lead to the crowding out of social resources, lowering the efficiency of production and resource utilisation and thus inhibiting the green transformation of the regional economy. In the market economy system, excessive local government dependence on land finance will distort land prices and cause a waste of resources. On the one hand, the current sharp reduction in the price of industrial land and bullish expectations for the future will increase the investment of industrial enterprises in non-development land, resulting in more land not being fully utilised or even being left idle. This will not only reduce the resource utilisation and production efficiency of the enterprises themselves but also reduce the possibility of other industries using the land, which will have a negative impact on regional development. The lack of overall economic development momentum will inevitably affect the ability and willingness of cities to transform greenly, thereby inhibiting the green transformation of the economy. On the other hand, local governments under greater financial pressure are more motivated to obtain funding income through land finance. Real estate is also favoured by local governments due to its high rate of return on investment, forming a closed-loop development of "land finance–real estate development". The tendency and preference of local governments to sell land for real estate development also means that resources that should have flowed into industries other than real estate are being squeezed out, the operating costs of the real economy are rising and the innovation and industrial upgrading of real enterprises are being hindered. In addition to being transmitted among enterprises, the negative effects of cost and resource allocation are also spread among regions, which in turn inhibits the green transformation of the urban economy.

Therefore, based on the above theoretical analysis, Hypothesis 1 is proposed.

H1. *The expansion of land finance will inhibit the green transformation of the economy.*

3.2. The Impact of Local Government Debt on Land Finance and Economic Green Transformation

Under the guidance of the "4 trillion" economic stimulus policy, local governments have begun to choose new debt financing models to raise funds for infrastructure construction, including issuing local government bonds and raising funds through financing platforms. While meeting the government's financial needs, these financing methods have also created a large scale of local government debt. The investment areas and repayment guarantee methods of funds raised by local governments through debt will have an impact on the way land finance affects the green transformation of the regional economy. The specific impact mechanisms include the following three points.

First, the impact of land finance on the degree of environmental pollution and its role in inhibiting the green transformation of the economy is affected by local government debt. The scale of local government debt reflects the amount of funds currently available to local governments. Therefore, the larger the scale of local government debt, the greater the potential motivation and financial support for local governments to accelerate urbanisation and achieve scale expansion under the "GDP-only" official assessment system. In this context, local governments will also be more inclined to use funds for infrastructure projects in non-environmental protection areas. Infrastructure investment projects are often medium-sized or large-scale projects with long time cycles and high resource consumption, which inevitably has a negative impact on the urban ecological environment and atmospheric environment during the construction process. In addition, during the rapid expansion of the city, local governments and officials have weak awareness of environmental protection, and they fail to take effective measures to curb environmental pollution during the construction period, which will further expand the degree of environmental pollution under the land finance system and further increase the inhibitory effect on the green transformation of the urban economy. Therefore, the larger the scale of local government debt, the stronger the motivation for local governments to urbanise land use, the higher the probability that land finance will increase the degree of environmental pollution, and the greater the inhibiting effect on the green transformation of the regional economy.

Second, the impact of land finance on the green transformation of the regional economy is mediated by local government debt. The interactive effect of local government debt and the scale of land finance will further promote the formation of land finance dependence and push up the scale of land finance. On the one hand, the expansion of the scale of local government debt issuance will encourage the government to conduct land sales more frequently to serve as collateral for debt or to repay the principal, thereby deepening its dependence on land finance policies. On the other hand, the expansion of the scale of local government debt will increase government investment in infrastructure. In the short term, infrastructure investment can be significantly capitalised into land prices, and rising land prices can alleviate the budget constraints faced by local governments through two channels: land mortgage borrowing and land transfer income. This will lead to more land mortgage loans and increase local government debt. The expansion of the scale of land finance will further increase the distortion of land resource allocation, further slowing down the process of industrial restructuring and upgrading, and the efficiency of resource allocation and utilisation among industries will continue to be low, ultimately inhibiting the green transformation of the urban economy. Therefore, the larger the scale of local government debt, the higher the dependence of local governments on land sales for revenue, the stronger the role of land finance in hindering industrial restructuring and the greater the inhibition of the green transformation of the regional economy.

Third, the impact of land finance on the green transformation of the regional economy is affected by local government debt. There are a positive interaction and interest coupling between local government debt and land finance, which makes local economic growth driven by local government debt unsustainable [51]. Local governments issue local government debt with land transfer income in the future as collateral. Due to the existence of a positive feedback mechanism between real estate prices and public investment, it will cause price fluctuations in the real estate market [52]. The uncertainty of income caused by housing price fluctuations will further expand the resource crowding-out effect of real estate development on other industries and increase the inhibitory effect of land finance on the green transformation of the regional economy. In addition, financial institutions prefer to provide credit to local governments with higher credit ratings and lower default risks. When the financing needs of local governments increase and the scale of their debt expands, it will also crowd out the financial resources of enterprises and individuals, which is not conducive to the development of green innovation projects by enterprises, and will further inhibit the green transformation of the urban economy. Therefore, the larger the scale of local government debt, the higher the degree of crowding out of social resources by land finance, and the more obvious the inhibiting effect on the green transformation of the urban economy.

Therefore, based on the above theoretical analysis, the following hypothesis is proposed.

H2. *Local government debt has a positive effect on the suppression of the green transformation of the economy by land finance.*

The theoretical analysis logical framework diagram is shown in Figure 1.

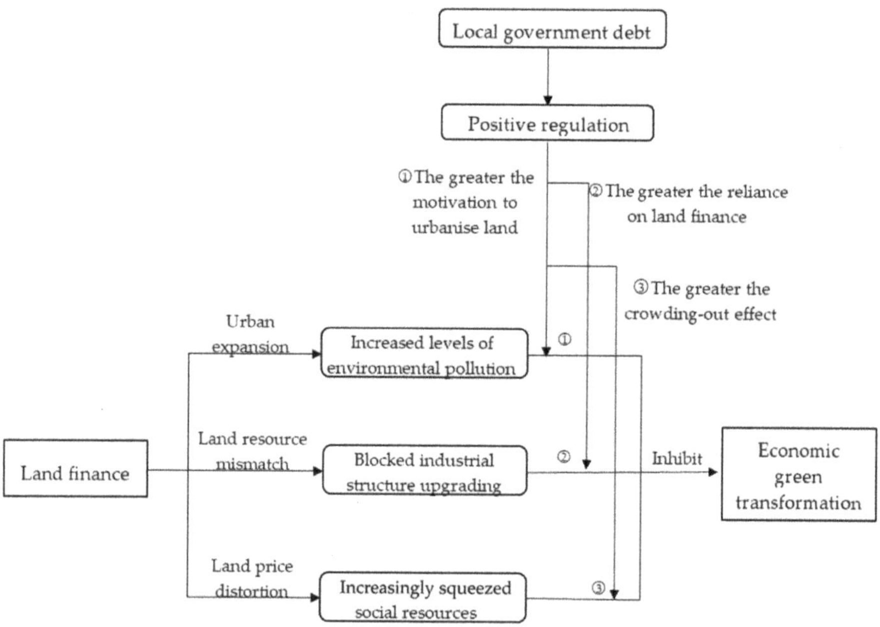

Figure 1. The conceptual framework for the impact of land finance and local government debt on the green transformation of the economy.

4. Empirical Design

4.1. Sample Selection and Data Sources

In this paper, the data sources for constructing the green GDP index are *China City Statistical Yearbook* and *China Science and Technology Statistical Yearbook*; the data for land transfer income are from *China Land & Resources Almanac*; the data for local government debt are from the Wind database; and the remaining control variables are from *China City Statistical Yearbook* and *China Science and Technology Statistical Yearbook*.

This paper takes all 293 prefecture-level cities in China into consideration, but due to the availability and accuracy of the data, the prefecture-level city samples in Xinjiang and Tibet are excluded, and the prefecture-level city samples with more missing data are also excluded. Finally, the extreme values are trimmed, and some missing data are supplemented by interpolation.

4.2. Variable Definition

4.2.1. Explained Variable

Green economic transformation. The measurement of economic green transformation adopts the methodology proposed by Wang (2014) [53], employing the DEA model to assess green economic efficiency. The process involves several steps: Initially, the entropy weight method is utilised to select industrial waste data, constructing a comprehensive environmental pollution index. Subsequently, this index is used to establish green GDP output indicators. Inputs including labour (L), capital (K) and technological development (T) are paired with the green GDP indicator as outputs. DEAP2.1 software is then employed to compute the green economic efficiency (TFP), which serves as the primary explanatory variable. The green GDP defined in this paper is the GDP/environmental pollution index, where GDP is the real GDP after removing the impact of local price indices, and the data come from *China Statistical Yearbook*. Industrial waste refers to the three main pollutants: industrial wastewater discharge, industrial sulphur dioxide emissions and industrial smoke emissions; the data come from *China City Statistical Yearbook*. The input indicator labour (L) is defined as the number of employees at the end of each region, and the data source is *China City Statistical Yearbook*; the technology input indicator is defined as the number of patent applications and authorisations at the end of each region as a substitute, and the data source is *China Science and Technology Statistical Yearbook*; the capital stock data are estimated based on the relevant research methods and research results of Wang (2017) [54], using the perpetual inventory method for estimation, and the fixed asset investment data are sourced from *China City Statistical Yearbook*, while the data required for calculating the price index and depreciation rate are sourced from *China Statistical Yearbook*.

4.2.2. Key Explanatory Variable

Land financial scale (LF). Drawing upon the research conducted by Zou (2015) [55], land finance is gauged by using local land transfer revenue (LF) as a proxy variable. This metric aptly mirrors the resource endowment and financial capability of local governments, playing a pivotal role as a primary explanatory variable.

4.2.3. Moderating Effect Variable

Local government debt. Local governments predominantly procure funding by issuing urban investment bonds through local financing platforms and other instruments. Moreover, the transparency regarding local government debt in China remains limited, resulting in the absence of precise data on its scale. Therefore, given the availability of data, this paper uses the total amount of municipal investment bonds as a proxy variable for the size of local government debt (LGD).

4.2.4. Control Variables

Since the green transformation of the economy is also affected by factors such as the level of economic development, industrial structure, technological innovation and environmental policy, in order to eliminate the interference of other factors on the model and control the factors that may affect the green transformation of the economy, the following control variables are selected in this paper: per capita real output, degree of openness to the outside world, industrial structure, per capita expenditure on science and technology education, fiscal decentralisation, environmental regulation and financial level. The principal variables are delineated in exhaustive detail and presented in Table 1.

Per capita real output. The level of regional economic development is an important factor affecting the green transformation of the economy. In this study, the level of economic development is used as a control variable, and only the impact of its first-order term on green development efficiency is considered.

Table 1. Variable names and definitions.

Category	Variable Name	Symbol	Definition
Explained variable	Green economic efficiency	TFP	Green economic efficiency measured by data envelopment method
Explanatory variable	Land finance	LF	Local land transfer income in the current year
Moderating effect variable	Local government debt	LGD	Estimating the scale of urban investment bond stock
Control variables	Real output per capita	pgdp	GDP/population
	Foreign direct investment accounts for GDP share	fdi	Foreign direct investment/GDP
	Proportion of tertiary industry	pti	Gross output value of tertiary industry/GDP
	Financial institution loan balance at end of year Share of GDP	ptlf	Financial institution loan balance/GDP
	Per capita expenditure on science and technology education	pse	Total expenditure on science and technology education/population
	Environmental regulation	er	The entropy weight method is used for the calculation
	Degree of fiscal decentralisation	fis	Fiscal revenue/fiscal expenditure

Degree of openness. The degree of openness of a region also affects its green economic efficiency. The main reasons include the technology diffusion effect [56] and the "pollution haven" hypothesis [57]. This paper selects the actual amount of foreign direct investment in each province and converts it into CNY units according to the intermediate exchange rate announced by the National Bureau of Statistics. The ratio of the converted FDI to the GDP of each region in the same year is used to measure the degree of openness of each prefecture-level city.

Industrial structure. Pender (2003) believes that the flow of factor resources from low-productivity sectors or low-growth sectors to high-productivity sectors or high-growth sectors will bring about a "structural dividend" effect. In addition, according to the law of Pareto–Clark, resources such as labour will first flow from agriculture to the secondary industry and then from the secondary industry to the service industry. This process will bring about a structural dividend. Therefore, the increase in the proportion of the tertiary industry in the city will promote green economic efficiency [58]. This paper uses the proportion of the tertiary industry in GDP as an important control variable.

Per capita expenditure on science and technology education. The level of scientific and technological innovation is also an important control variable. According to the endogenous economic growth theory, the higher the R&D investment of enterprises, the more conducive it is to promoting technological innovation and technological progress [59], accordingly improving enterprise production efficiency and green economic efficiency. This paper uses per capita expenditure on science and technology education to measure the level of scientific and technological innovation.

Fiscal decentralisation. Fiscal decentralisation can promote local governments to optimise resource allocation and engage in healthy competition, thereby laying the foundation for industrial upgrading and improving the soft environment for industrial upgrading, which is conducive to promoting industrial upgrading and thus promoting the green transformation of the economy [60]. This paper uses the fiscal autonomy index to measure fiscal decentralisation, expressed as budgeted fiscal revenue/budgeted fiscal expenditure [61].

Environmental regulation. The Porter hypothesis holds that appropriate environmental regulation will encourage enterprises to carry out innovative activities, thereby enhancing product competitiveness and thus promoting the efficiency of green development. However, some studies have found that environmental regulation will increase production costs and reduce R&D investment, which is not conducive to the efficiency of green development [62]. This paper draws on research by scholars such as Lin [63] to first standardise industrial wastewater discharge, industrial SO_2 emissions and industrial smoke (dust) emissions per unit of economic output. It then employs the entropy method to obtain weights, which are used to calculate the environmental regulation intensity index.

The level of financial development is also an important factor affecting the green transformation of the economy in various regions. In this paper, the proportion of the balance of loans from financial institutions at the end of the year to GDP is used to measure the level of financial development in various prefecture-level cities. In addition, environmental policies are also among the factors affecting the green transformation of the economy. As an important environmental policy in China, the low-carbon city pilot has been carried out in three batches, with 81 pilots. This paper adds the low-carbon city pilot as a control variable; however, as it has little impact on the empirical analysis and the results of this paper, it is not considered.

4.3. Model Setting

This paper is based on relevant economic panel data of 215 prefecture-level cities from 2015 to 2019. It selects the land transfer income of each prefecture-level city as the core explanatory variable and multiple control variables on a theoretical basis. Since the research sample contains short panel data of multiple prefecture-level cities, there are individual characteristics such as urban area and population that do not change or change little over time in each prefecture-level city. Individual effects are added to the model to control the omitted variable problem caused by individual changes in cities. At the same time, under the requirements of high-quality economic development, the process of green transformation of the economy in each city also has a trend of continuous change over time, which is controlled by adding a time fixed effect. For example, China's annual macroeconomic conditions and monetary policy also have an impact on the local economy. Through a two-way fixed effects model, the impact of the explanatory variable on the dependent variable can be more accurately identified, and unobserved individual heterogeneity and time heterogeneity can be controlled. The following benchmark model is established to test the impact of land finance on the green transformation of the economy:

$$TFP_{it} = \beta_0 + \beta_1 LF_{it} + \beta \sum controls + \mu_{it} + \delta_{it} + \varepsilon_{it} \quad (1)$$

In order to better understand how local government debt affects the hindrance of economic green transformation through land financing, a regulatory effect mechanism model, shown in (2) and (3), is constructed for future investigation. The empirical testing process involving moderating effects entails several steps: Initially, a baseline regression, based on model (2), examines the inhibitory impact of land finance on economic green transformation. Subsequently, if the regression coefficient of the core explanatory variable, β_1, proves significant, the analysis proceeds to test model (3). In the event that the regression coefficient, β_3, related to its interaction term displays significance, it establishes the presence of a moderating effect, signifying that local government debt moderates the influence of land finance on economic green transformation.

The specific model is as follows:

$$TFP_{it} = \beta_{01} + \beta_1 LF_{it} + \beta \sum controls + \mu_{it} + \delta_{it} + \varepsilon_{it} \quad (2)$$

$$TFP_{it} = \beta_{02} + \beta_1 LF_{it} + \beta_2 LGD_{it} + \beta_3 LF_{it} \times LGD_{it} + \beta \sum controls + \mu_{it} + \delta_{it} + \varepsilon_{it}. \quad (3)$$

Above, TFP_{it} is the measured green economic efficiency of city i in year t, LF_{it} is the land transfer income of city i in year t, LGD_{it} is the adjusting variable local government debt, $LF_{it} \times LGD_{it}$ is the interaction term between the explanatory variable land finance LF_{it} and local government debt LGD_{it}, controls is the control variable, μ_{it} and δ_{it} are time fixed effects and individual fixed effects and ε_{it} is the random interference error term.

5. Empirical Analysis

5.1. Descriptive Statistical Analysis

The table (Table 2) displays the mean, standard deviation, maximum and minimum values of the selected variables. All statistical values pertaining to the data indicate a reasonable level, and there are no outliers observed for the subsequent empirical analysis.

Table 2. Descriptive statistics.

Variable	N	Mean	sd	Min	Max
TFP	1046	1.017	0.062	0.617	1.492
LF	1046	0.017	0.027	0.001	0.193
LGD	1046	0.199	0.467	0.004	4.730
pgdp	1046	1.677	0.517	0.411	6.465
fdi	1046	200.973	262.512	0.024	1978.282
pse	1046	7.054	0.822	2.843	10.151
fis	1046	0.487	0.214	0.071	1.413
pti	1046	0.534	1.815	0.174	59.090
ptlf	1046	1.377	0.998	0.247	9.012
er	1046	0.0094	0.0274	0.0003	0.7725

The correlation analysis conducted on the research variables revealed Pearson coefficients at a reasonable level, indicating a low likelihood of covariance issues among the selected variables. In order to further investigate multicollinearity, the Variance Inflation Factor (VIF) test was utilised. The results indicate that the explanatory variable possesses VIF coefficients below 5 in all cases, suggesting the absence of multicollinearity among the chosen variables. This validation substantiates the earlier assertion.

5.2. Basic Regression Analysis

A panel regression model was used to conduct a multiple regression analysis on the sample data to verify the impact of land finance on the green transformation of the economy. By using Stata 18.0, the F-test p-value was 0, rejecting the null hypothesis and concluding that the fixed effects model was superior to the mixed effects model. The LM test and Hausman test were then conducted, and the p-value was 0, indicating that the random effects model was not selected for analysis and did not conform to the characteristics of the data. Based on the above test steps and results, the panel fixed effects model was finally selected as the benchmark regression model to verify the impact of land finance on the green transformation of the economy.

Table 3 lists the estimated results under different model settings. Column (1) is the baseline model in the research design without control variables, controlling for individual fixed effects, and the results are not significant. Column (2) adds the control variable of per capita real output and controls for individual fixed effects. It was found that when other conditions are constant, the explanatory variable LF is significantly negative, indicating that when other conditions are constant, the larger the scale of land finance, the stronger the inhibitory effect on the green transformation of the economy. Column (3) adds other control variables to (2), and the explanatory variable LF is still significantly negative, with the absolute value reduced. Finally, column (4) adds all control variables and controls for individual and time fixed effects. The results show that the impact coefficient of land finance on the green transformation of the economy is −2.184 from the perspective of the entire sample, which is significant at the 5% level, proving that land finance has an inhibiting effect on the green transformation of the economy as a whole, which is consistent with research hypothesis H1 in this paper.

Table 3. Land finance and economic green transition: panel fixed effects regression.

	(1)	(2)	(3)	(4)
LF	−0.850	−1.710 **	−1.668 **	−2.184 **
	(−1.20)	(−2.22)	(−1.99)	(−2.50)
Constant	1.089 ***	1.037 ***	1.249 ***	1.406 ***
	(78.09)	(44.41)	(4.50)	(3.84)
Controls	N	Y	Y	Y
Time	N	N	N	Y
Individual	Y	Y	Y	Y
Observations	1045	1045	1045	1045
R-squared	0.134	0.142	0.143	0.174

Notes: The t-value is presented in parentheses. Y for Yes, N for No. *** $p < 0.01$, ** $p < 0.05$.

5.3. Heterogeneity Analysis

5.3.1. Consideration of Regional Differences

Considering the disparities in economic development across China, regional financial disparities and varying reliance on and sensitivity to land finance are evident. Geographically, China's economic landscape broadly delineates stronger economic development in eastern cities compared with those in central and western regions. To analyse this, the sample was partitioned into three distinct geographic zones: the eastern, central, and western areas, comprising 85, 68, and 62 cities, respectively.

In Table 4, columns (5)–(7) report the regional heterogeneity. The coefficient of the impact of urban land finance on the green transformation of the economy in the eastern region is −3.583, and the result is significant at the 1% level. Conversely, in central and western cities, while the land finance coefficient shows negativity in relation to the environmental aspects, the significance level is less pronounced. The results show that the inhibiting effect of land finance on the green transformation of the economy is mainly found in cities in economically developed regions, and the inhibiting effect of land finance on the green transformation of the economy is not obvious in cities in central and western regions. Moreover, the spatial characteristics of regional debt scales reveal higher debt levels in China's eastern cities than in the central and western counterparts, proportional to the city's economic size. The reason for this is that the more economically advanced cities and regions have the capacity to service their debt in a way that matches the size of their debt. Consequently, the average government debt in eastern cities surpasses that of central and western regions. The significant inhibitory influence of land finance on economic transformation in the eastern region indicates a regulatory correlation between the adverse effects of land finance and the magnitude of government debt.

Table 4. Heterogeneity analysis.

	(5)	(6)	(7)	(8)	(9)	(10)	(11)
	Eastern Cities	**Central Cities**	**Western Cities**	**Tier 1 Cities**	**Tier 2 Cities**	**Tier 3 Cities**	**Tier 4 Cities**
LF	−3.583 ***	−0.618	−1.766	−2.132 **	−0.174	−0.463	0.425
	(−2.60)	(−0.36)	(−0.94)	(−2.02)	(−0.66)	(−0.62)	(0.56)
Constant	1.860 **	1.076	1.267 ***	2.281 *	1.453 ***	1.208 ***	0.989 ***
	(2.55)	(1.46)	(3.09)	(1.79)	(4.76)	(4.86)	(8.30)
Controls	Y	Y	Y	Y	Y	Y	Y
Time	Y	Y	Y	Y	Y	Y	Y
Individual	Y	Y	Y	Y	Y	Y	Y
Observations	419	330	297	73	125	215	632
R-squared	0.083	0.039	0.076	0.345	0.315	0.176	0.179

Notes: The t-value is presented in parentheses. Y for Yes. *** $p < 0.01$, ** $p < 0.05$ and * $p < 0.1$.

5.3.2. Consideration of Differences in Economic Output

To further corroborate these findings on the influence of land finance restraining environmental conservation and low-carbon economic transformation, particularly in economically prosperous cities, the sample was categorised based on the 2019 national city economic output rankings. The cities were classified into Tier 1, Tier 2, Tier 3 and Tier 4 groups, where Tier 1 comprises the top 20 cities in GDP ranking, Tier 2 spans from the 21st to the 50th, Tier 3 from the 51st to the 100th and the remaining cities fall into Tier 4.

In Table 4, columns (8)–(11) report the heterogeneity of the differences in economic output. As shown in Table 4 column (8), the main regression coefficient for the first-tier sample is −2.132, which is significant at the 5% level, indicating that land finance has a more significant negative effect on the green transformation of the economy in economically strong cities. Columns (9) to (11) show that the main regression coefficients for the second-, third- and fourth-tier city samples are not significant. The regression coefficients for the second- and third-tier cities are still negative. The findings indicate that for cities with a relatively middle ranking in terms of economic development and economic scale, the inhibiting effect of land finance on the green transformation of the economy is not yet apparent. However, there is a discernible trend of negative effects. In contrast, the coefficient for the fourth tier of cities is positive, suggesting that for regions and cities with lagging economic development, land finance has not yet demonstrated an inhibiting effect on economic development. Consequently, there is still the possibility of promoting economic development.

This heterogeneity results from the considerable geographical variations in China's economic development as well as the various economic development levels found in different parts of the country. Regional variation exists with regard to the consequences of local government debt on land financing to impede the green transformation of the economy due to differences in the efficiency of using debt funds, the investment of funds, the ability to bear debt and the ability to repay debt.

5.4. The Role of Local Government Debt as a Constraint on the Impact of Land Finance on the Green Transformation of the Economy

5.4.1. Examination of Local Government Debt's Moderating Effect

In order to further investigate the impact mechanism of local government debt and land finance on the green transformation of the economy, the regulatory effect of local government debt is studied. So, the interaction term of local government debt and land finance is introduced in order to study the moderating effect of local government debt.

In Table 5, columns (12) and (13) present the findings of the moderation test for the full sample. As illustrated in Table 5 (13), the coefficient of the interaction term $LF_{it} \times LGD_{it}$ is −1.338, which is statistically significant at the 0.05 level. The results of the moderation test indicate that an increase in the scale of local government debt is associated with a stronger inhibiting effect of land finance on the green transformation of the economy. The results of the moderation test indicate that local government debt exerts a moderating effect on the impact of land finance on the green transformation of the economy. This finding is consistent with research hypothesis H2 presented in this paper.

In Table 5, columns (14) and (15) and columns (16) and (17) report the moderating effect test for the eastern region urban sample and the first-tier urban sample, respectively, and indicate that there is land finance inhibition of the green transition of the economy in the sample of eastern cities and first-tier cities and that there is a moderating role for the land finance inhibition of economic green transition in local government debt.

The aforementioned analysis results are consistent with earlier theoretical hypotheses: public investment will crowd out private investment and lower overall investment efficiency; additionally, local government debt has a risk aggregation effect and interacts with land finance to further inhibit economic green transition. The aforementioned variables collectively contribute to the local government debt playing a reinforcing role in the utilisation of land financing for the purpose of promoting the green transformation of the economy.

Table 5. Moderating effect test.

	(12)	(13)	(14)	(15)	(16)	(17)
	Full Sample		Eastern Cities		Tier 1 Cities	
LF	−2.184 **	−2.362 ***	−1.249 ***	−0.947 *	−2.132 **	−1.328 *
	(−2.50)	(−2.66)	(−2.81)	(−1.79)	(−2.02)	(−1.66)
LGD		0.265 **		0.170 **		0.444 **
		(2.00)		(1.98)		(2.52)
LF× LGD		−1.338 **		−0.735 **		−1.630 **
		(2.30)		(−2.12)		(−2.24)
Constant	1.279 ***	1.139 ***	1.029 ***	0.958 ***	2.281 *	2.395
	(3.28)	(4.04)	(4.17)	(3.86)	(1.79)	(1.47)
Controls	Y	Y	Y	Y	Y	Y
Time	Y	Y	Y	Y	Y	Y
Individual	Y	Y	Y	Y	Y	Y
Observations	1045	1045	419	419	73	73
R-squared	0.175	0.167	0.189	0.201	0.345	0.488

Notes: The t-value is presented in parentheses. Y for Yes, N for No. *** $p < 0.01$, ** $p < 0.05$ and * $p < 0.1$.

5.4.2. Group Regression by Size of Local Government Debt

The government debt ratio is a measure of the overall government debt level of prefecture-level cities, calculated by comparing the scale of local government debt to the scale of local economic activity. The chosen samples were further categorised into high and low groups, with the median being used as the benchmark for regression analysis. This was performed to enhance the presentation of the cause-and-effect link between land finance and economic green transformation, as well as the impact of local government debt adjustments. We examined the differences in the impact of land finance on the green transition of the economy when the size of government debt varies through the above methodology. The results are shown in Table 6.

Table 6. Regression with sample grouping.

	(18)	(19)	(20)	(21)	(22)	(23)
	Full Sample, High Debt	Full Sample, Low Debt	Eastern Cities, High Debt	Eastern Cities, Low Debt	Tier 1 Cities, High Debt	Tier 1 Cities, Low Debt
LF	−2.014 **	−0.285	−4.649 ***	−8.608	−6.154 *	−1.247
	(−2.39)	(−0.94)	(−3.56)	(−1.63)	(−2.02)	(−0.14)
Constant	1.946 ***	1.045 ***	1.787 *	1.188	2.053	0.810 ***
	(3.21)	(21.58)	(1.75)	(0.99)	(0.47)	(5.16)
Controls	Y	Y	Y	Y	Y	Y
Time	Y	Y	Y	Y	Y	Y
Individual	Y	Y	Y	Y	Y	Y
Observations	459	523	227	192	50	23
R-squared	0.245	0.062	0.175	0.088	0.487	0.708

Notes: The t-value is presented in parentheses. Y for Yes.*** $p < 0.01$, ** $p < 0.05$ and * $p < 0.1$.

Regression of Full-Sample Subgroups by Local Government Debt Size

In Table 6, columns (18) and (19) report the results of the regression analysis for the full sample. Due to disparities in economic development among various provinces and municipalities across regions, assessing each locality's debt scale solely through the local debt balance stock indicator lacks comparability. Therefore, a more justifiable approach involves comparing it against comprehensive financial strength to derive the local government debt ratio. This ratio signifies the regional economy's growth capacity relative to the magnitude of local government debt and serves as a measure for each locality's debt scale. Column (18) shows that in the high-scale group of local government debt, land finance has a significant inhibitory effect on the green transformation of the economy, with a coefficient of −2.014, which is significant at the 5% level. Column (19) shows that in the low-scale group of local government debt, although the coefficient of land finance on the green transformation of the economy is still negative, indicating an inhibitory effect, it is not statistically significant. The regression findings within this subgroup highlight that the inhibitory impact of real estate finance on the economic green transformation is more pronounced in prefecture-level cities grappling with higher local government debt.

Regressions Grouped by Size of Local Government Debt in Eastern Cities

In Table 6, columns (20) and (21) show the regression results for the subgroup of eastern cities. In order to further explore the aforementioned conclusions and examine the moderating influence of local government debt, the magnitude of local government debt in the central and eastern cities of the sample was divided into high and low groups by using the median as a dividing threshold. Cities exceeding the median local government debt size were classified as the high-debt group (assigned a value of 1), while those below the median form the low-debt group (assigned a value of 0).

As shown in Table 6 (20), the substantial debt burden in eastern cities poses a significant obstacle to the progress of the green transformation of the economy, with a coefficient of −4.649, which is statistically significant at the 1% level. Simultaneously, land finance in the low-debt group in eastern cities has a consistent direction of the effect of the green transition of the economy, with a coefficient that is still negative but not statistically significant. The aforementioned group regression results demonstrate how the intensity of the land finance-induced economic green transition inhibitory effect is influenced by both high and low levels of government debt; in other words, local government debt amplifies this effect, thereby supporting research hypothesis H2.

Regressions Grouped by Size of Local Government Debt in First-Tier Cities

In Table 6, columns (22) and (23) report the regression results by the size of local government debt in first-tier cities. The local government debt burden ratio of the first-tier cities was divided into high and low groups for group regression based on the median, and the cities with local government debt size larger than the median were defined as the high-debt group and assigned the value of 1, while the cities with local government debt size smaller than the median were defined as the low-debt group and assigned the value of 0.

As shown in columns (22) and (23), with a coefficient of −6.154, which is significant at the 10% level, land finance has a substantial inhibiting impact on the green transition of the economy in major cities with large levels of debt; in the low-debt group, the coefficient of land finance's effect on the economy's green transition is still negative, indicating that it has an inhibitory effect but is not statistically significant.

The group regression results presented above indicate that local government debt plays a moderating role in the relationship between land finance and economic green transition, confirming hypothesis H2. Furthermore, the results suggest that as local government debt increases, the inhibitory effect of land finance on economic green transition becomes stronger.

5.5. Robustness Testing

The analysis in this paper employs various grouping regressions to examine the impact mechanism. These analyses consistently highlight the amplified moderating effect of local government debt on the relationship between land finance and the transition to a greener economy. Following this, a benchmark regression was conducted to further scrutinise the robustness of these findings. The results are shown in Table 7.

Table 7. Base regression robustness analysis.

	(24)	(25)	(26)	(27)	(28)
	TFP-SBM	**TFP-SBM**	**TFP-SBM**	**TFP**	**TFP**
LF	−0.278	−0.597 **	−0.582 **	−1.781 **	−3.774 ***
	(−1.12)	(−2.27)	(−2.19)	(−1.99)	(−3.43)
Constant	1.022 ***	0.995 ***	1.040 ***	1.354 ***	1.245 ***
	(214.24)	(110.27)	(9.36)	(3.80)	(2.73)
Controls	N	Y	Y	Y	Y
Time	N	N	Y	Y	Y
Individual	Y	Y	Y	Y	Y
Observations	1045	1045	1045	1032	839
R-squared	0.145	0.158	0.164	0.169	0.241

Notes: The t-value is presented in parentheses. Y for Yes, N for No. *** $p < 0.01$, ** $p < 0.05$.

5.5.1. Robustness Analysis with Replacement of Explanatory Variable

In Table 7, columns (24)–(26) report the results of the robustness analysis for the replacement of the explained variables. To assess the continued suppressive impact of land finance on environmental protection and low-carbon economic transformation, alongside the moderating role of local government debt, the primary explanatory variable undergoes substitution in the robustness test section. The actual analysis using the DEA method might introduce measurement bias due to potential random errors, making it challenging to authentically evaluate the impact of random errors and internal management variables. To address this, the super-efficient SBM model proposed by Tone (2001) [64] was adopted to recalibrate the economic efficiency of the green sector. Following the methodology outlined by Yuan (2020) [65], non-expected outputs were integrated into the super-efficient SBM model to gauge the new green economic efficiency, termed TFP-SBM. Subsequently, the model undergoes re-regression for further analysis. The findings demonstrate that the suppressive impact of land financing on the economic green transformation is robust.

5.5.2. Robustness Analysis with Partial Sample Exclusion

Table 7 (27) reports the results of a robustness analysis that excludes some of the samples. This study focuses on data of prefecture-level cities, a dataset notably comparable to provincial data and significantly larger than that of ordinary prefectures and cities. Consequently, the data from the four municipalities and certain samples from western Chinese provinces, where efforts are concentrated on developing their capital cities, like Chengdu and Xi'an, were excluded for robustness analysis. And the key explanatory variable continued to be statistically significant.

Furthermore, this paper examines the impact of local government debt reform. As illustrated in Table 7 (28), due to the 2015 local government debt reform, local government debt is subject to quota management and classified into budget management. The data on local government debt fluctuate greatly, so the 2015 sample was excluded. After excluding some samples, the impact of land finance on the green transformation of the economy is still significantly negative. The conclusions are consistent with the benchmark regression, indicating that the inhibiting effect of land finance on the green transformation of the economy is robust.

5.5.3. Robustness Analysis of Moderating Role of Local Government Debt

In Table 8, columns (29) to (34) report the results of the robustness analysis of the local government debt reconciliation. In columns (29) and (30), the moderating effect model replaces the dependent variable with TFP-SBM and uses a subsample after exclusion to conduct a robustness test of the moderating effect. The results show that the interaction term is still significantly negative, indicating that local government debt, as a moderating variable, has a robust moderating effect on the positive effect of land finance on the green transformation of the economy. In columns (31) to (34) of Table 8, the moderating effect of local government debt on the green transformation of the economy is robust to the replacement of the dependent variable in the moderating effect model for eastern cities and first-tier cities. The moderating effect is still significant after replacing the dependent variable with TFP-SBM, which once again confirms the robustness of the positive moderating effect of local government debt as a moderating variable on the impact of land finance on the green transformation of the economy.

Table 8. Robustness analysis of moderating effects.

	(29)	(30)	(31)	(32)	(33)	(34)
	Full Sample		Eastern Cities		Tier 1 Cities	
LF	−0.582 **	−0.612 **	−3.825 ***	−3.570 **	−6.938 **	−4.230 *
	(−2.19)	(−2.33)	(−2.69)	(−2.11)	(−2.29)	(−1.92)
LGD		0.093 *		0.604 **		1.453 ***
		(1.93)		(2.20)		(3.10)
LF × LGD		−0.579 **		−1.965*		−5.211 **
		(−2.57)		(−1.77)		(−2.69)
Constant	1.040 ***	1.056 ***	1.539 *	1.336 *	2.949	4.593
	(9.36)	(10.17)	(1.95)	(1.68)	(0.63)	(1.06)
Controls	Y	Y	Y	Y	Y	Y
Time	Y	Y	Y	Y	Y	Y
Individual	Y	Y	Y	Y	Y	Y
Observations	839	839	410	410	65	65
R-squared	0.164	0.162	0.205	0.217	0.437	0.553

Notes: The t-value is presented in parentheses. Y for Yes.*** $p < 0.01$, ** $p < 0.05$ and * $p < 0.1$.

5.5.4. Robustness Analysis of Regressions Grouped by Local Government Debt Size

In Table 9, columns (36)–(39) present the results of the robustness analysis of the regression, which was grouped according to the size of local government debt. The explanatory variable TFP-SBM was still utilised, and a robustness test was conducted by performing a grouped regression on the excluded sample. This was performed to verify the robustness of the grouped regression results on the size of local government debt. The regression results show that in both first-tier and eastern region cities, the impact of land financing on impeding the green transition of the economy becomes more pronounced as the level of local government debt increases, demonstrating the robustness of the grouping regression results.

Table 9. Subgroup regression robustness analysis.

	(35)	(36)	(37)	(38)	(39)
	Full Sample	Eastern Cities, High Debt	Eastern Cities, Low Debt	Tier 1 Cities, High Debt	Tier 1 Cities, Low Debt
LF	−3.774 ***	−2.863 ***	−0.076	−3.909 **	−3.774
	(−3.43)	(−2.96)	(−0.06)	(−2.20)	(−0.37)
Constant	1.245 ***	2.098 **	0.863 **	1.706 **	1.245 ***
	(2.73)	(2.35)	(2.13)	(2.41)	(2.73)
Controls	Y	Y	Y	Y	Y
Time	Y	Y	Y	Y	Y
Individual	Y	Y	Y	Y	Y
Observations	839	141	269	23	42
R-squared	0.241	0.315	0.054	0.496	0.241

Notes: The t-value is presented in parentheses. Y for Yes. *** $p < 0.01$and ** $p < 0.05$.

6. Conclusions and Discussion

6.1. Conclusions

This study utilises panel data from 215 prominent prefecture-level cities, covering the period from 2015 to 2019. Its aim is to empirically scrutinise the influence of land finance on economic green transformation, exploring the role and impact direction of local government debt as a regulatory variable. The research outcomes reveal the findings reported below.

First, a larger scale of land finance correlates with lower green economic efficiency, suggesting that land finance hampers urban economic green transformation.

Second, the influence of local government debt on the effect of land finance on the green transformation of the economy is substantial. With the escalation of local government debt, land finance's inhibitory influence on the green metamorphosis of the economy becomes more pronounced, thereby demonstrating a positive moderating effect.

Third, the impact of land finance and local government debt on the green transformation of the economy varies across different economic output scales and urban geographic regions. Empirical research conducted on subgroups reveals that within the eastern region and in prefecture-level cities with a larger economic output scale, there exists a correlation wherein a greater scale of local government debt is associated with more pronounced inhibition of economic green transformation due to land finance policies.

6.2. Discussion

The above empirical results confirm that the more serious the dependence on land finance and the higher the scale of government debt, the stronger the inhibiting effect on the green transformation of the economy in cities and regions. This reflects the more urgent need for economically developed cities to transform their economies and upgrade their industrial structure, as well as the greater need for governments to break away from their dependence on land finance and debt financing and find new and stable sources of revenue.

This paper shows that land finance has a significant inhibitory effect on the green transformation of the economy, which is consistent with Wang (2021) [66] considering the environmental effects of land finance. Land finance has an impact on economic growth as well as other aspects, so incorporating land finance and the green transformation of the economy into a unified research framework can help to expand research on the impact of land finance. On this basis, this paper also introduces the local government debt variable and empirically examines the moderating effect of local government debt on the impact of land finance on the green transformation of the economy, enriching the research content and perspective of land use policy and the green transformation of the economy.

In the world, the green transformation of the economy and green growth are ideal models for developing countries to achieve sustainable development [67]. The green transformation of the economy can be seen as a synergistic effect of simultaneous economic and ecological development and is the result of the combined effects of many factors, such as fiscal decentralisation, digitalisation, green innovation, green finance, etc. In addition to the above factors, this paper incorporates land finance and local government debt into the theoretical framework of green economic transformation and finds that the behaviour of local governments in making up for the funding gap through land finance and local government debt has a significant inhibitory effect on green economic transformation. Therefore, when local governments use land and debt to develop the economy, they should rationally exercise their right to allocate land resources and limit local government debt to a reasonable level.

7. Policy Implications

In 2008, a global financial crisis occurred. In order to boost the economy, the Chinese government proposed the "4 trillion" plan to stimulate domestic demand and promote development, of which 2.8 trillion, accounting for 70% of the total, was to be provided by local governments. Restricted by the reform of the tax-sharing system and the budget law, local governments began to rely on land finance and local financing platforms to increase local government revenue and raise related investment funds. Land finance and local debt have become important ways and means to solve the difficulties of local government revenue and expenditure and raise related investment funds. However, with the rapid development of the Chinese economy and the increase in the total economic volume, the effectiveness of land finance and local debt in promoting economic development has begun to weaken, and their negative effects have begun to appear, even showing signs of hindering economic development and inhibiting economic transformation.

Through an analysis of the interplay among land finance, local government debt and the green transition of the Chinese economy, this paper can provide valuable insights into the experiences of a wide range of developing countries around the world, which can help policymakers to better formulate more inclusive, equitable and sustainable development paths that utilise land resources and alleviate government debt in order to achieve a green transformation of the economy.

Combined with the above findings, this paper focuses on the following policy recommendations.

First, local governments should promote the transformation of land finance to reduce the dependence of local governments on land finance. This paper has shown that land finance has a significant inhibitory effect on the progress of the green transformation of the economy. Therefore, the government should accelerate land policy reform, improve the fiscal and taxation system and actively seek new sources of fiscal and taxation revenue, in addition to enhancing the sustainability of land-related fiscal revenue, thereby reducing the dependence on land finance. Secondly, according to the heterogeneous conclusions of the study, the current inhibitory effect of land finance on the green transformation of the economy is mainly reflected in the more economically developed regions. Therefore, according to the different economic development status of China's eastern, central and western regions, different land financing policies need to be implemented according to local conditions. While reducing the reliance on land financing in economically developed regions, it is necessary to strengthen supervision and intervention in land financing in economically underdeveloped regions to minimise the inhibiting effect of land finance on the green transformation of the economy.

Second, local government should optimise the structure and scale of debt and upgrade and adjust the industrial structure. Initially, by implementing legislation, rules and other appropriate high-level strategies, the government aims to regulate and limit its borrowing activities, ensuring that the amount of debt incurred by local governments remains within a reasonable and effective range. Secondly, we should optimise the debt structure, change the current financial dependence status quo and abandon the "financial expansion–economic development" inherent thinking, so that the scale and structure of government debt is at the best level to promote the development of economic green transformation. Finally, it is necessary to eliminate the traditional land finance development model and the development path of local government financing projects, increase investment in and support for basic research and promote the transformation of industries into high-technology and high-value-added industries with low-energy consumption and low-pollution areas, so as to promote the healthy development of the economy.

Thirdly, local governments should improve the efficiency of their debt capital allocation by adapting to local conditions and implementing policies that are appropriate for their cities. For the eastern region, local government debt should be invested in public services that effectively improve environmental quality and infrastructure that helps to innovate green development, so as to lead innovation and high-quality supply and create new demand. For the central and western regions, it is necessary to strengthen the performance evaluation of local government debt, pay attention to the long-term mechanism of debt-financed operations and actively expand effective social investment while avoiding crowding out private sector credit.

Author Contributions: Y.Z.: Conceptualisation, Methodology and Supervision. S.P.: Model analyses, Data curation and Writing—original draft and editing. Q.Z., Y.W., C.G. and X.L.: Writing—original draft. All authors have read and agreed to the published version of the manuscript.

Funding: This research was supported by National Social Science Fund of China (grant No. 21BJY128), National Natural Science Foundation of China (grant No. 72371178), Key project of the Social Science

Fund of Sichuan Province (grant No. SCJJ23ND38) and Major Project of the National Social Science Fund of China (grant No. 20&ZD081).

Data Availability Statement: The raw data supporting the conclusions of this article will be made available by the authors upon request.

Conflicts of Interest: The authors declare no conflicts of interest.

References

1. Wu, Q.; Li, Y.; Siqi, Y. The incentives of China's urban land finance. *Land Use Policy* **2015**, *42*, 432–442.
2. Guo, Z.; Gu, N. Institutional Change, Land Finance and Extended Urban Expansion—A New Perspective to Explain the Phenomenon of Inflated Urbanization and Industrial Structure in China. *Soc. Sci. Res.* **2013**, 8–14. [CrossRef]
3. Zhang, L.; Wang, X.; Xu, X. Financial incentives, promotion incentives and land grant behavior of local officials. *China Ind. Econ.* **2011**, *129*, 35–43.
4. Wu, P.; Sun, C.; Zhao, B. Local government financial pressure, official promotion incentives and land finance behavior: Theoretical analysis and empirical test. *Mod. Financ. Econ.-J. Tianjin Univ. Financ. Econ.* **2019**, *39*, 95–113.
5. Mo, J. Land financing and economic growth: Evidence from Chinese counties. *China Econ. Rev.* **2018**, *50*, 218–239. [CrossRef]
6. Zheng, S.; Sun, W.; Wu, J.; Wu, Y. "Generating Wealth with Land, Nurturing Land with Wealth"—A Study on the Investment and Financing Model of Urban Construction with Chinese Characteristics. *Econ. Res. J.* **2014**, *49*, 14–27.
7. Su, Y.; Wu, Y.; Choguill, L.C.; Luo, J.; Yu, X. Reflections on TOD in China: From land finance to inclusive growth. *J. Urban Manag.* **2024**, *13*, 294–307. [CrossRef]
8. Chen, L.; Chen, S. Institutional rethinking and optimization of regional fiscal balance—With the background of the construction of functional zones in municipalities directly under the central government. *Mod. Econ. Res.* **2024**, 11–19. [CrossRef]
9. Yang, J.; Yang, Q.; Liu, K. Land-based financing and debt growth-an empirical study based on prefecture-level city panel data. *Financ. Trade Econ.* **2018**, *39*, 52–68+117. [CrossRef]
10. Chen, Z.; He, Z.; Liu, C. The financing of local government in China: Stimulus loan wanes and shadow banking waxes. *J. Financ. Econ.* **2020**, *137*, 42–71. [CrossRef]
11. Puente-Ajovín, M.; Sanso-Navarro, M. Granger causality between debt and growth: Evidence from OECD countries. *Int. Rev. Econ. Financ.* **2001**, *35*, 66–77. [CrossRef]
12. Xu, C.; Cheng, L.; Zhuang, J. The impact and mechanism of local debt on regional economic growth: An analysis based on panel quantile model. *Economist* **2016**, *5*, 77–86.
13. Guo, Y.; Sun, X.; He, Y. Research on the incentive mechanism, growth performance and adjustment orientation of local fiscal leverage. *Econ. Res. J.* **2017**, *52*, 169–182.
14. Lyu, J. Analysis of the impact of local debt on economic growth: A liquidity perspective. *China Ind. Econ.* **2015**, 16–31.
15. Yang, C.; Lu, Y. Research on the Current Situation, Causes and Prevention Measures of Local Government Debt Risk. *Public Financ. Res.* **2013**, *11*, 58–60.
16. Cheng, Y.; Jia, S.; Meng, H. Fiscal policy choices of local governments in China: Land finance or local government debt? *Int. Rev. Econ. Financ.* **2022**, *80*, 294–308. [CrossRef]
17. Mao, R.; Liu, N.; Liu, R. Trigger mechanism of local government debt expansion and systemic financial risk. *China Ind. Econ.* **2018**, *4*, 19–38.
18. Tsui, K. China's infrastructure investment boom and local debt crisis. *Eurasian Geogr. Econ.* **2011**, *52*, 686–711. [CrossRef]
19. Liu, H.; Yang, G.; Yin, L. Government debt, economic growth and nonlinear effects. *Stat. Res.* **2014**, *31*, 29–38.
20. Chen, J. Analysis of the threshold effect of local government debt on economic growth in China. *Contemp. Financ. Econ.* **2018**, *10*, 33–44.
21. Jiang, J. Sustainable digital era: High-quality integration and development of digital economy and green economy. *Enterp. Econ.* **2021**, *40*, 23–30+161. [CrossRef]
22. Zhao, A.; Wu, C. Systematic evaluation of China's economic green transformation posture-based on entropy-OWA operator and gray correlation improvement TOPSIS. *Tech. Econ.* **2018**, *37*, 99–106. [CrossRef]
23. Chen, T.; Chen, K.; Li, K. Research on evaluation index system of green transformation of regional economy. *Stat. Decis. Mak.* **2019**, *35*, 59–62. [CrossRef]
24. Wai, X.; Huang, S. The impact of FDI preference on the green transformation of China's regional economy. *J. Cap. Univ. Econ. Bus.* **2018**, *20*, 62–72. [CrossRef]
25. Gao, X.; He, H. Cleaner production, green transformation and product quality upgrading of enterprises. *Stat. Res.* **2021**, *38*, 64–75. [CrossRef]
26. Yu, L.; Bi, X.; Zhang, W. Construction of green transformation evaluation system for industrial enterprises. *Stat. Decis.* **2019**, *35*, 186–188.
27. Xie, X.; Han, Y. How do local manufacturing enterprises realize "magnificent transformation" in green innovation?—A multi-case study based on the attention-based view. *Manag. World* **2022**, *38*, 76–106.
28. Zhang, Y.; Do, R.; Bai, Y. Measurement of the construction level of China's green, low-carbon and recycling development economic system. *J. Quant. Tech. Econ.* **2020**, *37*, 83–102.

29. Nie, L.; Wang, Y.; Zhang, J. Evaluation of green transition performance of resource-based cities—A test from 114 prefecture-level cities in China. *Tech. Econ.* **2022**, *41*, 141–152. [CrossRef]
30. Chen, T.; Hu, Q.; Yan, H. Research on the impact of financial agglomeration on the efficiency of the green economy: An empirical analysis based on the Super-DEA model and the SGMM model. *J. Tech. Econ. Manag.* **2020**, *4*, 9–14.
31. Sun, J.; Liu, W.; Zhou, Y. China's opening up, industrial structure and green economic growth: An empirical test based on inter-provincial panel data. *Manag. World* **2014**, 172–173.
32. Wu, J.; Hu, J. Environmental regulation, industrial restructuring and green economic growth: An empirical test based on provincial panel data in China. *Inq. Econ. Issues* **2018**, 7–17.
33. Candau, F.; Dienesch, E. Pollution Haven and Corruption Paradise. *J. Environ. Econ. Manag.* **2017**, *85*, 171–192. [CrossRef]
34. Khachoo, Q.; Sharma, R. FDI and Innovation: An Investigation into Intra- and Inter-industry Effects. *Glob. Econ. Rev.* **2016**, *45*, 311–330. [CrossRef]
35. Zhang, Z.; Li, T.; Ma, Q. Threshold effect of financial agglomeration on the efficiency of urban green economy: Based on the statistical data of nine national central cities in China. *Res. Technol. Econ. Manag.* **2020**, *284*, 98–102.
36. Hu, A.; Guo, A.; Zhong, F. Can the agglomeration of high-tech industries improve the efficiency of the regional green economy? *China Popul. Resour. Environ.* **2018**, *28*, 93–101.
37. Li, X.; Hong, G.; Huang, L. The mystery of China's land finance growth: The strategic reform of the tax system and the growth of land finance. *China Econ. Q.* **2013**, *12*, 1141–1160.
38. Fan, Z. The root of land finance: Financial pressure or investment impulse. *China Ind. Econ.* **2015**, 18–31.
39. Wang, L.O.; Wu, H.; Hao, Y. How does China's land finance affect its carbon emissions? *Struct. Chang. Econ. Dyn.* **2020**, *54*, 267–281. [CrossRef]
40. Xu, R.; Wu, Y. Interaction of carbon emission, energy consumption and economic growth in the Yangtze River Delta urban agglomeration-Empirical evidence based on panel linkage equation model. *Ecol. Econ.* **2016**, *32*, 32–38. [CrossRef]
41. Wang, D.; Tang, M. How does land urbanization affect ecological environment quality?—An analysis based on dynamic optimization and spatially adaptive semiparametric model. *Econ. Res.* **2019**, *54*, 72–85.
42. Li, Q.; Wang, Y.; Lu, K. Local government debt, banking structure and green credit level. *Gansu Financ.* **2020**, *12*, 49–54.
43. Zhang, Z.; Deng, W. Research on the effect and path of local government debt affecting corporate ESG. *Mod. Econ. Res.* **2022**, 10–21.
44. Tang, Y.; Su, X.; Liu, Q. Local Fiscal Pressure and Weak Environmental Regulation-Empirical Evidence of Behavioral Choice of Environmental Opportunism. *Theory Pract. Financ. Econ.* **2023**, *44*, 82–91. [CrossRef]
45. Ma, E.; Yang, X. Fiscal Pressure, Energy-intensive Enterprises and Carbon Emissions-A Quasi-Natural Experiment Based on the Reform of Education Rights. *Financ. Trade Econ.* **2022**.
46. Guo, Y.; Xue, J. Research on the nonlinear environmental effects of local government debt. *Stat. Res.* **2021**, *38*, 105–117.
47. Xiong, H. Study on the economic growth effect of local government debt--Based on the dual perspective of quantity and quality. *J. Macro-Qual. Res.* **2021**, *9*, 113–128.
48. Hou, S.; Song, L.; Wang, J.; Ali, S. How Land Finance Affects Green Economic Growth in Chinese Cities. *Land* **2021**, *10*, 819. [CrossRef]
49. Tang, P.; Chen, Y.; Xiao, J. Research on the mechanism and spatial effect of land finance affecting economic growth. *Land Econ. Res.* **2019**, *2*, 17–31.
50. Wu, Y.; Li, L. Performance evaluation of China's development zone policies: Based on the perspective of enterprise innovation capabilities. *J. Financ. Res.* **2017**.
51. Wu, J. Local debt, land finance and economic growth—An empirical analysis based on the linkage equation model. *Bus. Res.* **2018**, *8*, 38–44. [CrossRef]
52. Gao, R.; Zhu, Z.; Chen, C. Local debt and China's economic fluctuations: Analysis of the financial accelerator mechanism. *Econ. Res. J.* **2022**, *57*, 83–100.
53. Wang, J.; Geng, J. Calculation and empirical analysis of China's green economic efficiency. *Econ. Probl.* **2014**, *4*, 52–55.
54. Wang, H. GDP data revision and capital stock estimation in China: 1952–2015. *Econ. Sci.* **2017**, *6*, 16–30.
55. Zou, W.; Liu, H. Is land finance "drinking poison to quench thirst"—A spatio-temporal dynamic space panel analysis based on prefecture-level cities in China. *Economist* **2015**, *9*, 21–32.
56. Taskin, F.; Zaim, O. The role of international trade on environmental efficiency: A DEA approach. *Econ. Model.* **2001**, *18*, 1–17. [CrossRef]
57. Zeng, D.; Zhao, L. Pollution havens and industrial agglomeration. *J. Environ. Econ. Manag.* **2008**, *58*, 141–153. [CrossRef]
58. Peneder, M. Industrial structure and aggregate growth. *Struct. Chang. Econ. Dyn.* **2003**, *14*, 427–448. [CrossRef]
59. Chen, G.; Lan, B. Research on R&D expenditure, competition and regional innovation capability in China: An empirical analysis based on domestic patent application data and R&D data from 1998 to 2004. *Economist* **2007**, *3*, 101–106.
60. Huang, Q.; Li, G. Does fiscal decentralisation promote industrial upgrading in China? *Reform Econ. Syst.* **2019**, *5*, 129–135.
61. Liao, G.; Qin, J. Can green technological innovation effectively improve environmental quality? Based on the perspective of fiscal decentralisation. *Technol. Econ.* **2022**, *41*, 17–29.
62. Ren, S.; Li, X.; Yuan, B.; Li, D.; Chen, X. The effects of three types of environmental regulation on eco-efficiency: A cross-region analysis in China. *J. Clean. Prod.* **2018**, *173*, 245–255. [CrossRef]

63. Lin, C.; Gong, X.; Kong, F. Environmental regulation and green technological progress: Promotion or inhibition—Based on the perspective of spatial effects. *Macroeconomics* **2019**, *11*, 131–142.
64. Tone, K. A slacks-based measure of efficiency in data envelopment analysis. *Eur. J. Oper. Res.* **2001**, *130*, 498–509. [CrossRef]
65. Yuan, X.; Wang, J.; Zhang, J. Evaluation of urban efficiency under high-quality development—Empirical research from 19 cities at the sub-provincial level and above. *Urban Dev. Stud.* **2020**, *27*, 62–70.
66. Wang, G.; Salman, M. Understanding the spatial spillover effect of land finance on China's green development: Does the moderating role of industrial structure matter? *Environ. Sci. Pollut. Res. Int.* **2023**, *30*, 95959–95974. [CrossRef]
67. Ferreira, J.J.; Lopes, J.M.; Gomes, S.; Dias, C. Diverging or converging to a green world? Impact of green growth measures on countries' economic performance. *Environ. Dev. Sustain.* **2023**, *1*, 1–19. [CrossRef] [PubMed]

land

MDPI

Article

Promoting Urban Innovation through Smart Cities: Evidence from a Quasi-Natural Experiment in China

Mingbo Ji [1], Mengyun Jin [2], Lingyun Chen [3], Yuwei Liu [4] and Yihao Tian [2,*]

1 School of Economics, Sichuan University, Chengdu 610065, China; cdmypx@scu.edu.cn
2 School of Public Administration, Sichuan University, Chengdu 610065, China; jinmengyun@stu.scu.edu.cn
3 Administrative Services Center, Sichuan Provincial Water Resources Department, Chengdu 610017, China; chenlingyun988@gmail.com
4 School of Business, Chengdu University, Chengdu 610106, China; liuyuwei@cdu.edu.cn
* Correspondence: yihaotian@scu.edu.cn

Abstract: It is of great theoretical and practical significance to investigate the influence of intelligent city construction on urban innovation. Based on the data of 238 cities in China from 2006 to 2019, this paper utilizes the staggered difference-in-differences (staggered DID) model and the mediating effect model to examine the impact and mechanisms of smart city construction on urban innovation. We find that China's smart city pilot policies significantly promote urban innovation. Mechanism analysis shows that this innovation promotion effect acts through improving urban informatization, government financial expenditure on science and technology, and the upgrading of the city's industrial structure. Further analysis shows that smart city construction has a stronger promoting effect on innovation in cities of a larger scale, that located in the eastern region, and have a lower level of science and education. Overall, our findings provide new insights into urban innovation and objectively assess the impact of smart city construction in China.

Keywords: smart city; urban innovation; informatization; difference-in-differences; China

1. Introduction

Innovation is the primary driving force driving the transformation and development of the economy. The theory of endogenous growth proposes that the economy cannot rely on external forces to achieve sustained growth and that its determining factor is endogenous technological progress [1,2]. Economists have since studied the endogenous origins of economic growth more thoroughly, and a large body of literature provides evidence of a causal relationship between innovation and economic growth [3,4]. Cities are the spatial carriers of innovation activities [5], the gathering place of innovation resources and elements, and important places for knowledge creation and application [6]. Therefore, building an innovative country cannot be separated from urban innovation. Releasing the vigor of urban innovation and improving the level of urban innovation are vital to cities playing basic and supportive roles in the construction of a national innovation system. As cities are innovation machines, exploring how to enhance urban innovation capability is of non-negligible meaning for countries around the world seeking to improve their innovation capability and enhance sustainable economic development [7].

Smart cities underpinned by information and communications technology (ICT) are now an advanced form of urban development [8,9]. With the rapid progress of information technology, such as the Internet of Things (IoT), cloud computing and the Internet, smart cities have emerged, and countries around the world such as Switzerland, Norway, Australia, the United Kingdom, and China are committed to the construction of smart cities and to making them a sustainable and essential policy tool to improve urban resilience [10–12]. A large body of research on this "booming" phenomenon has yielded rich outputs [13].

Some studies have attempted to clarify the concept and characteristics of smart cities, pointing out that smart cities cannot simply be equated with technology but should be composed of a series of elements such as land, citizens, technology, and governance [14,15], as well as the idea that the smart city is innovation composed of technology, management, and policy [16]. However, the city is a complex and dynamic assemblage of relationships and technologies, and the smart city is also a dynamic concept, a fully consistent definition of which has not yet been established. Many scholars focus on the functions and performance of smart cities, exploring how they can be used to support higher quality urban spaces, better public service delivery, and citizen well-being. For example, smart cities enable the public sector to more effectively utilize information technology infrastructure and smart devices, which can increase the value of public services to communities [17]. Smart cities also help to address the challenges of social exclusion and environmental pollution faced in urban development, strengthen the scientific and technological level of government and the efficiency of resource allocation, and improving the regional environment [18,19]. In addition, others are concerned with the participation and interaction of stakeholders in the process of smart city construction and development [20–22], and have endeavored to enrich the theoretical framework of smart city research [23]. Still others have shifted their research focus to practices inside cities, shedding light on the different paths taken by different cities around the world in interpreting and applying the notion of the smart city [24]. In summary, these studies have emphasized the use of ICT and the importance of effectively integrating different city management systems, sharing information resources and promoting operational synergies among city systems [25,26]. Smart cities advocate intelligent city management and services, enhance city operations and management, improve public services, and contribute to the welfare of residents, with the ultimate goal of achieving the sustainable development of innovative cities. Exploring the causal relationship and mechanisms between smart cities and urban innovation is essential for advancing urban innovation, promoting national economic development, and planning for the future construction of smart cities [27].

The concept of urban innovation may vary according to different research perspectives. There are studies in the field of public administration that classify urban innovation into four dimensions including the agenda, process, product, and symbolic innovation of different policies [28]. In the economic sphere, cities have become key units of innovative activity, bringing together various elements such as enterprises, talent, capital and institutions, and innovation refers primarily to scientific and technological progress [29]. Urban innovation means the processes and products of innovative activities in cities that are primarily centered on scientific and technological progress. Urban innovation is characterized by significant temporal and spatial structure from the point of view of urban innovation networks or urban innovation systems [30,31]. The determined factors of urban innovation have long been in the spotlight, the early literature has highlighted the role of intra-city factors on innovation, such as city size [32], public infrastructure [33,34], educational resources [35,36], and fiscal decentralization [37]. With the advancement of research methods such as quasi-natural experiments, studies have begun to extend the boundaries of factors influencing urban innovation beyond the city to explore the role of macroeconomic policies and institutions [38]. A body of literature has empirically examined the promotional effects of macro-environmental regulation on urban innovation [39,40], some have found that innovative city policies stimulate urban innovation by raising the level of government fiscal expenditures, the degree of urban industrial agglomeration, and the level of human capital [41], and others have provided evidence that the innovative growth effect of such policies has significant spillover effects on neighboring cities [42]. However, only limited research has investigated the effects of smart cities on urban innovation as an important urban development paradigm. A study based on data from 309 European metropolitan areas proved that smart city development has a positive impact on urban innovation [43]. Some evidence from China that there is heterogeneity in the impact of smart city policies on urban innovation depending on geographic location [44]. Nonetheless, the mechanism by which smart cities affect urban innovation is still unclear and needs to be supplemented by further quantitative analyses and empirical evidence, especially as evidence from developing countries is scarce.

China offers unique conditions for empirically analyzing how smart cities affect urban innovation. First, in order to explore scientific approaches to the construction, operation, management, services, and development of smart cities, China began a national smart city pilot program in 2012 and expanded the scope of the pilot program in 2013 and 2014. The gradual implementation of smart city pilot policies has been characterized as a "quasi-natural experiment", creating an opportunity to open the black box of causality between smart city development and urban innovation [45]. Second, compared with studies based on cities in other countries, China has a large sample of cities, with as many as 293 prefecture-level cities, which makes a rich sample for the study. Therefore, based on China's smart city pilot policy, this paper establishes a staggered DID model to examine the impact of smart city construction on urban innovation and its specific mechanisms, utilizing data from China's city yearbooks from 2006 to 2019.

This study contributes to related research in three ways. First, we use data from Chinese cities to demonstrate the extent to which smart city construction promotes urban innovation. Our findings add evidence from developing countries to existing studies and are relevant to urban development in other developing countries around the world. Second, we contribute to the existing literature by analyzing in depth the mechanisms through which smart city construction affects urban innovation. We find that smart city pilot policies promote urban innovation through improving urban informatization, government financial expenditure on science and technology, and industrial structure upgrading. Third, we present a rich heterogeneity of results that help deepen our knowledge of how smart city policies affect urban innovation.

The rest of the paper is organized as follows. Section 2 introduces the institutional background of smart city construction in China and develops the research hypotheses. Section 3 presents the research design, including data sources, variables, and empirical models. Results are provided in Section 4. Section 5 presents conclusions and policy implications.

2. Institutional Background and Research Hypotheses

2.1. Institutional Background

The smart city is an urban form supported by a new generation of information technology and a knowledge society based on a next-generation innovation environment, stressing the role of ICT in improving the functioning of the urban system, facilitating knowledge transfer, constructing innovation networks [46], etc. In order to drive the construction of new urbanization and enhance the management capacity and service level of cities, the Ministry of Housing and Urban-Rural Development (MoHURD) of the People's Republic of China officially launched a notice on the development of national smart city pilot projects in November 2012, and issued the "Interim Measures for the Management of National Pilot Smart Cities", which guided and encouraged the construction of smart cities across the country, and identified 90 national pilot smart cities. In March 2014, China released the National New Urbanization Plan (2014–2020), which explicitly put forward "promoting smart cities" as one of the three major objectives of promoting the construction of new cities. In August of the same year, China's National Development and Reform Commission and eight other departments jointly formulated the guiding opinions on the healthy development of smart cities, proposing that by 2020, China should build a number of smart cities with distinctive features, significantly improve the comprehensive competitive advantages of cities, and achieve remarkable results in terms of livelihood services, innovative management, and cybersecurity. In November 2015, China finalized the framework of the smart city standard system and evaluation index system. In March 2016, China's 13th Five-Year Plan clearly states that it will make full use of big data and modern information technology to build a batch of demonstration-type smart cities featuring smart infrastructure, convenient public services, and fine-grained social governance. In 2018, China successively released national standards on top-level design, information technology operation, and information security guarantees for smart cities, which are used to standardize the order of smart cities and advance the higher-level of smart cities. In general, the scope of smart cities in China has been expanding, with China's MoHURD and Ministry of Science and Technology (MOST) announcing three batches of smart city pilot lists in 2012, 2013, and 2014, respectively, comprising a total of 290 pilot districts (including pilot municipalities, prefectures, counties, and cities at the county level).

From the perspective of relevant policies, the objectives of China's pilot smart cities are mainly in the areas of raising the efficiency of the supply of public services, improving the administrative efficiency of the government and the level of urban management, improving the urban living environment, upgrading the level of intelligence of public infrastructure, and developing a sophisticated system of urban safety and security networks. The main contents of China's smart city pilot policy can be summarized as follows: First, in the area of public services, the policy calls for the establishment of a modern information service system covering all people, the improvement of government administrative efficacy by means of smart applications, and the provision of convenient, efficient, and personalized smart medical care, education, and other public services. Secondly, in the area of social management, the policy requires the creation of a smart application system, strengthening data integration, information sharing, and business synergy, strengthening the ability of urban operation monitoring and intelligent security and emergency response, and ensuring the safe and efficient operation of the city. Third, in terms of the industrial system, the policy aims to give full play to the advantages of a new generation information and communication technologies, promote the digital, networked, and intelligent transformation of existing industries, develop new business forms such as smart logistics, smart agriculture, smart tourism, etc., and make industries more competitive.

2.2. Hypothesis Development

From the main objectives and contents of China's smart city pilot policy, the government's adoption of the smart city pilot policy should theoretically provide a significant boost to the innovation level of the pilot cities, which may be achieved by upgrading the level of informatization, increasing financial support, and optimizing the industrial structure.

First of all, the smart city pilot policy helps to bring the effects of information technology to cities, thereby promoting urban innovation. First, smart city construction places an emphasis on the use of a new generation of information technology, such as the IoT, cloud computing, big data, and spatial geographic information integration, to improve urban infrastructure, accelerate innovation in urban public services, promote the government's use of intelligent means to optimize the allocation of resources, improve the efficiency of resource utilization and the level of urban governance, and enhance urban resilience, which contributes to the creation of a favorable innovation environment for enterprises, colleges and universities, and other innovation bodies [47]. Second, the construction of smart cities is accompanied by the open sharing of substantial information resources, which reduces the information asymmetry, lowers transaction costs, and boosts the efficiency of innovation [48]. Third, in the process of smart city development, enterprises and other interested parties will accelerate the creation of new technologies and products through the application of ICT, and enhance the overall innovation capacity of the city.

Secondly, the smart city pilot policy will make the government increase its investment in innovation activities in science and technology and enhance the level of urban innovation. On the one hand, the financial expenditure of the government is an important tool for the construction of smart cities, and the pilot policy stresses that local governments should pay attention to funding planning and financial security when carrying out top-level design and planning for smart cities and, at the same time, they should include in their industrial planning the innovation costs of industrial transformation and upgrading and the aggregation of industrial elements, which means that the government's scientific and technological expenditures will grow as a result. On the other hand, the government's increased investment in R&D may assist enterprises in reducing R&D costs and avoiding R&D risks, stimulate enterprises and other organizations to strengthen long-term R&D, and encourage more investment in innovation, thus raising the level of urban innovation [49,50].

Finally, the smart city pilot policy places great emphasis on the development of smart industries and the economy, promotes the upgrading of existing industries and the development of new industries in the city, optimizes the industrial structure, and is conducive to promoting the gathering of all kinds of innovative elements in the city and stimulating the vitality of innovation in the city.

Based on the above analysis, this paper proposes the following research hypotheses:

Hypothesis 1. *Smart city construction can significantly promote urban innovation.*

Hypothesis 2. *Smart city construction can promote urban innovation by enhancing the level of informationization.*

Hypothesis 3. *Smart city construction can promote urban innovation by prompting governmental investment in science and technology.*

Hypothesis 4. *Smart city construction can enhance urban innovation by contributing to the upgrading of the industrial structure.*

3. Data and Methods

3.1. Variables

3.1.1. Explained Variable

The explanatory variable in this paper is the level of urban innovation. Patent data are an important indicator of a region's inventive capacity and are characterized by their accessibility and timeliness, so plenty of studies have used them to measure a region's innovation level [51]. However, as there are differences among different cities in various aspects of innovation, such as input of innovation factors, allocation of innovation resources, and the outputs of innovation, measuring the level of innovation of a city only by the number of patents fails to take into account the problem of heterogeneity between cities, and may make it difficult to reflect the true value of the city's innovation activities. Unlike the number of patents, some scholars and organizations are committed to constructing more objective and comprehensive measures of innovation. The Index of Regional Innovation and Entrepreneurship in China (IRIEC) is led by Peking University's Enterprise Big Data Research Center and jointly developed by Peking University's National Development Research Institute and Longxin Data Research Institute to examine the actual outputs of innovation and entrepreneurship activities, rather than the inputs. The index covers all industries and enterprises of different sizes in a region, and not only incorporates patent data into consideration, but also integrates and standardizes innovation data from different fields, such as technology and capital, given that innovation and entrepreneurship are complementary to each other. Hence, we adopt the IRIEC to measure the level of urban innovation.

3.1.2. Explanatory Variable

The smart city pilot policy is the core explanatory variable, which is set as a dummy variable in this paper, assigned a value of 1 if a city belongs to the smart city pilot as the treatment group subject to policy shock; conversely, other cities serve as the control group and are assigned a value of 0.

3.1.3. Control Variables

In order to reduce the endogeneity problem caused by omitted variables, this paper controls for the following variables: (1) The level of urban infrastructure. Urban infrastructure is an indispensable foundation for the development of urban economic and social undertakings, and is the basic condition for the survival and innovative development of modern cities [38]. The construction of urban roads is the basis for the city's economy and open interaction with the outside world, and to a certain extent, can reflect the city's level of development and innovation, so we adopt the ratio of urban road area to land area to reflect the level of local infrastructure. (2) Level of foreign direct investment (FDI). FDI can stimulate the technological innovation effect of the host country and has an obvious technology spillover effect [52]. We use the ratio of the actual amount of utilized FDI to the regional GDP as an indicator of the level of foreign investment, and because of the large differences between the ratios, the data are logarithmized. (3) Financial development level. Innovation cannot be separated from financial support, which plays a pivotal role in reducing financing costs and promoting risk management [53]. This variable is measured using the logarithm of the ratio of the balance of deposits of financial institutions to regional GDP. (4) Population size, measured using the logarithm of the number of permanent urban residents. (5) Level of economic development, measured using the logarithm of urban GDP.

3.1.4. Mediating Variables

The mediating variables of concern mainly include the level of urban informatization, the level of advanced and rationalized urban industrial structure, and the degree of government financial support.

The promotion and application of ICT represented by the Internet has become an effective way for policy makers to enhance innovation [54], so we adopt the ratio of the number of Internet broadband access subscribers to the resident population to measure the level of urban informatization.

The financial support of the Chinese government acts as an integral part in fostering the development of high-tech industries and strategic emerging industries, which is favorable to urban innovation activities. In this paper, the science and technology expenditure component of the government's financial expenditure is selected to measure the degree of government support for innovation, and the indicator is logarithmized.

Optimization of the industrial structure refers to the process of transforming and upgrading an economy to more high-end, highvalue-added, and high-technology industries in the process of economic development, consisting of an advanced industrial structure and industrial structure rationalization as two key dimensions [55,56]. An advanced industrial structure refers to the process of gradually increasing the proportion of non-agricultural industries in an economy, which generally implies a shift in the economy's industries from traditional agriculture and light industry to higher-end, high-technology, high-value-added manufacturing and services, as measured by the coefficient of the hierarchy of the industrial structure. Rationalization of the industrial structure, meaning the dynamic process of strengthening the coordination capacity and increasing the level of association between different industries, reflects the degree of effective utilization of resources and the degree of coordination between the input and output structures of factors among different industries, and is measured by the Theil index of industrial structure.

3.2. *Data*

First, the data on urban innovation came from the IRIEC published by Peking University's Open Research Data Platform, which contains the innovation and entrepreneurship index for 292 cities in China. Second, the data on smart city pilots were established based on the list of smart city pilots published by MoHURD and MOST in 2012, 2013, and 2014. Since we are mainly studying the impact of this policy on urban innovation, the four municipalities (Beijing, Shanghai, Tianjin, and Chongqing are at the same administrative level as the provinces) and counties in the list were deleted, and the final sample consisted of 238 prefectural-level cities, of which 84 cities were in the experimental group and the remaining 154 were in the control group. Third, all other data were obtained from the China Urban Statistical Yearbook. Eventually, after combining, cleaning, and standardizing all the data, we arrived at a total of 3332 valid samples from 2006 to 2019, and the descriptive statistics of each variable are shown in Table 1.

Table 1. Descriptive statistics.

Variables	N	Mean	S.D.	Min	Max
City Innovation Index	3332	0.555	0.272	0.02	1
Infrastructure Level	3332	1.205	1.363	0.009	12.707
Foreign Investment Level	3332	−3.629	2.909	−13.337	8.288
Financial Development Level	3332	0.713	3.137	−9.82	10.69
Population Size	3332	5.912	0.613	3.85	7.413
Economic Development Level	3332	14.829	2.677	4.163	19.374
Informatization Level	3332	1.608	1.239	0.001	14.074
Government Investment	3332	9.889	1.539	4.205	15.529
Degree of Advanced Industrial Structure	3332	6.448	0.35	5.517	7.836
Rationalization of Industrial Structure	3332	−1.704	1.06	−8.321	5.332

3.3. Empirical Model

The difference-in-differences method can simultaneously control for group effects (experimental and control groups) and time effects (before and after the pilot), and identify the net effect from exogenous policy shocks as much as possible. Considering that smart city construction is carried out in batches, this paper constructs the following staggered DID model [45]:

$$cityinno_{it} = \alpha + \beta smartcity_{it} + \sum \delta x_{it} + \mu_i + \eta_t + \varepsilon_{it} \quad (1)$$

In the above equation, i represents the city and t represents the year. $cityinno_{it}$ is the explanatory variable, representing the city's innovation level. $smartcity$ is a dummy variable representing smart city construction and $smartcity_{it} = treat_i * post_{it}$. $treat_i = 1$ if city i is a pilot smart city, $treat_i = 0$ otherwise, and $post_{it} = 1$ if city i becomes a pilot city during the policy implementation period, and $post_{it} = 0$ otherwise. Thus, $smartcity_{it} = 1$ means that city i was a pilot smart city in year t, $smartcity_{it} = 0$ means that city i was not a pilot smart city in year t. β is the estimated coefficient of smart city construction, which is the core parameter of interest in this paper. If the obtained estimate $\hat{\beta} > 0$, it indicates that there is a significant positive impact of smart city construction on urban innovation. If $\hat{\beta} < 0$, it indicates that smart city construction has a negative impact on urban innovation, which is manifested as unfavorable to urban innovation. x_{it} is a set of control variables. μ_i is the city-fixed effect to control for inherent characteristics that do not change over time at the city level. η_t is the time-fixed effect to control for time-level factors that do not vary with area. ε_{it} is the random error term.

Further, in order to test the mechanism of smart city construction acting on urban innovation, we also constructed the following model:

$$MV_{it} = \alpha_1 + \beta_1 smartcity_{it} + \sum \delta_1 x_{it} + \mu_i + \eta_t + \varepsilon_{it} \quad (2)$$

$$cityinno_{it} = \alpha + \beta_2 smartcity_{it} + \beta_3 MV_{it} + \sum \delta_2 x_{it} + \mu_i + \eta_t + \varepsilon_{it} \quad (3)$$

In Equation (2), MV_{it} is the mediating variable, if the estimated $\hat{\beta}_1$ of the coefficient β_1 passes the significance test at the 10% level, then it suggests that there is a significant effect of smart city construction on the mediating variable.

In Equation (3), if the estimates of coefficients β_2 and β_3 both pass the significance test at the 10% level and coefficient β_2 turns out to be smaller, it explains that the mediating variable MV_{it} is an important transmission mechanism for smart cities to influence urban innovation. The mediating effect is an indirect effect, i.e., $\beta_1 * \beta_3$; β_2 is a direct effect. The relationship between mediating effect and total effect should satisfy the equation: $\beta = \beta_2 + \beta_1 * \beta_3$.

4. Results and Discussion

4.1. Baseline Estimated Results

Benchmark regression results are reported in Table 2, Column (1) demonstrates the results with no control variables and fixed effects added, Column (2) includes time-fixed effects and city-fixed effects, and Column (3) continues to include control variables. It can be seen that after incorporating fixed effects and control variables, the regression result coefficient of column (3) is 0.0307, which is still statistically significant at 1%, indicating that the construction of smart cities has a significant positive impact on urban innovation, and that Hypothesis 1 has been verified, which is also consistent with the conclusions of existing studies [43,57,58].

Table 2. Effects of smart city construction on urban innovation.

	(1)	(2)	(3)
smartcity	0.0436 ***	0.0527 ***	0.0307 ***
	(3.5537)	(4.5997)	(4.3235)
Infrastructure Level			0.0163 ***
			(7.6541)
Foreign Investment Level			0.0317 ***
			(17.6369)
Financial Development Level			0.0353 ***
			(11.2886)
Population Size			0.1560 ***
			(30.6473)
Economic Development Level			0.0973 ***
			(34.6467)
Constant	0.5471 ***	0.5455 ***	−1.7462 ***
	(105.3784)	(125.6951)	(−42.8866)
City-fixed effect	No	Yes	Yes
Time-fixed effect	No	Yes	Yes
N	3332	3332	3332
R-squared	0.0035	0.3413	0.7498

Notes: *** indicates significance at the 1% level. Numbers in parentheses are t-values.

From the results of the control variables, the estimated coefficient of the level of urban infrastructure is significantly positive at the 1% level, indicating that the construction of urban roads significantly affects the level of urban innovation, and the government should give attention to urban infrastructure. The estimated coefficient of the level of foreign investment is significantly positive at the 1% level, which means that investment attraction is conducive to the development of urban innovation, and the government should deregulate, increase the level of openness of the other side of the city, and emphasize on outreach and interoperability to promote the development of urban innovation. The regression coefficients of the level of financial development, population size and economic development are all significantly positive at the 1% level, suggesting that the government should emphasize the development of urban finance and economy on the basis of securing the population base, and promote innovation and urban innovative development by economy.

4.2. Parallel Trend Test

The key premise in using staggered DID is to satisfy the parallel trend assumption, that is, the change in the level of innovation in the cities in the treatment and control groups was not significantly different before the announcement of the list of smart city pilots. To test this trend, we constructed an event analysis model as follows [45]:

$$cityinno_{it} = \alpha + \sum_{t=-8}^{7} \lambda_t D_{it} + \sum \delta x_{it} + \mu_i + \eta_t + \varepsilon_{it} \quad (4)$$

In Equation (4), D_{it} is a set of dummy variables that takes the value of 1 if city i becomes a pilot smart city in year t, and vice versa takes it the value of 0. λ_t is the coefficient of interest, reflecting the difference in innovation between the cities that have carried out smart city construction and those that have not, in year t when the list of pilot smart cities was issued.

We use data from 2006 to 2019, and the smart city pilots include three phases in 2012, 2013, and 2014, so the time difference before and after the policy pilots is 8 years before the policy and 7 years after the policy. In graphing, we summarize data from 5–8 years before the pilot list was announced to 4 years before the policy presentation, and data from 7 years after the pilot list was announced to 6 years after the policy presentation. Figure 1 shows that the coefficient estimates of the smart city construction effect are not significant in the pre-implementation periods, demonstrating that there is no significant difference in innovation levels between the cities that have carried out smart city construction and those that have not before the announcement of the pilot list, and thus the sample passes the parallel trend test.

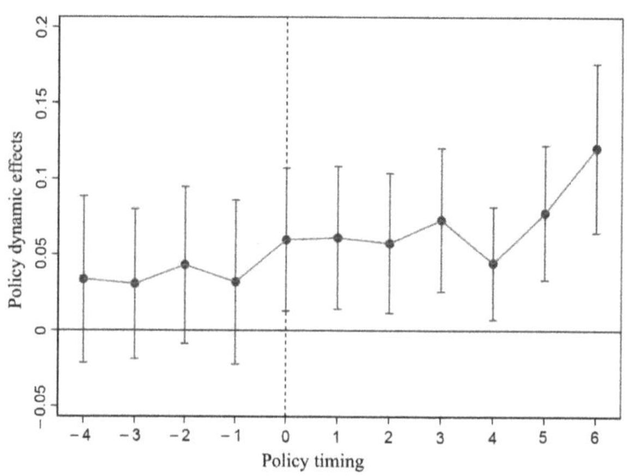

Figure 1. Parallel trend test.

4.3. Placebo Test

4.3.1. Time Placebo Test

This study employs a time placebo test to rule out the effect of time variation on the differences in urban innovation between the treatment and control groups [59]. Specifically, we constructed spurious exogenous shock times, denoted by $smartcity_{t-3}$ and $smartcity_{t+3}$, for the announcement of the smart city pilot list with a three-year lead and a three-year lag, respectively, and brought them into Equation (1) for regression analysis.

The regression results are shown in Table 3, where the estimated coefficient of $smartcity_{t-3}$ fails the significance test, which shows that there is no systematic difference in the time trend between the treatment and control groups, which is further evidence of the facilitating effect of smart city construction on urban innovation. The estimated coefficient of $smartcity_{t+3}$ is significantly positive at the 1% statistical level, indicating that the promoting effect of smart city construction on urban innovation is still significant three years after execution (although the regression coefficient is reduced compared to the baseline estimation), which may be attributed to the government's other priorities in terms of its smart city construction goals, such as putting more emphasis on the construction of smart communities and the provision of smart public management and services, among others.

Table 3. Time placebo test results.

	(1)	(2)
$smartcity_{t-3}$	0.0027 (0.4165)	
$smartcity_{t+3}$		0.0282 *** (2.7346)
Control variables	Yes	Yes
City-fixed effect	Yes	Yes
Time-fixed effect	Yes	Yes
N	3332	3332
R-squared	0.9006	0.7489

Notes: *** indicates significance at the 1% level. Numbers in parentheses are *t*-values.

4.3.2. Urban Placebo Test

We continue to employ a commonly used placebo test to ensure the reliability of the baseline regression results. Referring to Cai et al. [60], we randomly selected 84 cities in our sample as the false treatment group and the rest of the cities as the false control group in order to assess the impact of smart city construction as a placebo on urban innovation. After repeating the above process 500 times, we obtained the corresponding regression coefficients and *p*-values and plotted the kernel density distribution of the 500 estimated coefficients (See Figure 2). The baseline regression coefficient estimates, on the other hand, are located in the distal position of the tail of the distribution of spurious regression coefficients, which is a small probability event in the urban placebo test, illustrating that our baseline estimation results are significant rather than due to unobservable factors, which further validates and ensures the reliability and credibility of the findings.

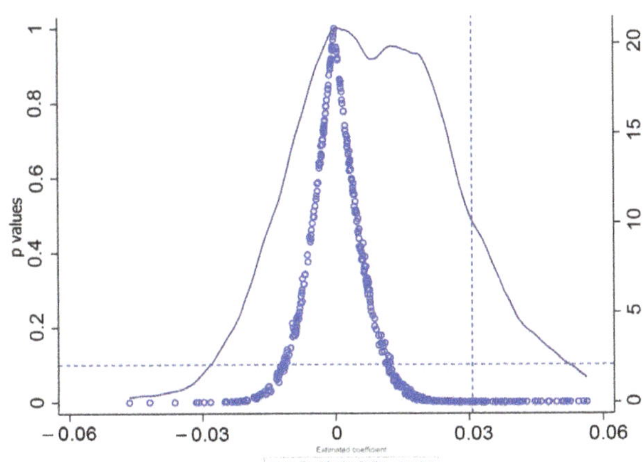

Figure 2. Urban placebo test results.

4.4. Robustness Test

The previous analysis has shown that smart city construction significantly promotes urban innovation. To further ensure the robustness of the estimation results, we continue the regression analysis in terms of the two dimensions of screening the sample data and replacing the explanatory variables.

Sample adjustment. In order to avoid the impact of extreme values on the benchmark regression results, we winsorized the sample by 1% up and down and re-adopted Equation (1) for the regression. The regression results are reported in Column (1) of Table 4, and we can see that the estimated coefficients of the smart city pilot policy are still statistically significantly positive at 1% after excluding extreme values, which is basically consistent with the benchmark estimation results.

Table 4. Results of robustness tests.

	Sample Adjustment	Replacement of Explained Variables	
	(1)	(2)	(3)
smartcity	0.0281 ***	0.0360 ***	0.0423 ***
	(3.9750)	(4.4447)	(6.3214)
Control variables	Yes	Yes	Yes
City-fixed effect	Yes	Yes	Yes
Time-fixed effect	Yes	Yes	Yes
N	3332	3332	3332
R-squared	0.7513	0.6836	0.7729

Notes: *** indicates significance at the 1% level. Numbers in parentheses are *t*-values.

Replacement of explained variables. The Peking University Open Research Data Platform also released the per capita innovation index and the per unit area innovation index of cities. We incorporate these two indexes into the benchmark regression model to further examine the impact of smart cities on urban innovation, and the regression results are reported in Column (1) and Column (2) of Table 4, respectively. The estimated coefficients of the pilot smart city policies are all significantly positive at the 1% significance level, further verifying the positive contribution of smart city construction to urban innovation.

4.5. Mechanism Analysis

4.5.1. Effect of Informatization Level

Smart city construction includes advancing the application of modern information and communication technologies such as big data, cloud computing, and IoT, which contributes to the enhancement of the urban information infrastructure and the process of urban informatization [61,62], which is crucial for facilitating the agglomeration of urban innovation factors, stimulating urban innovation activities, and bolstering the city's innovation capacity. The popularization and application of information technology can change the economic structure, increase science and technology-intensive industries, and improve the technological level and competitiveness of cities [63]. We introduced the city informatization level into the mediation effect model for analysis, and the regression results are reported in Columns (1) and (2) of Table 5. Column (1) reports the effect of smart city construction on the level of city informatization, and the regression coefficient of smart city construction is 0.161 and passes the significance test of 1%, which indicates that smart city construction significantly promotes the construction of a city's informatization. The results in Column (2) indicate that smart city construction and city informatization level have a significant positive impact on urban innovation, suggesting that the mediating effect is significant and that improvements in smart information facilities resulting from the pilot policy deliver innovative benefits to the city [57]. The direct effect of smart city construction affecting urban innovation is 0.0259, and the mediating effect through urban informatization construction is 0.0048, and Hypothesis 2 is verified.

Table 5. Mechanism analysis.

	Effect of the Level of Informatization		Effect of Government Science and Technology Investment		Effect of Industrial Structure Upgrading			
	(1)	(2)	(3)	(4)	(5)	(6)	(7)	(8)
smartcity	0.1610 *** (4.4975)	0.0259 *** (3.6784)	0.1240 *** (3.6614)	0.0204 *** (3.1222)	0.0138 *** (1.4801)	0.0278 *** (4.0701)	−0.1192 *** (−3.1171)	0.0276 *** (3.9149)
Informatization Level		0.0298 *** (8.7037)						
Government Investment				0.0831 *** (24.7348)				
Degree of Advanced Industrial Structure						0.2136 *** (16.7007)		
Rationalization of Industrial Structure								−0.0263 *** (−8.2094)
Control variables	Yes	Yes	Yes	Yes	Yes	Yes	Yes	Yes
City-fixed effect	Yes	Yes	Yes	Yes	Yes	Yes	Yes	Yes
Time-fixed effect	Yes	Yes	Yes	Yes	Yes	Yes	Yes	Yes
N	3332	3332	3332	3332	3332	3332	3332	3332
R-squared	0.6938	0.7554	0.8222	0.7890	0.7401	0.7693	0.5225	0.7548

Notes: *** indicates significance at the 1% level. Numbers in parentheses are *t*-values.

4.5.2. Effect of Government Investment in Science and Technology

We introduce governmental science and technology investment into the mediated effect model for analysis, and the regression results are reported in Columns (3) and (4) of Table 5. Column (3) reports the effect of smart city construction on governmental science and technology investment, and the results show that the estimated value is 0.124 and is significant at a 1% statistical level, showing that the implementation of the smart city pilot policy leads to an increase in the city's science and technology expenditure. The results in Column (4) indicate that smart city construction and urban science and technology investment can both significantly promote urban innovation, which means that the mediating effect is significant. Specifically, the direct effect of smart city construction affecting urban innovation is 0.0209, and the mediating effect through governmental science and technology investment is 0.0103, which verifies Hypothesis 3. Pilot smart city policies have led governments to increase financial investment in science and technology, injecting more financial support for urban innovation. The finding also somewhat adds to the body of related research that explores the relationship between government subsidies and innovation performance [64]. In China, the government is the main driver of smart city development. When a city is established as a national pilot smart city, local governments tend to invest in the city's R&D activities, especially in science and technology R&D activities related to making the city smarter, in ways such as establishing smart city R&D centers and providing tax incentives for enterprises, which serve to boost the transformation and application of scientific and technological achievements, and to enhance the city's scientific and technological level and innovation capacity [65].

4.5.3. Effect of Industrial Structural Upgrading

We analyze the mediating effect of industrial structure upgrading from the dimensions of industrial structure advancement and industrial structure rationalization, respectively. The results of the analysis of the mediation effect of an advanced industrial structure are shown in Columns (5) and (6) of Table 5, and the results show that the regression coefficient of smart city construction on the level of industrial structure advancement is significantly positive, and at the same time the regression coefficients of smart city construction, and advanced industrial structure in Column (6) pass the test of significance at the level of 1%, which indicates that the mediation effect of an advanced industrial structure is established, and the coefficient of the mediation effect is 0.0029. Columns (7) and (8) are tests of the mediating effect of industrial structure rationalization, and the results show that the regression coefficient of smart city construction on the level of rationalization of industrial institutions is significantly negative, which means that the Theil index of industrial structure is reduced, illustrating that the construction of a smart city helps to optimize the allocation of resources, improve the efficiency of resource utilization, accelerate the flow of information between industries, contribute to the fusion of industries, and coordinate the development of various industries in the city [56], all of which are advantageous to enhancing the overall innovation competence of the city. Consequently, Hypothesis 4 is verified, expanding the existing knowledge of the channels through which smart city construction contributes to urban innovation.

4.6. Heterogeneity Analysis

4.6.1. City Size

City size refers to the number of people in a city or the size of the city's area, which to a certain extent reflect the condition of the city's resources, such as population, economy, finance, and infrastructure; important elements affecting urban innovation [32]. Therefore, there may be differences in the effect of smart city construction on urban innovation among cities of different sizes. We divide the sample into four groups: small- and medium-sized cities with a resident population of less than 1 million in the urban area, large cities with a resident population of more than 1 million and less than 5 million, supercities with a resident population of more than 5 million and less than 10 million, and megacities with a resident population of more than 10 million, based on the criteria for categorizing the size of cities published by the State Council of China. The regression results (1)–(4) in Table 6 show that the innovation effect of smart city construction is greater for larger cities, which is more obvious for large cities and supercities. However, when the size of the city is too large, the coefficient of the impact of smart city construction on urban innovation decreases, which may be due to the fact that after a certain degree of city size, the cost of urban innovation will gradually increase with the expansion of the size, while the innovation effect of smart city implementation also gradually decreases. For one thing, megacities require a higher level of infrastructure and public services to support urban development, which can entail significant investment and management costs and reduce the cost-effectiveness of innovation. In addition, the increased complexity and diversity faced by megacities in the course of their development increases the difficulty of urban management and decision-making, as well as the risk and uncertainty of innovation, which may reduce the benefits of innovation brought about by the construction of smart cities.

Table 6. Heterogeneity analysis.

	City Size				City Location		Science and Education Level	
	Small and Medium-Sized Cities	Large Cities	Supercities	Megacities	East	Midwest	High	Low
	(1)	(2)	(3)	(4)	(5)	(6)	(7)	(8)
smartcity	−0.0168	0.0278 ***	0.0297 ***	0.0212 **	0.0436 ***	0.0159 *	0.0228 ***	0.0259 ***
	(−0.6474)	(3.2251)	(2.6405)	(2.0224)	(3.7889)	(1.9286)	(3.0230)	(3.3657)
Control variables	Yes	Yes	Yes	Yes	Yes	Yes	Yes	Yes
City-fixed effect	Yes	Yes	Yes	Yes	Yes	Yes	Yes	Yes
Time-fixed effect	Yes	Yes	Yes	Yes	Yes	Yes	Yes	Yes
N	68	2185	969	109	1274	2058	490	2842
R-squared	0.9227	0.7150	0.7758	0.9554	0.7122	0.7456	0.9258	0.7081

Notes: *, ** and *** indicate significance at the 10%, 5%, and 1% levels, respectively. Numbers in parentheses are t-values.

4.6.2. City Location

The development of cities is often shaped by their geographical location and the resulting resource endowment [66]. Compared to the central and western regions, cities in eastern China, with convenient transportation, abundant resources, and relatively high levels of human capital and foreign investment, have unique advantages regarding the agglomeration of innovation factors such as population, economy, finance, and infrastructure, and this difference in resources may cause heterogeneity in the effects of smart cities on urban innovation. We divide the sample into two groups of eastern developed cities and central and western less developed cities for regression. The regression results (5) and (6) in Table 6 show that compared with the central and western regions, the urban innovation effect brought about by smart city construction in the eastern region is more significant, suggesting that there are regional differences in the impact of smart city construction on urban innovation, and that smart city construction should be implemented according to the local conditions by taking full account of the geographic location and resource conditions of the cities, so as to better utilize its facilitating effect on urban innovation.

4.6.3. Level of Urban Science and Education

Smart city construction depends on high-tech practitioners, and colleges and universities are important bases for cultivating and nurturing scientific and technological innovation talent [35]. Examining whether the innovation effect of smart city construction varies according to the level of science and education in the city is helpful for us to clarify the focus of resource investment in smart city construction. Since 2017, China has formally proposed the construction of "double first-class" universities, which often reflect the higher quality of higher education resources and scientific research level of the cities they are located in. According to the possession of "double first-class" universities in a city, we classify the city as a city with a higher level of scientific and educational development with such universities and a city with a lower level of scientific and educational development. According to the results in Columns (7) and (8) of Table 6, there is no significant difference in the impact of smart city construction on urban innovation in different levels of science and education, and all of them show that they significantly promote urban innovation. From the regression coefficients, the promotion effect of smart city construction on urban innovation is slightly larger in cities with lower levels of science and education than in cities with higher levels of science and education, which is likely to explain why the marginal innovation effect brought about by smart city construction is relatively small in cities with "double first-class" universities, where human capital and innovation factors are fully utilized, and the marginal innovation effect is relatively small in cities with "double first-class" universities.

5. Conclusions and Policy Implications

Taking advantage of the exogenous shock brought by China's smart city pilot policy, this paper empirically analyzes the impact of smart city construction on urban innovation and its mechanism based on the panel data of 238 prefectural-level cities in China from 2006 to 2019, and mainly draws the following conclusions: First, the construction of smart cities can significantly enhance the level of urban innovation. Second, the mediating effect analysis shows that smart city construction can promote urban innovation development by improving the level of urban informatization, promoting government investment in science and technology, and optimizing industrial structure. Third, the heterogeneity analysis finds that the effect of smart cities on urban innovation varies according to the differences in city size, regional location, and science and education level. In particular, the innovation effect of smart city construction is more obvious in larger-scale cities, cities in the eastern region, and cities with lower levels of science and education. Overall, our main conclusions are in line with the findings of studies based on different cities internationally [43] and have the potential for broader implications. By elucidating the mechanisms through which smart city construction enhances urban innovation and identifying heterogeneity in the effects, our study offers valuable insights for international debate. Our conclusions provide a foundation for cross-national comparative analysis and policy exchange, facilitating a deeper understanding of the role of smart cities in fostering urban innovation globally, although the need for policy considerations tailored to the specific city characteristics of different countries must also be underscored.

Based on the conclusions of the study, there is a need to focus on the potential of smart city construction as a sustainable policy tool and to develop a policy mix that comprises the areas of smart infrastructure, industrial optimization, fiscal investment, and public governance. There are policy implications as follows:

First, the construction of modern information infrastructure in the region should be further strengthened and consolidated. The government can increase investment in information infrastructure such as data centers, cloud computing platforms, and high-speed Internet, so as to ensure that the level of informatization in the city matches the needs of the smart city.

Second, the structure of government investment in science and technology should be further optimized. The government should appropriately invest more in smart city-related scientific and technological research and development and innovation projects according to the needs of urban development, facilitating sustained breakthroughs in key technological areas. At the same time, policy measures such as financial subsidies and tax incentives can be used to encourage enterprises to increase their investment in science and technology R&D and promote scientific and technological innovation and transformation of products. However, an assessment mechanism should be established to mitigate the adverse effects of government intervention [67].

Besides, the role of modern information and communication technologies in optimizing industrial structure should be given full play. The industrial layout should be rationally planned according to the development needs and resource advantages of the city to promote industrial agglomeration and the improvement of the industrial chain, and to bring about a synergistic effect and overall competitiveness of the industries. On the one hand, the government should encourage the integration of digital technologies with traditional industries and explore intelligent paths for the development of traditional industries. On the other hand, it should actively cultivate new industries, boost the in-depth integration of modern ICT with new energy development, intelligent manufacturing, and other industries, and develop green, intelligent, and high value-added industries.

Finally, differentiated smart city construction strategies should be implemented for different types of cities according to local conditions. For example, for the more developed large-scale cities and cities in the eastern region, the main priority can be to advance the intelligent development of high-end manufacturing, modern service industries, and other fields, enhancing the intelligence level of urban governance, and accelerating the agglomeration of innovation factors. For cities with a lower level of science and education, further input into science and technology education can be increased on the basis of full consideration of their own development advantages, and scientific and technological innovation and talent cultivation can be strengthened. Future research is needed to understand alternative socioeconomic impacts of smart city construction, how governments can design smart city policies to reap benefits and avoid pitfalls, and the sustainability and scalability of smart city initiatives in a changing technological environment, in addition to investigating the role of stakeholders such as citizens, businesses, and others in shaping the effectiveness of smart city strategies.

Author Contributions: Conceptualization, M.J. (Mingbo Ji) and Y.T.; methodology, M.J. (Mengyun Jin) and Y.T.; software, M.J. (Mengyun Jin) and L.C.; validation, M.J. (Mingbo Ji), M.J. (Mengyun Jin), and Y.T.; formal analysis, M.J. (Mengyun Jin), L.C. and Y.L.; investigation, L.C. and Y.L.; resources, L.C., Y.L. and Y.T.; data curation, M.J. (Mingbo Ji) and M.J. (Mengyun Jin); writing—original draft preparation, M.J. (Mingbo Ji); writing—review and editing, M.J. (Mingbo Ji), M.J. (Mengyun Jin) and Y.T.; visualization, L.C. and Y.L.; supervision, Y.T.; project administration, M.J. (Mingbo Ji); funding acquisition, Y.T. All authors have read and agreed to the published version of the manuscript.

Funding: This research is supported by Sichuan University's "From 0 to 1" Project entitled "Study on the Formation Mechanisms and Socioeconomic Impacts of Health Literacy in Ethnic Areas: A Survey from Pastoral Areas in Northwest Sichuan" (Grant No. 2022CX04).

Data Availability Statement: Publicly available datasets were analyzed in this study. These data can be found here: https://opendata.pku.edu.cn/ (Peking University Open Research Data), accessed on 4 March 2023; https://data.stats.gov.cn/easyquery.htm?cn=C01 (National Bureau of Statistics of China), accessed on 7 March 2023.

Conflicts of Interest: The authors declare no conflicts of interest.

References

1. Lucas, R.E. On the Mechanics of Economic Development. *J. Monet. Econ.* **1988**, *22*, 3–42. [CrossRef]
2. Romer, P.M. Increasing Returns and Long-Run Growth. *J. Polit. Econ.* **1986**, *94*, 1002–1037. [CrossRef]
3. Arkolakis, C.; Ramondo, N.; Rodríguez-Clare, A.; Yeaple, S. Innovation and Production in the Global Economy. *Am. Econ. Rev.* **2018**, *108*, 2128–2173. [CrossRef]
4. Kogan, L.; Papanikolaou, D.; Seru, A.; Stoffman, N. Technological Innovation, Resource Allocation, and Growth. *Q. J. Econ.* **2017**, *132*, 665–712. [CrossRef]
5. Anttiroiko, A.-V. City-as-a-Platform: The Rise of Participatory Innovation Platforms in Finnish Cities. *Sustainability* **2016**, *8*, 922. [CrossRef]
6. Feldman, M.P.; Audretsch, D.B. Innovation in Cities: Science-Based Diversity, Specialization and Localized Competition. *Eur. Econ. Rev.* **1999**, *43*, 409–429. [CrossRef]
7. Turok, I.; Bailey, D.; Clark, J.; Du, J.; Fratesi, U.; Fritsch, M.; Harrison, J.; Kemeny, T.; Kogler, D.; Lagendijk, A.; et al. (Eds.) *Transitions in Regional Economic Development*, 1st ed.; Routledge: London, UK, 2018; ISBN 978-1-315-14373-6.
8. Khan, H.H.; Malik, M.N.; Zafar, R.; Goni, F.A.; Chofreh, A.G.; Klemeš, J.J.; Alotaibi, Y. Challenges for Sustainable Smart City Development: A Conceptual Framework. *Sustain. Dev.* **2020**, *28*, 1507–1518. [CrossRef]
9. Laufs, J.; Borrion, H.; Bradford, B. Security and the Smart City: A Systematic Review. *Sustain. Cities Soc.* **2020**, *55*, 102023. [CrossRef]
10. Duygan, M.; Fischer, M.; Pärli, R.; Ingold, K. Where Do Smart Cities Grow? The Spatial and Socio-Economic Configurations of Smart City Development. *Sustain. Cities Soc.* **2022**, *77*, 103578. [CrossRef]
11. Shamsuzzoha, A.; Nieminen, J.; Piya, S.; Rutledge, K. Smart City for Sustainable Environment: A Comparison of Participatory Strategies from Helsinki, Singapore and London. *Cities* **2021**, *114*, 103194. [CrossRef]
12. Zhu, S.; Li, D.; Feng, H. Is Smart City Resilient? Evidence from China. *Sustain. Cities Soc.* **2019**, *50*, 101636. [CrossRef]
13. Anthopoulos, L.G. Understanding the Smart City Domain: A Literature Review. In *Transforming City Governments for Successful Smart Cities*; Rodríguez-Bolívar, M.P., Ed.; Public Administration and Information Technology; Springer International Publishing: Cham, Switzerland, 2015; pp. 9–21. ISBN 978-3-319-03167-5.
14. Dameri, R. Searching for Smart City Definition: A Comprehensive Proposal. *Int. J. Comput. Technol.* **2013**, *11*, 2544. [CrossRef]
15. Zubizarreta, I.; Seravalli, A.; Arrizabalaga, S. Smart City Concept: What It Is and What It Should Be. *J. Urban Plan. Dev.* **2016**, *142*, 04015005. [CrossRef]
16. Nam, T.; Pardo, T.A. Smart City as Urban Innovation: Focusing on Management, Policy, and Context. In Proceedings of the Proceedings of the 5th International Conference on Theory and Practice of Electronic Governance, Tallinn, Estonia, 26–28 September 2011; ACM: Tallinn, Estonia, 2011; pp. 185–194.
17. Bartelt, V.L.; Urbaczewski, A.; Mueller, A.G.; Sarker, S. Enabling Collaboration and Innovation in Denver's Smart City through a Living Lab: A Social Capital Perspective. *Eur. J. Inf. Syst.* **2020**, *29*, 369–387. [CrossRef]
18. March, H.; Ribera-Fumaz, R. Smart Contradictions: The Politics of Making Barcelona a Self-Sufficient City. *Eur. Urban Reg. Stud.* **2016**, *23*, 816–830. [CrossRef]
19. Chu, Z.; Cheng, M.; Yu, N.N. A Smart City Is a Less Polluted City. *Technol. Forecast. Soc. Chang.* **2021**, *172*, 121037. [CrossRef]
20. Yeh, H. The Effects of Successful ICT-Based Smart City Services: From Citizens' Perspectives. *Gov. Inf. Q.* **2017**, *34*, 556–565. [CrossRef]
21. Paskaleva, K.; Cooper, I. Open Innovation and the Evaluation of Internet-Enabled Public Services in Smart Cities. *Technovation* **2018**, *78*, 4–14. [CrossRef]
22. Leite, E. Innovation Networks for Social Impact: An Empirical Study on Multi-Actor Collaboration in Projects for Smart Cities. *J. Bus. Res.* **2022**, *139*, 325–337. [CrossRef]
23. Chatfield, A.T.; Reddick, C.G. Smart City Implementation Through Shared Vision of Social Innovation for Environmental Sustainability: A Case Study of Kitakyushu, Japan. *Soc. Sci. Comput. Rev.* **2016**, *34*, 757–773. [CrossRef]
24. Karvonen, A.; Cugurullo, F.; Caprotti, F. *Inside Smart Cities: Place, Politics and Urban Innovation*; Routledge: London, UK, 2018; ISBN 978-1-351-16618-8.
25. Kirimtat, A.; Krejcar, O.; Kertesz, A.; Tasgetiren, M.F. Future Trends and Current State of Smart City Concepts: A Survey. *IEEE Access* **2020**, *8*, 86448–86467. [CrossRef]
26. Zhao, F.; Fashola, O.I.; Olarewaju, T.I.; Onwumere, I. Smart City Research: A Holistic and State-of-the-Art Literature Review. *Cities* **2021**, *119*, 103406. [CrossRef]
27. Abu-Rayash, A.; Dincer, I. Development of Integrated Sustainability Performance Indicators for Better Management of Smart Cities. *Sustain. Cities Soc.* **2021**, *67*, 102704. [CrossRef]

28. Dente, B.; Coletti, P. Measuring Governance in Urban Innovation. *Local Gov. Stud.* **2011**, *37*, 43–56. [CrossRef]
29. Florida, R.; Adler, P.; Mellander, C. The City as Innovation Machine. *Reg. Stud.* **2017**, *51*, 86–96. [CrossRef]
30. Li, D.; Wei, Y.D.; Wang, T. Spatial and Temporal Evolution of Urban Innovation Network in China. *Habitat Int.* **2015**, *49*, 484–496. [CrossRef]
31. Lau, A.K.W.; Lo, W. Regional Innovation System, Absorptive Capacity and Innovation Performance: An Empirical Study. *Technol. Forecast. Soc. Chang.* **2015**, *92*, 99–114. [CrossRef]
32. Gao, K.; Yuan, Y. The Effect of Innovation-Driven Development on Pollution Reduction: Empirical Evidence from a Quasi-Natural Experiment in China. *Technol. Forecast. Soc. Chang.* **2021**, *172*, 121047. [CrossRef]
33. Hou, Y. Agglomeration Spillover, Accessibility by High-Speed Rail, and Urban Innovation in China: A Focus on the Electronic Information Industry. *Habitat Int.* **2022**, *126*, 102618. [CrossRef]
34. Yigitcanlar, T.; Degirmenci, K.; Butler, L.; Desouza, K.C. What Are the Key Factors Affecting Smart City Transformation Readiness? Evidence from Australian Cities. *Cities* **2022**, *120*, 103434. [CrossRef]
35. Ferraris, A.; Belyaeva, Z.; Bresciani, S. The Role of Universities in the Smart City Innovation: Multistakeholder Integration and Engagement Perspectives. *J. Bus. Res.* **2020**, *119*, 163–171. [CrossRef]
36. Xiong, Y.; Li, S. Can the Establishment of University Science and Technology Parks Promote Urban Innovation? Evidence from China. *Sustainability* **2022**, *14*, 10707. [CrossRef]
37. Yang, S.; Li, Z.; Li, J. Fiscal Decentralization, Preference for Government Innovation and City Innovation: Evidence from China. *Chin. Manag. Stud.* **2020**, *14*, 391–409. [CrossRef]
38. Zou, C.; Huang, Y.; Wu, S.; Hu, S. Does "Low-Carbon City" Accelerate Urban Innovation? Evidence from China. *Sustain. Cities Soc.* **2022**, *83*, 103954. [CrossRef]
39. Fan, F.; Lian, H.; Liu, X.; Wang, X. Can Environmental Regulation Promote Urban Green Innovation Efficiency? An Empirical Study Based on Chinese Cities. *J. Clean. Prod.* **2021**, *287*, 125060. [CrossRef]
40. Li, J.; Du, Y. Spatial Effect of Environmental Regulation on Green Innovation Efficiency: Evidence from Prefectural-Level Cities in China. *J. Clean. Prod.* **2021**, *286*, 125032. [CrossRef]
41. Zhou, Y.; Li, S. Can the Innovative-City-Pilot Policy Promote Urban Innovation? An Empirical Analysis from China. *J. Urban Aff.* **2023**, *45*, 1679–1697. [CrossRef]
42. Gao, K.; Yuan, Y. Government Intervention, Spillover Effect and Urban Innovation Performance: Empirical Evidence from National Innovative City Pilot Policy in China. *Technol. Soc.* **2022**, *70*, 102035. [CrossRef]
43. Caragliu, A.; Del Bo, C.F. Smart Innovative Cities: The Impact of Smart City Policies on Urban Innovation. *Technol. Forecast. Soc. Chang.* **2019**, *142*, 373–383. [CrossRef]
44. Wang, B.; Loo, B.P.Y.; Huang, G. Becoming Smarter through Smart City Pilot Projects: Experiences and Lessons from China since 2013. *J. Urban Technol.* **2022**, *29*, 3–24. [CrossRef]
45. Beck, T.; Levine, R.; Levkov, A. Big Bad Banks? The Winners and Losers from Bank Deregulation in the United States. *J. Financ.* **2010**, *65*, 1637–1667. [CrossRef]
46. Angelidou, M. The Role of Smart City Characteristics in the Plans of Fifteen Cities. *J. Urban Technol.* **2017**, *24*, 3–28. [CrossRef]
47. Nie, C.; Zhong, Z.; Feng, Y. Can Digital Infrastructure Induce Urban Green Innovation? New Insights from China. *Clean Technol. Environ. Policy* **2023**, *25*, 3419–3436. [CrossRef]
48. Putra, Z.D.W.; van der Knaap, W. Urban Innovation System and the Role of an Open Web-Based Platform: The Case of Amsterdam Smart City. *J. Reg. City Plan.* **2018**, *29*, 234–249. [CrossRef]
49. Lach, S. Do R&D Subsidies Stimulate or Displace Private R&D? Evidence from Israel. *J. Ind. Econ.* **2002**, *50*, 369–390. [CrossRef]
50. Czarnitzki, D.; Ebersberger, B.; Fier, A. The Relationship between R&D Collaboration, Subsidies and R&D Performance: Empirical Evidence from Finland and Germany. *J. Appl. Econom.* **2007**, *22*, 1347–1366. [CrossRef]
51. Pfister, C.; Koomen, M.; Harhoff, D.; Backes-Gellner, U. Regional Innovation Effects of Applied Research Institutions. *Res. Policy* **2021**, *50*, 104197. [CrossRef]
52. Yam, R.C.M.; Lo, W.; Tang, E.P.Y.; Lau, A.K.W. Analysis of Sources of Innovation, Technological Innovation Capabilities, and Performance: An Empirical Study of Hong Kong Manufacturing Industries. *Res. Policy* **2011**, *40*, 391–402. [CrossRef]
53. Gao, D.; Zhou, X.; Mo, X.; Liu, X. Unlocking Sustainable Growth: Exploring the Catalytic Role of Green Finance in Firms' Green Total Factor Productivity. *Environ. Sci. Pollut. Res.* **2024**, *31*, 14762–14774. [CrossRef]
54. Paunov, C.; Rollo, V. Has the Internet Fostered Inclusive Innovation in the Developing World? *World Dev.* **2016**, *78*, 587–609. [CrossRef]
55. Zhu, B.; Zhang, M.; Zhou, Y.; Wang, P.; Sheng, J.; He, K.; Wei, Y.-M.; Xie, R. Exploring the Effect of Industrial Structure Adjustment on Interprovincial Green Development Efficiency in China: A Novel Integrated Approach. *Energy Policy* **2019**, *134*, 110946. [CrossRef]
56. Hao, X.; Li, Y.; Ren, S.; Wu, H.; Hao, Y. The Role of Digitalization on Green Economic Growth: Does Industrial Structure Optimization and Green Innovation Matter? *J. Environ. Manag.* **2023**, *325*, 116504. [CrossRef]
57. Wang, J.; Deng, K. Impact and Mechanism Analysis of Smart City Policy on Urban Innovation: Evidence from China. *Econ. Anal. Policy* **2022**, *73*, 574–587. [CrossRef]
58. Xu, N.; Ding, Y.; Guo, J. Do Smart City Policies Make Cities More Innovative: Evidence from China. *J. Asian Public Policy* **2022**, *15*, 1–17. [CrossRef]

59. Chen, K.; Li, Q.; Shoaib, M.; Ameer, W.; Jiang, T. Does Improved Digital Governance in Government Promote Natural Resource Management? Quasi-Natural Experiments Based on Smart City Pilots. *Resour. Policy* **2024**, *90*, 104721. [CrossRef]
60. Cai, X.; Lu, Y.; Wu, M.; Yu, L. Does Environmental Regulation Drive Away Inbound Foreign Direct Investment? Evidence from a Quasi-Natural Experiment in China. *J. Dev. Econ.* **2016**, *123*, 73–85. [CrossRef]
61. Mehmood, Y.; Ahmad, F.; Yaqoob, I.; Adnane, A.; Imran, M.; Guizani, S. Internet-of-Things-Based Smart Cities: Recent Advances and Challenges. *IEEE Commun. Mag.* **2017**, *55*, 16–24. [CrossRef]
62. Jiang, D. The Construction of Smart City Information System Based on the Internet of Things and Cloud Computing. *Comput. Commun.* **2020**, *150*, 158–166. [CrossRef]
63. Ke, X.; Chen, H.; Hong, Y.; Hsiao, C. Do China's High-Speed-Rail Projects Promote Local Economy?—New Evidence from a Panel Data Approach. *China Econ. Rev.* **2017**, *44*, 203–226. [CrossRef]
64. Guerzoni, M.; Raiteri, E. Demand-Side vs. Supply-Side Technology Policies: Hidden Treatment and New Empirical Evidence on the Policy Mix. *Res. Policy* **2015**, *44*, 726–747. [CrossRef]
65. Szczygielski, K.; Grabowski, W.; Pamukcu, M.T.; Tandogan, V.S. Does Government Support for Private Innovation Matter? Firm-Level Evidence from Two Catching-up Countries. *Res. Policy* **2017**, *46*, 219–237. [CrossRef]
66. Gao, D.; Tan, L.; Mo, X.; Xiong, R. Blue Sky Defense for Carbon Emission Trading Policies: A Perspective on the Spatial Spillover Effects of Total Factor Carbon Efficiency. *Systems* **2023**, *11*, 382. [CrossRef]
67. Gao, D.; Li, Y.; Tan, L. Can Environmental Regulation Break the Political Resource Curse: Evidence from Heavy Polluting Private Listed Companies in China. *J. Environ. Plan. Manag.* **2023**, 1–27. [CrossRef]

land

MDPI

Article

Urban Modeling in the Global South and Sustainable Socio-Territorial Development: Case of Puebla-San Andrés Cholula Urban Binomial, Mexico

Anne K. Kurjenoja

InLAB, Territory and Urban Artifact Research Group, Ex. Hda. Santa Catarina Mártir, Cholula 72810, Puebla, Mexico; annek.kurjenoja@udlap.mx; Tel.: +52-2228035934

Abstract: To understand urbanization across the Global South, it is indispensable to consider situated heterogeneous urban situations shaped by global and local forces and their intersections. In the case of Mexico, the political and economic desire for globalization has extended beyond the great metropolis of Mexico City to mid-size cities triggering the formulation of their own urban strategies to become global. This paper explores the connections between neoliberal public policy, globalization, urban modeling, and socio-territorial sustainability in the territorial binomial of the city of Puebla and the municipality of San Andrés Cholula in the period of 2011–2017 and its current consequences, addressed by the local planning route map, and informed by the Orange Economy-guide of the Inter-American Development Bank (IDB), using relevant bibliographical sources, mapping, interviews of stakeholders, and fieldwork. The ultimate purpose of the research project described herein is to provide a multi-dimensional analysis of the development of the territorial situation of the Puebla-San Andrés Cholula-urban binomial, and of its current and potential future consequences, offering supporting information for its urban planning. The research results exposed here reveal urban modeling processes informed by the Global North urban globalization theory, development of global cities in the Global South, local socio-territorial dynamics characterized by economic and political interests imprinted in the public policy, and socio-territorial patterns inherited from the colonial past, resulting in socio-economic and racial discrimination, population displacements, real estate speculation, and risking ecological and environmental sustainability.

Keywords: Global South; urban modeling; neoliberal public policy; globalization; socio-environmental sustainability; Mexico

1. Introduction

To study urbanization across the Global South, it is indispensable to consider situated heterogeneous urban situations shaped by global and local forces and their intersections [1,2]. In the case of Mexico, the political and economic desire for globalization has extended beyond the great metropolis of Mexico City to mid-size metropolitan areas triggering the formulation of their own urban strategies to become global.

In the postcolonial theory framework, Aihwa Ong in her *Powers of sovereignty: State, people, wealth, life* [1], has pointed out how the recent decades have marked a shift away from models of stabilized world arrangements, also in the case of studying urban development. Mouton and Shatkin [2] have mentioned that we should not follow the presumption of being able to explain the recent urbanization processes in the postcolonial developing world only through the Euro-American theoretical core of urban studies, as their local characteristics and conditions disrupt its all-embracing universal concepts and development patterns. Thus, lines of thought have emerged especially in Asia, to attend to the specificities and patterns of urbanization across the post-colony. As Ong [1], besides Mouton and Shatkin [2], observed that the demand for postcolonial thinking in urban studies has triggered the interest in the analysis of heterogeneous urban situations in the post-colony shaped by both global and deeply rooted local forces and their intersections. Thus, Ong [3] proposed the exploration of the global as a historically contingent concept shaped by the interactions between global and locally situated components characterized by varied political, economic, social, and cultural conditions. Mouton and Shatkin [2], citing Hart (2016) [4], have suggested that cities are not simple recipients of global processes, but agential territories and sites for their production juxtaposed with spatial-historical conjunctures. When analyzing their urban territories through a conjunctural approach, it is thought of as indispensable to study their specific spatio-temporal conditions, such as those related to a colonial past, to acquire a deep understanding of their histories of social, economic, and political change, and how these have been connected to larger structural forces, as those of globalization.

According to Shatkin [5], in the developing and postcolonial world the studies of urban globalization have been much concentrated on the megacities of the Global South and on the dramatic spatial expansion of their territory. This explosive growth has frequently meant even violent changes in the surrounding regions, transforming the habitability of large areas detonating human displacements and socio-territorial and environmental changes. The local policy informed by the global has triggered especially in Asia, the promotion of renewal plans to up-scale urban regions through the interaction between locally situated actors and the neoliberal global dynamics, giving a powerful example for other regions of the post-colony how to increment the economic and political importance of their cities in the globalizing world.

In Latin America, globalization appeared in the 1990s as a luring alternative for urban development, but the true start-up for this in Mexico began after 2007 triggered by ProMéxico, a trust of the Federal Government. Though, much of the research work done about urban transformation until now in Mexico has been concentrated on the biggest metropolitan area of the country, Mexico City, the political and economic desire for globalization reached various mid-size Mexican cities and their suburban areas since the beginning of the 21st century, reformulating their urban strategies now aiming at becoming global. Among them, there is the city of Puebla and its metropolitan area with 2,328,663 inhabitants [6], the capital of the federal state of the same name, located in the central plateau of Mexico. Under the state government of 2011–2017, Puebla aspired to be one of the Latin American global cities and as such, through an aggressive public policy, promoted urban territorial transformations and re-engineering of its image through urban modeling, expanding these interventions beyond its own limits to the territory of its Western neighbor, the rural municipality of San Andrés Cholula, thus giving origin to the development of an urban binomial.

In this context, this paper presents five initial statements it proposes to discuss, based on the 2014–2023 research work: (1) The urban development of Puebla-San Andrés Cholula-urban binomial has been influenced by the Global North and the Southeast Asian neoliberal development patterns, manifesting characteristics of both in its territorial management; (2) Regarding the creation of a global urban image, Southeastern Asian style urban and territorial modeling has been applied through public politics; (3) The local colonial history and the socio-racial discrimination has been inherited and transferred to the 21st century urban territorial management; (4) The juxtaposition of the above mentioned aspects have aggravated the already existing socio-territorial polarization observable in the urban inclusion-exclusion strategies reinforced by the urban planning processes of the 2011–2017 period, and lastly (5) The territorial processes triggered by neoliberal public policy of the mentioned period and the corresponding land management strategy have menaced the socio-territorial and environmental balance, quality of life and urban sustainability in the area of the Puebla-San Andrés Cholula-urban binomial. Thus, this article explores, at which point could we study urban binomial's urbanization processes from the scope of development patterns of the Global North. Furthermore, could the 21st-century Latin American urbanization–metropolization be analyzed in the light of experiences and research done in other parts of the post-colony as in Southeast Asia, exploring similarities? Finally, can the Global

North and the Southeast Asian influences be observed in the urban development of a Mexican mid-size metropolitan area, such as the urban binomial of the city of Puebla and the municipality of San Andrés Cholula, and juxtaposed with local conditions, public policy, and economic interests, and which have been their consequences considering the socio-territorial and environmental sustainability? As answers to these questions, this article shows how the Puebla-San Andrés Cholula urban binomial has been transformed into an agential territory in which spatial-historical conjunctures are juxtaposed with urbanization patterns informed by a neoliberal public policy, influenced by the Global North urban globalization theory and urban modeling in Asian global cities, putting thus at risk the social, ecological, and environmental sustainability.

2. Materials and Methods

This paper, informed by bibliographical research, field studies, and interviews, aims to explore which has been the role of different urban actors, such as real estate developers, state government, and local stakeholders, in the re-assemblage of urban territories and development processes in Puebla-San Andrés Cholula-urban binomial, but it also explores how the local rural population of the latter has been impacted by the change and how the territorial transformations have impacted the environmental conditions and sustainability.

The article is informed by the Global North and Global South postcolonial lines of thought, revising how these localized tendencies of development have influenced the binomial's own processes of urban transformations. Furthermore, in 2016 a series of interviews were conducted, first with the urbanists who participated in the 1990s planning processes that triggered the first stage of the neoliberal urbanization in the territory of the binomial being the basis for the second 2011–2017 stage of urban transformations. In the same year, more interviews were held with the architect, responsible for the design of various public spaces in the San Andrés Cholula territory, under the concept of urban modeling and staging.

Also, fieldwork was carried out in 2014–2016 through photographing the urban transformations in progress, besides mapping the territorial impacts. From 2018 onwards, the results and consequences of the 2011–2017 development have been observed, especially focused on the territorial conflicts and socio-environmental demands denounced by the San Andrés Cholula's Indigenous habitants. In these terms, more interviews have been conducted with the local communities, revision of the legal framework carried out, and more fieldwork undertaken to document the socio-territorial and environmental effects of the urban transformations. The following chart of Figure 1 describes the workflow carried out:

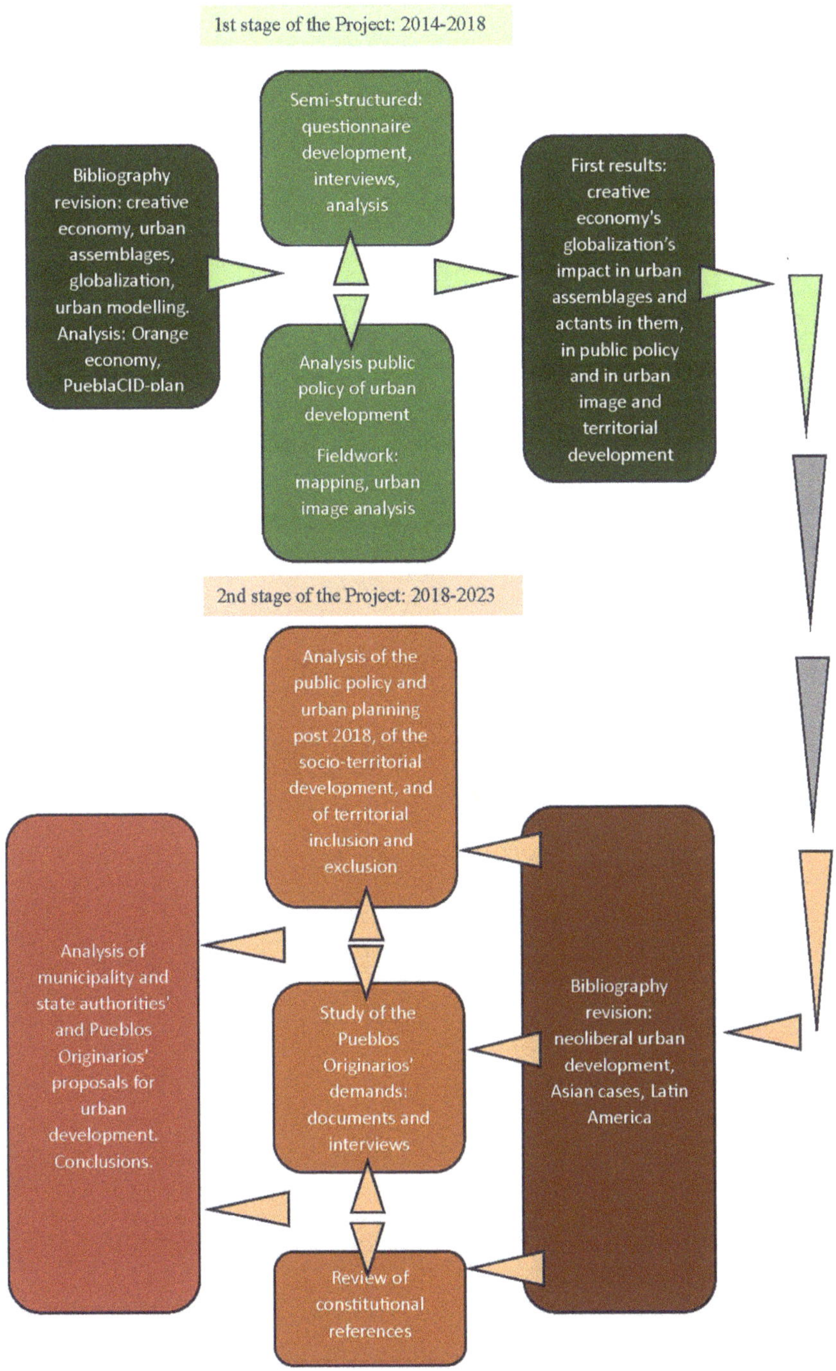

Figure 1. Research project workflow chart elaborated by Kurjenoja. (2023).

3. Case Study

The focus area of this case study, the metropolitan area of Puebla and the border zone between it and the municipality of San Andrés Cholula, is located in central Mexico, 106 km to the Southeast of Mexico City, as presented in the following map of Figure 2.

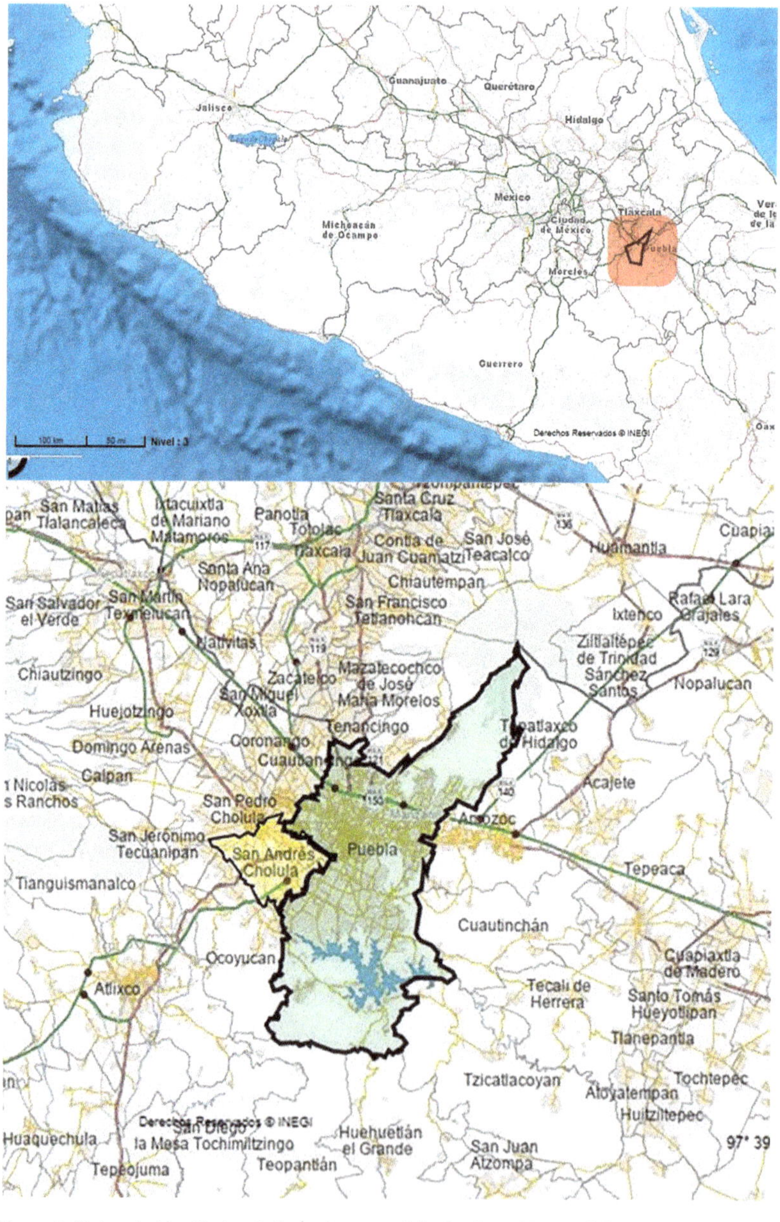

Figure 2. Elaborated by Kurjenoja from the map of the Instituto Nacional de Estadística y Geografía, INEGI, Mapa digital de México, 2023 [7]. (2023).

In the following sections, the origin of San Andrés Cholula and the colonial foundation of Puebla are described, as well as their urban development until current times.

3.1. Puebla-San Andrés Cholula: Foundation, Development and Globalization of an Urban Binomial

3.1.1. Cholula: The Ancient City

Ashwell [8,9] describes how the territorial development of Cholula, today consisting of two municipalities, San Pedro and San Andrés Cholula, has been the product of a long civilizing process and human presence dating back at least 20,000 years B.C. Between the year 300 B.C. and the beginning of the Christian era, Cholula quickly transformed into a large city. At this time, the oldest pyramidal structure of the Cholultecan ceremonial center of the period 200–100 BC, Tlachihualtepetl, 'the hill made by man' (see Figure 3) or the Great Pyramid of Cholula dedicated to the god Quetzalcoatl, was erected representing a natural hill associated with water, tule trees and flowers [8,9].

Figure 3. The Great Pyramid of Cholula. Photo taken by Carrera, C. (2023).

During the colony, and despite the territorial division into two separate communities, the two Cholulteca villages, San Pedro and San Andrés, continued to maintain their customs and habits without major changes until the mid-20th century, also preserving their definition as a living *altepetl*, that is, a socio-spatial organization of the territory similar to many others that gave rise to many Mesoamerican cities [9,10]. Both Cholulas are characterized by different fabrics juxtaposed in their territory such as geography, land use, landscape, and biocultural and socioeconomic aspects.

San Andrés Cholula has always been a rural municipality with the predominant characteristic of local smallholder agriculture [9,11]. Rural life in San Andrés Cholula continued its own course until 1988, when by the state expropriation decree, extensions of communal agricultural lands, *ejidos*, were assigned as a territorial reserve for the growth of the neighboring city of Puebla, detonating resistance movements among Cholula population [9].

3.1.2. Puebla: Foundation and Development of a Colonial City

Nancy Churchill [12] describes, how shortly afterward Hernando Cortés arrived in what is now Mexico in 1519, the Spanish Colony of New Spain started the construction of its new colonial bastions to guarantee territorial domination. One of the first to be established was the city of Puebla, founded in 1531, halfway between the port of Veracruz and what is now Mexico City, no man's land on the Eastern side of the ancient city of Cholula. The new city found an ideal site on this high plain crossed by several rivers adequate for agricultural and commercial activities soon converting it to the second-most important city of New Spain.

Loreto López [13] describes, how shortly after the city's foundation the urban settlement began to grow to have as its Eastern limit the San Francisco river, nurtured by numerous springs and creeks with their origin on the slopes of the nearby mountains. Further away to the West and outside the urban structure another local river, the Atoyac, was used to irrigate the agricultural production of the new settlement. From the very beginning, the two rivers were used for the production of hydraulic energy, needed for the first pre-industrial productive plants, such as mills, tanneries, soap factories, and cotton and wool workshops.

For Vélez Pliego [14], maps and plans of the colonial cities evidence how colonizers discredited the right of the native people to the land and to the hegemonic urban space through systems of ethnic and racial discrimination and control taking advantage of hydrological and topographical conditions of the terrain. These helped to divide the urban territory and to separate the Spanish cities from the Indian settlements that with time were transformed into mestizo and finally urban proletariat districts in the urban peripheries. Thus, as Nancy Churchill [12] describes, the Spanish city was constructed on the Western bank of the San Francisco River, leaving the Eastern bank to the Indigenous people brought to the city as the labor force. As Homi Bhabha [15] has said, the colonial representation of the political and cultural hegemony, and the social truth was translated to the urban form of colonial cities as Puebla (see Figure 4). The colonial idea of the Otherness triggered an ideology of the construction of socio-cultural barriers for different human existences to give protection to the epistemological edge of the West. This characteristic of the fragmentation of urban territories based on socio-ethnic and racial aspects of the population, as we can see further on, has survived until today and is strongly present in the contemporary patterns of urban development characterized by social inclusion and exclusion.

3.1.3. Modernization of a Colonial City and Emergence of the Urban Binomial

After its foundation, the city structure of Puebla suffered very few changes conserving its colonial limits until the 1940s, as Puebla's authorities had rejected the modernization and urban renewal, and the development was stagnant until 1993, when the state government of Manuel Bartlett Díaz launched the Angelópolis Plus-plan with the intension to trigger an urban reformation through integral planning and swinging doors wide open to welcome the globalized economy and foreign investment. When the period of his state government ended in 1999, the Angelópolis Plus plan was abandoned [16] until 2011, when the elected governor Rafael Moreno Valle revived this project, now having in mind its potential to insert Puebla as a Latin American node in the networks of the global economy. Contemporary lifestyles, urban security, cultural industries, historical patrimony and memory, smart urban environment, and renovated urban image were key concepts to catapult the city to international knowledge through nominations and certifications. In this framework, the Puebla CID plan [17], influenced by the Orange Economy model [18], promoted it in 2015 as a Smart City and later in the same year as a Creative City in Design by UNESCO [16].

The recovery of the ideas of the Angelópolis Plus plan in 2011 by the Moreno Valle state government gave special importance to the urban territory by the Atoyac river. Going back in time, for the year 1989 the urban growth of Puebla had reached its Eastern bank though it consisted mainly of irregular settlements that the city council years later had to formalize and provide with urban services. Besides them, along the river, popular housing complexes have been constructed by the Institute of the National Housing Fund for Workers (INFONAVIT). Thus, this area maintained its character as a low and middle-income district. On the other hand, the 1993 Angelópolis Plus-plan exposed the economic potential of the Western bank of the Atoyac for the urban expansion of the city of Puebla [19,20], with the ultimate inconvenience that, according to the colonial Royal Certificate and the official municipality limits, it belonged to the neighboring rural municipality of San Andrés Cholula. Puebla's aspirations to extend its territory beyond the Atoyac river triggered a political controversy and a legal struggle in federal courthouses over the ownership of this potentially lucrative urban land between these two municipalities. Insisting on expanding Puebla beyond the Atoyac river, the Angelópolis Plus plan of the state government promoted extensive expropriations of San Andrés Cholula's agricultural lands to give space for urban growth. These actions of appropriation were executed from 1993 onwards and especially from 2011 to 2017, as the state government started to carry out an organized transformation of this zone to a 'New Puebla' with luxury malls, hotels, entertainment, and cultural centers, universities and residential gated communities connected to extensive and exclusive green belts along the Atoyac-river [21], initiating territorial struggles between San Andrés Cholula's rural population and state authorities about the right to the land and to the natural resources, though finally having to accept the loss their rural lands.

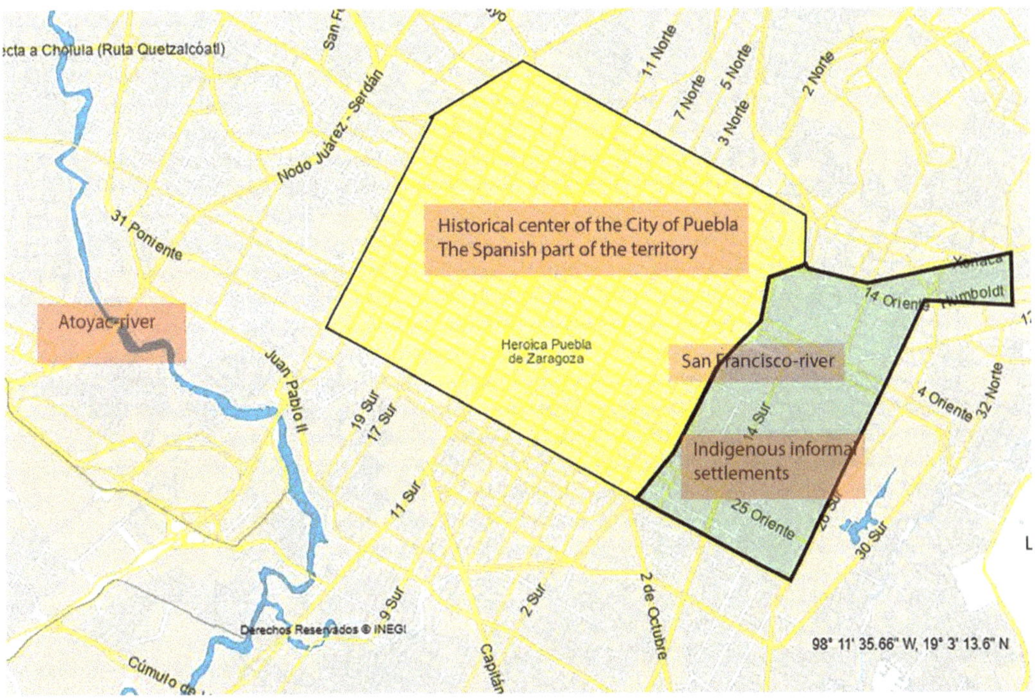

Figure 4. Historical center of Puebla. Elaborated by Kurjenoja from the map of the Instituto Nacional de Estadística y Geografía, INEGI, Mapa digital de México, 2023 [7]. (2023).

4. Results

4.1. *References from the Global North: Global Economy, Expulsions, and Environmental Crisis*

In 2010, Saskia Sassen [22] defined the metropolis as sites in which globalization has found wide opportunities to take root triggering development processes aiming at pinning them up as nodes for the management of global financial flows. The by-product of these processes, as Sassen observed, was urban polarization due to the increment of urban poverty triggered by macro-social tendencies related to cities' own economic, political, and cultural processes and practices. Harvey [23], on the other hand, observed an especially notable characteristic in these contemporary urban processes: their focus on the development of changes in lifestyles considering them as merchandise transforming the city itself to a consumer product offering conditions for temporal or permanent experiential life in the framework of the creative economy for those who could afford it. The resulting urban polarization caused by the unbalanced distribution of resources can then be seen in the urban image and structure through the emergence of gated communities, compact urban centers with spectacular skyscrapers and public spaces, and, on the other hand, misery belts in the urban peripheries as 'micro-states' of prosperity and poverty.

In her 2014 book *Expulsions. Brutality and Complexity in the Global Economy* [24], Sassen tackles the issue of 'logics of expulsion', mentioning how the last two decades have evidenced territorial expulsions of people due to the economic dynamics of our time, fracturing cities in a before unseen ways, making visible growing tendencies of inequality as, according to Sassen, a pathology of contemporary global capitalism. With this, she is paying attention to the augmented social fragmentation that is appearing as a growing urban and socio-territorial characteristic even in the contemporary cities of the Global North. Furthermore, she declares how advanced political economies have created realities in which financial, political, and legal complexities tend to produce elementary brutalities [24]; "[t]he globalization of capital and the sharp rise in technical capabilities have produced major scaling effects" [24] p. 3. Sassen points out the contemporary logic of speculation-driven finance to seek hyper-profits frequently triggering damaging consequences to people, places, and the environment. She explains quite interestingly how the territorially expelled, now living at long distances from the new and complex invasive and dominating power systems of persons, networks, and technology occupying renovated urban cores, still are indispensable for the social infrastructure of power of global cities [24]. This is especially true in a country such as Mexico, where the new urban elites, strictly wanting to live separated from the lower-class population, also need them to give essential maintenance, security, and housekeeping services to their homes, offices, and *gated communities*.

Finally, Sassen [24] draws our attention to the environmental damage caused by globalization and the commodification of water and land mentioning how the deterioration of the quality of land, water, and air has affected the urban population, but especially the marginalized urban populations triggering human displacements worldwide. Though, Sassen is not analyzing environmental problems from a detailed urban scope but concentrates her observations on the global scale of economics and politics. Related to this, the Latin American theorist Néstor García Canclini [25] has made considerations of what globalization has meant on the regional scale. He explains how at the beginning of the 1990s the tendency observable in Latin America was that of countries reordering their national economies to attract investments and to increase their competitiveness in the global market. Thus, globalization policies were accepted because they promised development and prosperity the people and governments of the developing countries badly needed, with the notion that not all groups were permitted to participate in the globalization processes. Thus, this differentiation triggered the emergence of inequality between those connected to global networks and those separated from them, reflecting at this point the inherited colonial difference, as can be further seen in the case of Puebla-San Andrés Cholula-urban binomial's socio-territorial development.

4.2. *References from the Global South: Local Scope of the Postcolonial Southeast Asia*

4.2.1. Cities and National Ambitions

According to Ong [1], the first decades of the 21st century meant the expansion of the global city idea to the Asian continent, revealing national aspirations reflected in local public policy focused on the 'worlding' of big cities. The rise price of the new Asian financial region attracted a great volume of international trading away from traditional Western financial centers. Added to this, investments in financial technologies incremented the pace and volume of trading and the scale of financial interactions especially in Southeast

Asia. As a consequence, the appearance of new global cities in Asia challenged the already established Global North concepts about urban globalization triggering the need to study the phenomenon from the postcolonial and local scope. What Ong observes is that the Asian environmental dynamics of financial reconfiguration emerged from specific assemblages of urban aspirations focused on the desire to create and promote globalized urban hotspots, in which the global was fused with the situated post-colonial realities, giving a totally new dimension and paradigm to what it means to be a global city in the post-colony, on one the hand tightly connected to a capitalist world but on the other, powerful visual icons for the emerging non-Western economic, financial and political powers [1].

In this context, as Aihwa Ong [3] pointed out how "emerging nations exercise their new power by assembling glass and steel towers to project particular visions of the world" (p. 1). Singapore, Shanghai, Seoul, and Hong Kong, as new powerful centers of finance, started transforming their urban landscape and skyline through cutting-edge innovations in urban architecture and design, to expose through their urban image their new role in the global economy reinventing, as Ong mentions, urban norms and paradigms that showcased their position as global cities. Thus, in Asia, cities started to embody nationalist ambitions of wealth, power, and international recognition. They became sites for instantiating their countries' claims to global significance. As such, their urban codes of city norms and accomplishments were not detachable from their situated politics related to their role as image icons in a global inter-city game of gaining international prestige.

Thus, in terms of strategies for the creation of novelty and speculative urban imaginaries, urban modeling tools used in Asia have spread beyond its borders, to other parts of the post-colony, showcasing national and local ambitions through discursive and material gestures, inspired by non-Western urban image icons. What modeling means here, is the creation of new urban concepts related to sustainability, habitability, or being world-class eagerly followed elsewhere in the post-colony hoping to be able to reproduce or surpass the already existing successful effects [3]. Mexico and its cities have not been exceptions in these 'worlding of tendencies by adopting Asian recipes of urban modeling. In these terms, the Puebla CID plan [17] and IDB's Orange Economy document [18] refer to Asian examples, highlighting the importance of urban design and architecture in the construction of an imaginary competitive global city distinguished by its creative, innovative, and productive capacity.

4.2.2. Territorial Order, Land Speculation, and Urban Modeling

Ong [3] points out how Asian neoliberal urban practices have rearticulated and reassembled as material, technological and discursive elements in the processes of remaking the urban territories by new alliances between political and economic actors. What has been notable in these processes, is how they have ignored the urban poor or other groups not fitting to the neoliberal global form, expulsing them from the most profitable urban lands to improvised settlements in the urban outskirts. Ong [3] highlights how the Western theory of globalization does not totally respond to the locally situated 'worlding' processes, urban resistances, and socio-territorial patterns of the developing post-colony and their consequences, which should be understood considering the realities emerging from the colonial roots. For Ong, the colonial past plays an important role in the present urban culture and territorial order: "Postcolonial theory is thus as much about how contemporary urban situations have been shaped by colonial legacies of injustice as by contemporary problems of urban underdevelopment" [3] (p. 9).

Ong argued that urban modeling is a political tool to change the social spirit of an urban environment exposing its symbolic values and desirable icons of 'world-class' amenities: luxury hotels, shopping malls, entertainment and convention facilities, cultural centers, international enclaves, and airports as urban attributes. Thus, cities aspire to be sites for the expansion of a new middle class offering attractive conditions for working and living. With urban modeling as a political and economic tool, strategies that Goldman [26] calls 'speculative urbanism' emerge: the urban land has become the key issue for planners and developers. Goldman mentions how in the East Asian context, the practice of selling land and displacing the original population to favor public and private goods and services has become one of the primary interests of governments with aspirations to re-engineer their cities to be globally competitive. Thus, the speculative government has become the new political rationality in which the primary relationship between the local people and their government is related to landowning.

Goldman [26] describes how in Asia there has been a vigorous promotion of what he calls the 'urban turn' in which regulatory changes by states and new investment strategies by private firms have triggered the creation of ambitious master plans to design worlding of urban landscapes. To achieve a successful urban transformation through modeling, local state agencies have privatized many of their public tasks transferring them to an expanding local and transnational elite. In Asia it has been sometimes difficult to find vast open areas to build upon and new land for urban expansion must have been acquired from the villages of the local rural population. Thus, as Goldman denounces, the previous agricultural lands have been transformed into high-value gated communities, world-class hotel complexes, 'medical tourism' hospitals, and business centers by developers converting the cheap public and undervalued private or communal lands into real estate skyrocketing their value and triggering so a massive displacement crisis. The 'urban turn' observed by Goldman in Asia, has also landed in Latin America. In the case of our urban binomial, the public policy since 1993 and all the regulatory changes related to landowning and urban norms have been decisive in triggering urban and territorial transformations focused on 'worlding'. Indeed, as in Asia, the 'speculative urbanism' executed here, as explained further in this article, has triggered considerable human displacements.

4.2.3. Becoming Global in Latin America?

As mentioned before, the economic development of the Asian countries, the 'Asian Tigers', influenced the Orange Economy model [18] for Latin America developed by Felipe Buitrago Restrepo and Ivan Duque Márquez, the model being sponsored by the Inter-American Development Bank (IDB), and based on the concept of the creative economy defined by John Howkins in 2007 [27]. Howkins focused his idea on the economic potential of talent and creativity that could be exploited through creative goods and services in the framework of architecture, arts and design, research, fashion, development of software, etc. Buitrago and Duque highlight the creative economy's potential to boost the social and economic development in Latin America and the Caribbean recognizing though also the dangers of its implementation: unemployment and migration of the technologically unskilled triggering social instability. On the other hand, it could motivate a socio-economic transformation for many Latin Americans, being able to experience the highest growth in wealth in their history, enjoying disposable income for the first time in their lives to personalize their consumption in which the technology is integrated into the consumption of contents.

In the Orange Economy framework, the focal task of cities is that of attraction, retention, and reproduction of the talented segment of the population through urban development plans of creative cities, based on the idea of an equilibrium between creativity, enjoyment, and environment, understanding this latter as amenities and lifestyle. This creative environment must be provided with high-quality infrastructure (public transportation system, Internet connections, highways, airports, etc.). The last step in the development of a creative city is obtaining global recognition as an internationally identified center of cultural excellence, attracting varied and abundant tourism for its architecture, festivals, museums, history, and nightlife [18]. We can argue that in the context of the socioeconomic realities of Latin America, the creative cities described in the document of Buitrago and Duque would be artificial 'bubbles' separated from the conflictive socio-environmental and socio-territorial realities of the region. Thus, as could be expected, the Puebla CID plan inspired by the Orange Economy idea, among other triggers, was from the beginning a potential detonator of territorially situated processes of inclusion and exclusion, reinforced by neoliberal urban public policy.

4.3. *Worlding the 'New Puebla'*

4.3.1. Urban Modeling and Creation of a Global Imaginary

For the 2011–2017 Puebla state government, the neoliberal public policy formed the main development framework for driving Puebla as an important Latin American node and benchmark in the global economy. At that point, the state government found federal support for its plans in ProMéxico, founded in 2007 and dissolved in 2019 by the leftist López Obrador federal government. ProMexico was a trust of the Federal Government—sectored into the Ministry of Economy—promoting international trade and investment aiming at consolidating Mexico as an attractive, safe, and competitive destination for foreign investment [28]. Thus, the previously mentioned implied generation of new paradigms for the planning, programming, and execution of a new urban reality based on providing an improved social environment and urban sustainability, distinguished by international recognitions and nominations, such as Smart City and World Capital in Design by the UNESCO, recognizing it as an attractive city to live [17].

To trigger the desired urban development, the 2011–2017 state government implemented aggressive strategies of public and economic policies that opened opportunities for urban developers and architects inspired by the urban modeling that had given an iconic form and skyline to Southeast Asian global cities [29]. The globalizing Puebla had to focus its efforts on structuring its socio-economically fragmented and historically polarized territory to adapt it to the new urban ideals established in the Puebla CID plan, among them creating an adequate global image [17].

The urban transformation of Puebla through the Puebla CID plan [17] was intended to make it a creative city as described in the Orange Economy model [18]. But the most part of its urban territory was not apt for that; the Northern, Eastern, and Southeastern parts of its territory were occupied by low and middle-income populations, people employed in commerce, services, or industry. The most potential areas for the development of the creative city idea were those on the Western and Southwestern sides of the city, mainly beyond the Atoyac River, with the abovementioned inconvenience of not being part of the urban land of the city of Puebla, but that of the municipality of San Andrés Cholula. Despite the legal disputes, the iconic 'New Puebla', the Angelópolis district, began its development in this area as the focus point to attract talent and business and to produce creative goods and services. Following the idea of an iconic urban image as that of the Southeast Asian global cities, a great number of residential and office towers with spectacular architecture and modern gated communities were designed, and experiential public spaces were created. With all that, the urban image of Angelópolis followed the ideals of the contemporary visual culture with its contained and aggressive aesthetics [21,29]. Certainly, the new urban structure was converted to an attractive area to live for the new groups of the population moving in from near and far, and with it, the number of habitants in the municipality of San Andrés Cholula in 2020, [20], increased by 53.8% compared with the year 2010 [30]. The new incomers changed the socio-economic profile of Cholula from rural to a conflicting combination of low-income and high-income populations living side by side. The logic behind all the development was that of business focused on the increment of the surplus value and productivity of land, and the real estate developers had the central role in setting the pace for the development in the Angelópolis district [21] incrementing the socio-spatial segregation very much reflecting the inherited colonial discrimination pattern based on, in the case of Cholula territory, creation of urban barriers to separate the Indigenous communities, the *Pueblos Originarios*, from the new higher middle-class dwellers of the Angelópolis district (See Figure 5 (below): blue—Angelópolis, yellow—Puebla's poor, irregular settlements, red—*Pueblos Originarios* communities).

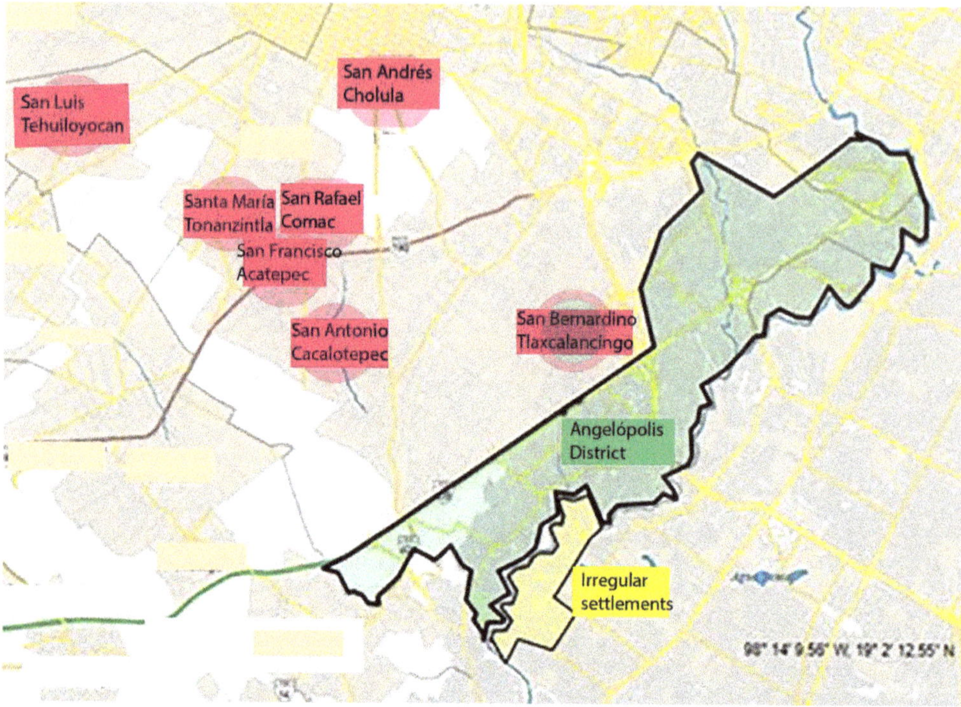

Figure 5. Development of the Atoyac river area. Elaborated by Kurjenoja from the map of the Instituto Nacional de Estadística y Geografía, INEGI, Mapa digital de México, 2023 [7]. (2023).

Finally, even the Atoyac River recovered its colonial role as a socio-economical barrier, now separating the irregular settlements on the Eastern side of it, still occupying the land legally belonging to the city of Puebla. As Guevara-Romero and Ramírez-Rosete [31] observe, those living on the Eastern bank perceive their opportunities to connect to the basic urban infrastructure and services as precarious compared with the real estate development on the Western bank, as shown in the aerial view of 2019 of the following Figure 6: above 1. Concentrations of poverty and informal settlements on the Eastern bank; 2. The Lomas de Angelópolis gated communities, 3. The La Vista gated community, 4. The Atoyac River is the border between Cholula (above the river) and the City of Puebla (below the river). Below, from the left to the right: Angelópolis residential towers in construction on the Western bank in 2019; irregular settlements on the Eastern bank of the Atoyac River in 2014; Angelópolis shopping center in 2014).

Figure 6. (**Above**) Elaborated by Soto Sandoval (2019) from Google Maps [32]. (**Below**) Photos taken by Romero Romero (2014).

4.3.2. Pueblos Originarios, Neoliberal Land Management, Territorial Disputes, and Socio-Racial Discrimination

The neoliberal urban development of the Angelópolis district has impacted especially the Indigenous population of the municipality of San Andrés Cholula, and its *Pueblos Originarios,* or First Peoples communities, as descendants of the original Indigenous population of the site. Cholula's territory has been inhabited for thousands of years, and thus recognized and protected by the Mexican Constitution [33–36]: it "recognizes Indigenous peoples' traditional ways of social, cultural, political and economic organization, and decrees their right to manage their habitat and preserve their land (Article 2)" [34] (p. 2), confirming through Article 27 that "laws must protect the integrity of Indigenous lands" [34] (p. 2). What is important here, is not the racial dimension of the sense of the law, as Article 2 of the Mexican Constitution establishes as the primary criteria to determine the indigenous identity, the self-awareness of Indigeneity as the basic criteria for being *Pueblo Originario*. The distinction thus links a person or a group of people to an Indigenous community acknowledged by the Mexican State, due to cultural, historical, political, linguistic, and other types of connections [34,37]. Also, the international law, through the United Nations Declaration on the Rights of Indigenous Peoples of 2007, in its Article 32, establishes that

> States shall consult with indigenous peoples through their own representative institutions to obtain their free prior informed consent prior to the approval of any project affecting their lands or territories and other resources, particularly in connection with the use of mineral, water or other resources. [38] (p. 12)

The previous reference to the Constitution is especially relevant, as the municipality of San Andrés Cholula and its population has been strongly impacted by the 1990s and 2011–2017 periods of neoliberal metropolitan development as 25% of its territory, until today, has been affected by expropriations and urbanization promoted by an aggressive real estate business and socio-environmental transformations [36]. This is relevant, as the identity of the *Pueblos Originarios* of Cholula is rooted in their territory, land, and traditional socio-spatial form of organization, the before mentioned *altepetl,* having a strong connection to their historical territory through a particular socio-spatial and bio-cultural relationship with land [34].

Despite the legal protection of the indigenous territories in San Andrés Cholula, a great number of hectares of rural lands have been converted to urbanized land under economic and political pressure by the state government and the real estate business triggering constant conflicts related to landowning: "San Andrés has been subject to constant conflicts linked to displacement, expropriation, touristic exploitation, social and urban exclusion, and disputes over land rights" [35] (p. 3), threatening the traditional land use [39].

4.4. *Key Findings*

4.4.1. Global Influences

Conceptually, the Puebla CID-plan adopted the Global North scope of the city as a commodity providing fashionable lifestyles and experiential life as important attractors for talent the local businesses and industry were looking for to boost them internationally [17]. It is important to highlight that urban transformation processes tend to follow the interests of the real estate business, the fact that Mexico constantly creates conflicts about the possession and occupation of land and problems related to the right to the city, with some similarities with the land management problems in many urban areas in Southeast Asia. If the globalization of cities in the Global North has meant an unequal distribution of wealth causing urban polarization, the same phenomenon is present in metropolitan areas as that of the Puebla-San Andrés Cholula-binomial, though the socio-economic distance between the different groups of urban habitants is enormously bigger than in any developed country. Here it is important to pay attention to how the colonial idea of the city as a tool for socio-territorial subordination is still prevailing.

In Southeast Asia, the image of cities has been created to promote each nation's potential as economic and financial nodes in global networks. In the case of Puebla, ProMexico's participation inspired by the Orange Economy model of the IDB and the development of the Puebla CID plan [17] focused on achieving international nominations and recognition in technology, culture, and design as the main attractors of international investments, and global actors as the UNESCO were drawn as driving forces for the urban development. What is certainly notable is the adaption of the Southeast Asian model of speculative urbanism and urban modeling to create an urban image as a cutting-edge differentiator concentrated in the Angelópolis district. Thus, it is there where the main pole of global significance of an economic potential was intentioned to emerge highlighting its profile as a smart and creative city and producer of vanguard culture and knowledge.

As mentioned before, the Latin American colonial past is present in Puebla's current urban realities despite all the influence and impact of the globalization processes. In general, social inclusion and exclusion are more than visible in the city, as well as subtle tools giving perpetuity to this separation, such as the urban connectivity through the public transportation system limiting the accessibility of certain population groups to exclusive lifestyle zones as Angelópolis district. (Figure 7 below describes the influences of the Global North theoretical core, Southeast Asian urban examples, and local, Latin American, and colonial realities in the urban development addressed in the Puebla CID plan).

4.4.2. Urban Planning, Land Use and Management: Social Sustainability, Discrimination, and Population Displacements

What has been observable from the beginning of the creation of the Territorial Reserves in the Cholula territory has been the presence of speculative practices by the real estate business backed up by state and municipal governments that, through expropriations and higher land tenure taxation, have reduced the extension of the rural land, triggering population displacements causing loss of identity and traditions, and finally also, accelerated population growth due to new inhabitants occupying the new habitational developments [36]. The urban planning 2011–2017 in the Angelópolis district, appropriating the previous rural lands of San Andrés Cholula and its Indigenous communities through planning tools, proposed to create a compact city with a dense land use and vertical edification, seeking to 'modernize' it, applying urban modeling strategies to create an attractive contemporary urban image and skyline. In terms of land use, land management dynamics did not recognize the limits of the *Pueblos Originarios*, thus making it easier to connect San Andrés Cholula territories to the urban development of Puebla.

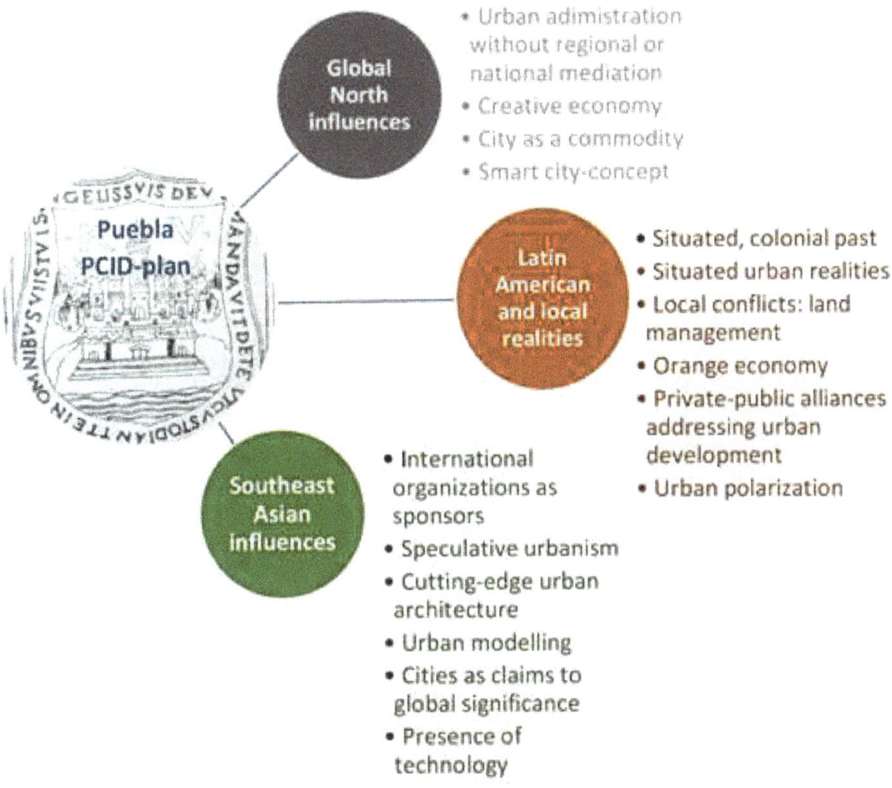

Figure 7. Diagram elaborated by Kurjenoja. (2021).

The 2011–2017 land management paved the way for the current clash between the state and municipal authorities, and the *Pueblos Originarios*, about the right to the land and to the auto-determination, as established in the Constitution. On the other hand, and very importantly, socio-ethnic disputes have emerged in the Angelópolis district, as instead of promoting integration and tolerance between the original population and newcomers, the planning has introduced the idea of a necessity of repelling the different 'Other' erecting urban barriers and socio-economic obstacles. The property management systems introduced by the real estate business have isolated the newcomers from the original population through walls and other restrictions, meanwhile the latter feel the first as invaders that have occupied their inherited lands by force. Urban modeling has created 'urban bubbles' of artificial realities in the name of modernization, intentionally disconnected from the surrounding urbanities. More grievously, the speculation has skyrocketed the value of the land, as also the land tenure taxation, triggering the expulsion and displacement of the original population without economic resources to cover the costs, putting in danger the socio-territorial sustainability through demographic integration.

4.4.3. Ecological and Environmental Sustainability

The dense urbanization in the Angelópolis district has severely affected the local ecological balance and agriculture especially due to diminishing water reservoirs (See Figure 8, presenting water resources of Central Mexico. Grey-no sufficient water resources; red-water resources available; blue-San Andrés Cholula; Yellow-Puebla). Traditionally, the water supply of Cholula communities has been based on families' own, particularly wells that now, due to the overexploitation of underground water supplies by new urban developments have left the local communities and their agriculture without sufficient irrigation. Also, since the 2011–2017 planning, no all-covering basic urban services such as potable water or sewer systems [35] were contemplated. For instance, wastewater and solid waste have been a serious problem, as much of them end up in the creeks and rivers of the area, polluting them and the surrounding lands. It would be indispensable to define all the surviving natural areas with special environmental values to be urgently protected to improve the catchment of water to permit the recovery of water resources and to guarantee the continuity of local agricultural activities. On the other hand, local endemic vegetation, today dangerously affected and disappearing due to the intense construction activity and lack of water, has traditionally provided food supplies for the local population that now are getting scarce. Thus, the territorial growth pattern powered in the 2011–2017 period, has put existing ecological thresholds at risk, as original environments have been partially modified by anthropogenic activities. Correspondingly, environmental restoration and preservation would be essential since they contribute to the environmental balance and maintenance of the territory's ecosystems [36]. The Political Constitution of the United Mexican States [33] declares, that to give order to human settlements and to define adequate provisions, reserves, and use of land, water, and forest, adequate measures should be sought to guarantee the construction of infrastructure, planning, and regulation of the new settlements including their maintenance and improvement, considering their future growth. The Constitution also demands the preservation and restoration of environmental balance and the division of large rural estates, taking into account the collective exploitation and organization of farming cooperatives and the development of small rural properties to stimulate agriculture, livestock farming, forestry, and other economic activities in rural communities. Finally, it prohibits the destruction of natural resources and damages against property to the detriment of society [34].

Finally, in terms of environmental sustainability, Article 104 of the Territorial Planning and Urban Development Law of the state of Puebla [40], establishes that urban densification is permissible when the capacity of water, drainage, electricity, or mobility services is not exceeded, considering the existing urban structure, and improving habitability. This, observing the results of the 2011–2017 urbanization process and its current consequences, has not happened, triggering a process of environmental deterioration that should be urgently attended to.

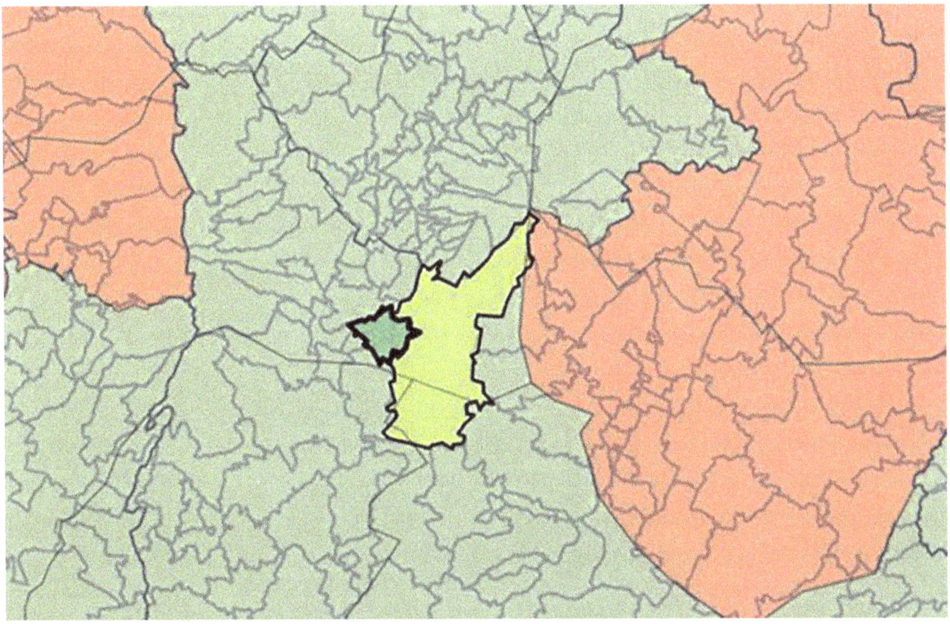

Figure 8. Water resources, central Mexico. Elaborated by Kurjenoja from the map of the Instituto Nacional de Estadística y Geografía, INEGI, Mapa digital de México, 2023 [7]. (2023).

5. Discussion

We have presented in the introduction of this article five initial statements regarding the urban modeling and its impact on the socio-territorial sustainability in the metropolitan binomial of Puebla-San Andrés Cholula we are going to discuss here. Our first statement declared that the urban development of the area has been influenced by the Global North and the Southeast Asian neoliberal lines of thought, manifesting characteristics of both in its territorial management. Indeed, after having studied the Orange Economy ideas [18], the Puebla CID documents [17], and the 2011–2017 urban development and its consequences, we can find coincidences with Saskia Sassen's [22] observations about globalization in the Global North and how it has triggered urban processes in which cities have been converted into products through localized or localizable urban image nuanced by globalized image patterns. Also, we can find in Puebla-San Andrés Cholula-binomial what David Harvey has mentioned: contemporary urban processes related to changes in lifestyles converted to commodities just as the cities themselves to products in a world of consumerism, tourism, and cultural industries as generators of experiences to those having the monetary resources to pay for them. Thus, we can see that even in the developed world, the polarization is triggering an unbalanced distribution of resources reflected in the urban image itself and in its spaces, fragmenting cities into parts forming 'micro-states' of well-being and those of poverty [23]. In the context of globalization, neoliberal public policy, and 'urban image construction', the application of ephemeral and distinctive concepts for cities such as 'being in trend' has been used as tools for their promotion as 'fashionable cities' triggering real estate developments, as can be seen also in the Puebla-San Andrés Cholula-binomial [41].

On the other hand, and as Aihwa Ong [1] highlights, the 21st century has meant the emergence of new economic powers competing with those of the Global North. In the case of Asia, the continent's big cities very soon began to reflect the national aspirations and thus were focus points for special 'worlding' strategies to be able to compete with the Western global cities. As a consequence, the appearance of new global cities in Asia challenged the established Global North concepts about urban globalization and triggered the need to observe the phenomenon from the postcolonial and local scope. In this sense, the colonial past nuances the present urban culture and territorial order due to legacies of injustice as those of peripheralization of the urban poor through the erection or reinforcement of urban barriers to separate the different and antagonistic urban groups from each other, or through expropriations of the land of the rural population for the urban development triggering human displacements and socio-territorial conflicts related to them [3]. What Ong mentions about 'redesigning' the city in the developing world related to highly speculative practices in land management, is also present in the Puebla-San Andrés Cholula urban development.

Secondly, we have presented the phenomenon of the creation of a global urban image, through urban modeling and redesigning of big cities especially in Southeast Asia. As Ong [1] points out, in the context of globalizing ex-colonies of Asia, the urban image of the great cities has been converted to the symbol of the emerging economic and political power in which cutting-edge urban, technological, tectonic, and architectural innovations are transmitting the idea of these nations' new role as powerful global actors demanding international recognition. In the framework of urban modeling, as Goldman [26] mentions, regulatory changes promoted by city administrations have made possible the new investment strategies and through them, master plans on the hands of private stakeholders to design 'worlding' urban landscapes, thus also Angelópolis-district. Here, as in Asian examples, the previous agricultural lands were now transformed into gated communities, private hospitals, malls, and business centers, triggering huge increments in land value. Besides the territorial re-design, the architectural patterns of Singapore, Shanghai, Hong Kong, and others, emerged as iconic examples to follow, emphasizing the attractiveness and efficiency of their compact city model and vertical skyline. During the 2011–2017 period and through the Puebla CID-plan, the Southeast Asian cities were mentioned as examples to follow, as icons of economic success and visibility on the global scale. From 2011 until today, we can clearly observe how the Puebla-San Andrés Cholula development concentrated on the Angelópolis-district has followed the strategies of urban modeling aiming at a world-class city skyline of glass towers and gravity-challenging architectural forms aiming at converting it to an icon for an emerging Latin American global city (See Figure 9 Exemplifying urban modeling in the Angelópolis-district).

Thirdly, we declared that the local colonial history and the socio-racial discrimination have been inherited and transferred to the 21st-century urban territorial management. Going back to the beginning of this article, the decade of the 1990s in Mexico meant a radical change in the public politics related to land management. As Schumacher et al. [42] mention, the 1992 Land Reform liberated the communal lands, '*ejidos*', that before that year could not be commercialized, but were now available for the real estate business triggering a massive urban sprawl outside urban cores and emergence of vast peri-urban areas. Besides this, in the case of the Puebla-San Andrés Cholula region and Angelópolis-district, the 1993 Angelópolis-plus-plan made possible the vast expropriations of rural lands of the *Pueblos Originarios* of San Andrés Cholula, to give space to the construction of the 'New Puebla'. 2011–2017 urban development based on the Puebla CID plan, seeking a global image, and providing an attractive urban environment for a higher middle-class experiential lifestyle, triggered the forced (expropriations) or silenced (high land tenure taxation) displacements of the Indigenous population. On the other hand, the real estate business promoted the idea of separating the urban newcomers from the *Pueblos Originarios* and other vulnerable groups through urban barriers, such as walls surrounding the new urban developments, surveillance systems controlling the entrances, and urban mobility favoring the use of private cars. Thus, the new higher middle-class population perceives the Indigenous communities and rural communities as threatful 'Others' and the latter see the first as illegal invaders. Lastly, colonial socio-racial discrimination has been present also in the planning processes in which, although having a constitutional right to be heard, the opinions of the Indigenous population have been only partially considered [36].

Figure 9. Photo taken by Carrera. X. (2023).

In the fourth place, we have shown how the combination of the mentioned aspects has aggravated the already existing socio-territorial polarization observable in the urban inclusion-exclusion strategies reinforced by the urban planning processes of the 2011–2017 period. The Angelópolis planning in the hands of private stakeholders of real estate business, and the state and municipality administrations supporting them, has triggered confrontations between them and the local Indigenous and rural communities about the right to the land and to the local traditions and culture related to it. The *Pueblos Originarios* have demanded the valorization of their cultural practices, conservation of their agricultural lands, as well as the traditional landscape, urban image, and character of the public spaces of their villages [36], meanwhile the state and municipality authorities have insisted on the necessity of 'modernization' of the Cholula territory through its re-engineering and urban modeling. Territorial controversies nuanced by socio-racial differences and economic and political interests that emerged since the 1990s and reinforced in 2011–2017 have not been resolved despite negotiations carried out.

Lastly, we stated that the territorial processes triggered by the neoliberal public policy of the 2011–2017 period and the corresponding land management strategy have menaced the socio-territorial and environmental balance and quality of life in the metropolitan area of Puebla, resulting today in an unsustainable situation. The municipality authorities and the Indigenous communities point out as the main problems related to the socio-territorial and environmental sustainability the following: chaotic urban development, real estate business speculation, overexploitation of the territory and its resources, loss of agricultural land, loss of identity, customs, and habits, and accelerated demographic growth and displacement of original habitants [36]. Regarding socio-territorial sustainability, the planning strategies should be focused on population integration and not socio-territorial segregation, as until now.

Meanwhile, Puebla's population has decreased from 2010 to 2020, from 1,834,936 to 1,724,831 habitants, and the population of the municipality of San Andrés Cholula has increased from 2010 to 2020 with 53.8%, meaning this from 115,976 inhabitants to 160,069 inhabitants [43] mainly due to new people moving into the locality from Puebla and other federal states, demanding for housing [36]. The accelerated demographic growth has affected, to begin with, the underground water supplies that for years have been overexploited, causing problems in the local agriculture and also a lack of water all over the urban territory. Augmented paved and constructed surfaces of the urban territory have been for long affecting aquifer recharge areas aggravating the water problem, and elevating temperatures in the locality. Finally, the new urban development requires amplified mobility systems, now proposed as wider streets and highways, augmenting the level of noise and air pollution [36]. Thus, thinking of urban densification and compact city model through the application of urban modeling seems not to provide sustainable solutions to the problems initiated in the 1990s and intensified in 2011–2017.

6. Conclusions

The National Council for Assessment of the Policy for Social Development, CONEVAL [44], reported that in 2020, in the city of Puebla 39.9% of its population lived in poverty, meanwhile in 2020 the percentage had grown to 41.4%. In the case of San Andrés Cholula, the same numbers were 68.3% (2010) and 49.71% (2020). Thus, public policy and public-private initiatives intended to trigger a new, sustainable urban economy seemed not to have given a solution to the basic, urban socio-economic problems in the state capital meanwhile San Andrés Cholula's percentages may be quite deceptive, as the decreasing numbers of the urban poor may be supposed to be due to the incoming wealthy, higher middle-class population occupying the new attractive gated communities and fashionable department towers. Indeed, the 2020 INEGI census shows, how the demographic growth in San Andrés Cholula has been that of 4.5% per year due to the new population moving in. As a consequence, the urban binomial of Puebla-San Andrés Cholula is today facing the danger of socio-territorial fracturing, making it socially unstable and unsustainable, added to the aggravated environmental problems related to water, air pollution, the disappearance of originally local vegetation and incrementing erosion of the soil that menaces its quality of habitability.

6.1. Suggestions for Public Policy

Though both the municipality and state authorities as the local rural population identify the same basic problems regarding the urban development of the Puebla-San Andrés Cholula-urban binomial, the approach they give to resolving them is contrasting. To face the challenge of solving these problems, it would be indispensable to create functional mechanisms of dialogue and negotiation to avoid confrontations between different interest groups and their stakeholders. The public policy and the corresponding urban planning should take into consideration, (1) not to conceptualize the city as a commodity and provider of lifestyles, nor carry out the urban and territorial transformation based on urban modeling, as has been happening since 1990s in the Angelópolis.district, (2) to treat the current needs and desired futures of all the occupants of the Puebla-San Andrés Cholula-urban binomial in a balanced way, taking into consideration the rural communities, habitants of the consolidated urban areas, but also the recent newcomers already established on the

Western bank of the Atoyac-river, despite racial, social, cultural or economic differences, and respecting their constitutional rights, detached from the inherited colonial thinking of socio-racial differentiation, through particular site-tailored development strategies adapted to local realities, (3) to give an organic delimitation for the rural communities' and *Pueblos Originarios'* territory, recognizing their natural dynamics of growth, through inclusive population consultation mechanisms and co-creation workshops, (4) to assure that the mentioned delimitations provide adequate zones for urban transition, not barriers, between the traditional land uses and new urban developments, (5) to control de accelerated urban growth caused by the uncontrolled real estate business through strategic public policy and planning tools, accompanied by effective mechanisms of urban and environmental monitoring carried out by independent experts, and (6) to carry out detailed environmental studies of the complete Puebla metropolitan area and extended to the neighboring state of Tlaxcala, as the basis for a regional environmental public policy and integral sustainable urban and community planning, to link the macro-scale urban development to environmental problematics and existing supplies of natural resources in the Puebla-Tlaxcala valley.

6.2. Limitations and Future Studies

The ultimate purpose of the studies and research project described herein is to provide a multi-dimensional analysis of the development of the territorial situation of the Puebla-San Andrés Cholula-urban binomial, and of its current and potential future consequences, to elaborate analyzed data as supporting information for urban planning of the municipality of San Andrés Cholula. These studies, now focused on and limited to the binomial's territory, should be amplified to consider all of the Puebla metropolitan area, as mentioned above, as many of the issues highlighted here, such as environmental questions and mobility network, extend beyond its limits to other municipalities and even to the territory of the neighboring federal state of Tlaxcala. The potential limitation to achieving functional, long-term sustainable planning and public policy through collaboration and bottom-up strategies between municipality, state and federal government, and local stakeholders and communities, is the disposition to a multi-level collaboration under the considerable political and economic pressures exercised by different interest groups.

Funding: 2014–2017: the Mexican National Council for Science and Technology, CONACYT, funding number 247777, for "Proyectos de Desarrollo Científico para Atender Problemas Nacionales" ("Scientific Development Projects to Address National Problems"). Since 2017, the research has had no external funding.

Data Availability Statement: No new data were created in this study. Data sharing is not applicable to this article.

Acknowledgments: Special thanks to the students Karla Sofía Soto Sandoval and Edna Alejandra Romero Romero, of the Department of Architecture, Universidad de las Américas Puebla (UDLAP) for their collaboration in this project, as well as to the Janina C. Carrera and the Xavier Carrera for their photos.

Conflicts of Interest: The author declares no conflict of interest.

References

1. Ong, A.H. Powers of sovereignty: State, people, wealth, life. *Focaal* **2012**, *64*, 24–35. Available online: http://citeseerx.ist.psu.edu/viewdoc/download?doi=10.1.1.1076.8435&rep=rep1&type=pdf (accessed on 7 October 2023). [CrossRef]
2. Mouton, M.; Shatkin, G. Strategizing the for-profit city: The state, developers, and urban production in Mega Manila. *Environ. Plan. A Econ. Space* **2020**, *52*, 403–422. Available online: https://journals.sagepub.com/doi/pdf/10.1177/0308518X19840365?casa_token=qUIUqAYABkMAAAAA:9xqHCghYVQmFnwaAUN_2mWgwnoeK1yPwKLTtwjFc2NgdQwl0LqATw2EjaHTxdYRX1q_V41TtILCD4cE (accessed on 7 October 2023). [CrossRef]
3. Ong, A.H. Introduction: Worlding cities, or the art of being global. In *Worlding Cities: Asian Experiments and the Art of Being Global*; Roy, A., Ong, A.H., Eds.; Blackwell Publishing Ltd.: London, UK, 2011; pp. 1–26.
4. Hart, G. Relational comparison revisited: Marxist postcolonial geographies in practice. *Progress in Human Geography* **2018**, *42*, 371–394. Available online: https://journals.sagepub.com/doi/abs/10.1177/0309132516681388. (accessed on 17 November 2023). [CrossRef]

5. Shatkin, G. Planning to forget: Informal settlements as 'forgotten places' in globalising Metro Manila. *Urban Stud.* **2004**, *41*, 2469–2484. Available online: https://journals.sagepub.com/doi/pdf/10.1080/00420980412331297636?casa_token=qEUTrRsrT8AAAAAA:XhQKvO6Ody7joUpAwsn7lq4XsY4Bu0zZZzoN29npLJwmDzCsq8Tp34Fj7b4HfEw89HqtSJ7ruIeWwdk (accessed on 7 October 2023). [CrossRef]
6. Instituto Nacional de Estadística y Geografía, INEGI. Síntesis de Resultados. Zona Metropolitana de Puebla-Tlaxcala. XII Censo General de Población y Vivienda. 2000. Available online: http://internet.contenidos.inegi.org.mx/contenidos/productos/prod_serv/contenidos/espanol/bvinegi/productos/historicos/2104/702825496609/702825496609_1.pdf (accessed on 7 October 2023).
7. Instituto Nacional de Estadística y Geografía, INEGI, Mapa digital de México. 2023. Available online: http://gaia.inegi.org.mx/mdm6/?v=bGF0OjE5LjA2NDEwLGxvbjotOTguNzYzNjEsejozLGw6YzExMXNlcnZpY2lvcw== (accessed on 18 November 2023).
8. Ashwell, A. Cholula: Su herencia en una red de agujeros. *Elem. Cienc. Cult.* **2004**, *11*, 3–11. Available online: https://elementos.buap.mx/directus/storage/uploads/00000002632.pdf (accessed on 7 October 2023).
9. Schumacher, M.; Kurjenoja, A.K. La figuración del paisaje de Cholula a través de la mirada de Alexander von Humboldt: Reflexiones sobre su transformación territorial y socio-cultural. *Tekopora Lat. Am. J. Environ. Humanit. Territ. Stud.* **2021**, *3*, 139–166. [CrossRef]
10. Reyes García, C. *Altepetl: Ciudad Indígena*; Cholula en el Siglo XVI; UNAM: Mexico City, Mexico, 1976.
11. Bonfil Batalla, G. *Cholula: La Ciudad Sagrada en la era Industrial*; UNAM: Mexico City, Mexico, 1973.
12. Churchill, N.E. Erasing Popular History: State Discourse of Cultural Patrimony in Puebla, Mexico. In Proceedings of the XXII International Congress of the Latin American Studies, Miami, FL, USA, 16–18 March 2000; pp. 1–10. Available online: https://citeseerx.ist.psu.edu/viewdoc/download?doi=10.1.1.552.5857&rep=rep1&type=pdf (accessed on 7 October 2023).
13. Loreto López, R. El microanálisis ambiental de una ciudad novohispana. Puebla de los Ángeles, 1777–1835. *Hist. Mex.* **2008**, *57*, 721–774. Available online: https://www.jstor.org/stable/25139804 (accessed on 7 October 2023).
14. Vélez Pliego, F. Puebla de Zaragoza, Antigua Ciudad de los Ángeles, Patrimonio Cultural de la Humanidad. *Rev. Soc. Ciudad. Territ.* **2011**, *1*, 1–47. Available online: https://issuu.com/ablmartilu/docs/pueblafvp (accessed on 7 October 2023).
15. Bhabha, H. *Location of Culture*; Routledge: London, UK; New York, NY, USA, 2004.
16. Kurjenoja, A.K.; Ismael, M.E. ¿Puebla, ciudad creativa ó inteligente? In *Ciudad, Capital y Cultura*; Hernández, A., Kurjenoja, A., Ismael, A.K., Comp, M.A., Eds.; Editorial Itaca: Mexico City, Mexico, 2018; pp. 21–36.
17. ProMéxico. *Mapa de Ruta. Puebla Capital de Innovación y Diseño*; ProMéxico: Mexico City, Mexico, 2015.
18. Buitrago Restrepo, F.; Duque Márquez, I. *Economía Naranja*; Inter-American Development Bank IDB: Washington, DC, USA, 2013; Available online: https://books.google.com.mx/books/about/La_econom%C3%ADa_naranja.html?id=I9N2DwAAQBAJ&printsec=frontcover&source=kp_read_button&hl=es-419&redir_esc=y#v=onepage&q&f=false (accessed on 7 October 2023).
19. Kurjenoja, A.K. Interview in person with a Local Urbanist. Personal communication, 2016.
20. de Puebla, G.E. *Programa Regional de Ordenamiento Territorial Angelópolis*; Puebla: Mexico City, Mexico, 1993.
21. Kurjenoja, A.K. Economía creativa y cultura en la transformación desarrollo de dos ciudades mexicanas: Puebla y Tijuana. In *Globalópolis (Provisional Name of the Book)*; Marchand, M., Morales, I.M., Meza, E., Eds.; Editorial Porrúa: Mexico City, Mexico, 2023; *in press*.
22. Sassen, S. *Territorio, Autoridad y Derechos de los Ensamblajes Medievales a los Ensamblajes Globales*; Katz: Madrid, Spain, 2010.
23. Harvey, D. *Rebel Cities. From the Right to the City to the Urban Revolution*; Verso: London, UK; New York, NY, USA, 2012.
24. Sassen, S. *Expulsions. Brutality and Complexity in the Global Economy*; The Belknap Press of Harvard University Press: Cambridge, MA, USA; London, UK, 2014.
25. García Canclini, N. *Imagined Globalization*; Duke University Press: Durham, NC, USA, 2014.
26. Goldman, M. Speculating on the next world city. In *Worlding Cities: Asian Experiments and the Art of Being Global*; Roy, A., Ong, A.H., Eds.; Blackwell Publishing Ltd.: London, UK, 2011; pp. 1–26.
27. Howkins, J. *Creative Economy*; Penguin Books: London, UK, 2007.
28. ProMéxico. Wikipedia. Available online: https://es.wikipedia.org/wiki/ProM%C3%A9xico (accessed on 7 October 2023).
29. Kurjenoja, A.K. Interview in person with the Architect of PCID Public Spaces. Personal communication, 2016.
30. DataMéxico. *San Andrés Cholula*; DataMéxico: Municipio de Puebla, Mexico, 2020.
31. Guevara-Romero, M.L.; Ramírez Rosete, N.L. Condiciones de movilidad en colonias marginadas. Unidad territorial Atlixcáyotl, Puebla. *Bitácora Urbano Territ.* **2019**, *29*, 31–38. [CrossRef]
32. Google Maps. Available online: https://www.google.com.mx/maps/@19.0753479,-98.3221855,13z?entry=ttu (accessed on 18 November 2023).
33. Honorable Congreso de la Unión. *Constitución Política de los Estados Unidos Mexicanos*; 25 Legislatura Cámara de Diputados: Mexico City, Mexico, 1992.
34. Schumacher, M.; Guizar Villalvazo, M.; Kurjenoja, A.K.; Durán-Díaz, P. The Writ of Amparo and Indigenous Consultation as Instruments to Enforce Inclusive Land Management in San Andrés Cholula, Mexico. *Land* **2023**, *12*, 9. [CrossRef]
35. Gutiérrez Juárez, E.G.; Kurjenoja, A.K.; Schumacher, M.; Guizar Villalvazo, M.; Gonzalez Meza, E.; Durán-Díaz, P. Neoliberal Urban Development vs. Rural Communities: Land Management Challenges in San Andrés Cholula, Mexico. *Land* **2022**, *11*, 1058. [CrossRef]
36. InLAB. Territorio and Urban Artifact Research Group. Interview in person with Cholultecas Unidos en Resistencia, CHUR. Personal communication, 2023.

37. Tetra Tech. Effective Engagement with Indigenous Peoples: USAID Democracy, Human Rights, and Governance Sector Guidance Document. United States Agency for International Development. 2020. Available online: https://www.climatelinks.org/sites/default/files/asset/document/2020_USAID_Effective-Engagement-with-Indigenous-Peoples-USAID-Sustainable-Landscapes-Sector-Guidance-Document.pdf (accessed on 7 October 2023).
38. United Nations. United Nations Declaration on the Rights of Indigenous Peoples. 2007. Available online: https://www.un.org/development/desa/indigenouspeoples/declaration-on-the-rights-of-indigenous-peoples.html (accessed on 15 December 2022).
39. United Nations and General Assembly. *Universal Declaration of Human Rights*; United Nations: Paris, France, 1948. Available online: https://www.un.org/en/about-us/universal-declaration-of-human-rights2 (accessed on 7 October 2023).
40. Ley de Ordenamiento Territorial y Desarrollo Urbano del estado de Puebla. 2017. Available online: https://www.congresopuebla.gob.mx/index.php?option=com_docman&task=doc_download&gid=12924&Itemid= (accessed on 7 October 2023).
41. Anderson, N. *Sociología de la Comunidad Urbana. Una Perspectiva Mundial*; Fondo de Cultura Económica: San Diego, CA, USA, 1993.
42. Schumacher, M.; Durán-Díaz, P.; Kurjenoja, A.K.; Gutiérrez-Juárez, E.; González-Rivas, D.A. Evolution and Collapse of *Ejidos* in Mexico—To What Extent Is Communal Land Used for Urban Development? *Land* **2019**, *8*, 146. [CrossRef]
43. Instituto Nacional de Estadística y Geografía, INEGI. *Panorama Sociodemográfico de Puebla. Censo de Población y Vivienda 2020*; INEGI: Aguascalientes, Mexico, 2021.
44. El Consejo Nacional de Evaluacion de la Politica de Desarrollo Social, CONEVAL. Estadísticas de Pobreza en Puebla. Available online: https://www.coneval.org.mx/coordinacion/entidades/Puebla/Paginas/principal.aspx. (accessed on 7 October 2023).

land

MDPI

Review

A Review on the Use of Geodesign Processes in Managing Flood Vulnerability

Bo Pang * and Brian Deal

Department of Landscape Architecture, University of Illinois at Urbana-Champaign, Champaign, IL 61801, USA; deal@illinois.edu
* Correspondence: bp4@illinois.edu

Abstract: The integration of geodesign with vulnerability analysis offers a novel approach to landscape, urban, and environmental planning challenges. The interdisciplinary nature, engaged processes, and advanced spatial analytics in geodesign provide important component pieces for identifying and addressing vulnerable places. A comparative case-study approach explores the geodesign process and its efficacy in addressing flooding issues in diverse settings. The objectives of this review include assessing geodesign processes in the development of vulnerability indices and assessments, exploring their effectiveness and limitations, and providing insights across different scales and contexts. The work seeks to contribute to the advancement of vulnerability assessment methodologies and their broader role in the fields of landscape design and urban and environmental planning. The findings of this study are expected to provide insights for researchers and practitioners looking to produce more effective, sustainable, and resilient planning and design outcomes.

Keywords: geodesign; vulnerability analysis; landscape design; urban and environmental planning; sustainable and resilient planning; interdisciplinary; flood vulnerability; comparative case study

1. Introduction

In the evolving landscape of urban planning and design, the multifaceted challenges of resilience and urban sustainability have come to the fore. Cities around the world are grappling with policy and design approaches for managing social, economic, and ecological infrastructure more sustainably while simultaneously trying to improve resilience against climate-driven disasters. One such common disaster, in particular, poses a significant threat to urban areas on a global scale. It has the potential to cause widespread damage to infrastructure, disrupt economic activity, and endanger numerous human lives: flooding. Flooding vulnerability is a problem that affects many urban areas around the world. It is influenced by a wide range of factors, including geographic location, socio-economic conditions, and the affected built environment. Assessing and mitigating flood vulnerability is crucial for developing effective strategies to enhance urban resilience and sustainability. In this paper, we argue that the integration of geodesign processes and vulnerability analysis represents a pioneering approach to assessing, planning, and designing for flood mitigation and adaptation. Geodesign and its associated processes for landscape analysis and design represent the merger of geographic, landscape, and urban design principles [1]. It provides an interdisciplinary framework for large-scale planning and design by incorporating advanced spatial analysis and geographic information systems (GIS) science. The methodology has been instrumental in bringing analytics to complex design solutions [2], assessing outcomes of various design interventions [3], and crafting sustainable solutions to environmental, social, and urban challenges [4]. Despite an increasing application across fields such as urban planning [5], landscape architecture [6], and environmental conservation [7,8], the use of geodesign processes in assessing the vulnerability of populations or geographies has been limited. The geodesign approach, however, leverages transdisciplinarity and spatial analytical approaches to facilitate a comprehensive understanding of the interplay between environmental, social, and built factors that contribute to flood vulnerability.

In terms of landscape design, urban planning, and environmental management, vulnerability can be characterized by the susceptibility of ecological, social, or infrastructural systems to stresses and shocks [9,10]. Cutter describes it as the degree to which a system is susceptible to a specific stress; this includes changes in environment and/or extreme events [9]. Similarly, Turner et al. [11] argue that vulnerability is not solely determined by exposure to hazards (perturbations and stresses) but also by the sensitivity and resilience of the system experiencing such hazards. In general, vulnerability refers to the propensity or predisposition to be adversely affected by external or internal factors, such as environmental hazards, economic shocks, social inequalities, or personal circumstances. It represents a state of susceptibility or exposure to harm, damage, or loss [12]. Vulnerability can be influenced by various factors, including socio-economic status, age, gender, disability, geographic location, and access to resources or support systems. For the most part, the literature emphasizes the importance of understanding vulnerability through a lens of exposure, sensitivity, and resilience.

While vulnerability highlights the potential for harm or negative consequences, resilience focuses on the capacity to overcome adversity and maintain or regain a state of equilibrium or functionality [13,14]. Individuals, communities, or systems with higher levels of resilience are better equipped to navigate through challenges, mitigate the impact of adverse events, and recover more quickly and effectively. If vulnerability is the condition of susceptibility, resilience is the ability to adapt, recover, and bounce back from adverse or susceptible situations, challenges, or disturbances.

It is important to note that vulnerability and resilience are not static or absolute states; rather, they exist on a continuum and can be influenced by various factors, including personal characteristics, social support systems, and access to resources [15]. Furthermore, vulnerability and resilience are interconnected, as high levels of vulnerability can diminish resilience while enhancing resilience can reduce vulnerability to future adversities.

According to Cutter et al. [16], understanding urban vulnerability (and its evolution toward resilience) requires a comprehensive and integrated approach that transcends traditional disciplinary boundaries. This is especially important within the realm of complex urban and landscape design problems [17]. Transdisciplinary processes, a notable component of geodesign, aim to create a holistic understanding of specific design issues by drawing insights from multiple disciplinary perspectives to produce a more holistic understanding of the interdependencies between natural and human systems [18]. Geodesign process provides a basis for transdisciplinary engagement, bridging the gap between data-driven analysis and creative problem-solving.

Geodesign principles can be effectively integrated into vulnerability assessments by leveraging the spatially explicit and transdisciplinary nature of the process. However, despite the growing recognition of geodesign's potential in various fields, there is a lack of research specifically focusing on its application in vulnerability assessments. There is evidence in the literature on the theoretical benefits of integrating geodesign principles with vulnerability analysis, but there is a dearth of empirical studies that evaluate the effectiveness and limitations of the theoretical model in real-world contexts. Moreover, the existing literature does not adequately address the challenges and best practices associated with implementing the geodesign framework in vulnerability assessments across different scales and contexts. We aim to address this by critically evaluating comparative case studies with a particular focus on flooding. By examining the strengths, weaknesses, and lessons learned from these case studies, we hope to provide insights into the potential benefits and challenges of integrating geodesign into flood vulnerability assessments and to identify best practices and recommendations for future research and practice.

The primary objectives of this research are to (a) investigate how geodesign's comprehensive and integrated approach can improve vulnerability indices and assessments; (b) explore the effectiveness and limitations of applying geodesign principles in real-world vulnerability assessment case studies; and (c) provide insights into the challenges and best practices associated with implementing the geodesign framework in vulnerability assessments across different scales and contexts. The work seeks to contribute to the advancement of vulnerability assessment methodologies and the broader fields of landscape design and urban and environmental planning. The findings of this study are expected to provide insights for researchers and practitioners looking to integrate geodesign principles into vulnerability assessment processes, ultimately leading to more effective, sustainable, and resilient planning and design outcomes.

This paper argues that the inclusion of vulnerability analysis within a geodesign process can enhance the way vulnerabilities are identified, assessed, and mitigated, leading to more informed, effective, and adaptive design strategies for reducing risk. To better understand the potential synergies between geodesign and vulnerability analysis and to identify the gaps in existing research, Section 2 provides a comprehensive review of the literature in these two fields. This review highlights the key contributions of previous studies and explores how geodesign can further inform vulnerability assessment and mitigation strategies, setting the stage for the present research. In Section 3, we outline our case study methodology, detailing parameters and evaluative framework used to examine the application of geodesign principles in addressing vulnerabilities with a specific focus on vulnerability to flooding. Section 4 describes our findings and discusses the impact of integrating geodesign on urban flood resilience. We discuss the broader implications of our study in Section 5, evaluating the utility of the methods described. The paper concludes with a summary of key insights and recommendations for advancing the application of geodesign in vulnerability assessments, along with a discussion of some next steps in Section 6.

2. Literature Review: Geodesign and Vulnerability

Geodesign. Geodesign, as defined by Steinitz [1], is a design and planning method that couples landscaped scaled design and design processes with impact simulations informed by analytically derived geographic contexts. Its interdisciplinary collaboration [6], stakeholder engagement [5], and spatial analysis [7] make it well-suited for assessing the multidimensional nature of vulnerability.

The literature on Geodesign suggests that geodesign principles and methods could enhance vulnerability assessments by providing a more integrated and spatially explicit approach [3], there is a lack of empirical studies demonstrating the effectiveness and limitations of this integration in practice. Geodesign principles can effectively enhance vulnerability assessments by providing a more integrated, spatially explicit, and transdisciplinary approach [3]. Traditional vulnerability assessment methods often struggle to capture the dynamic and multidimensional nature of the vulnerability, highlighting the need for more comprehensive and adaptive approaches [16]. Geodesign leverages the power of geospatial data, environmental factors, and stakeholder input to create a holistic understanding of the spatial distribution of vulnerability and identify targeted interventions [1]. By integrating advanced spatial analysis techniques, collaborative decision-making processes, and design thinking, geodesign offers a promising framework for addressing the complex challenges of vulnerability assessment and enhancing the resilience of urban systems [19]. This research aims to address this gap by critically evaluating the application of the geodesign framework in conducting vulnerability assessments through comparative case studies.

Vulnerability. Vulnerability in urban and environmental planning refers to the susceptibility of ecological, social, or infrastructural systems to stresses and shocks [9,10]. Key dimensions include physical exposure, socio-economic sensitivities, and adaptive capacity [12], with urbanization, climate change, and socio-economic disparities as major drivers [20,21]. Birkmann [12] suggests that it is essential to consider the physical, social, economic, and environmental dimensions of vulnerability when assessing the resilience of any urban systems [12]. Physical vulnerability refers to the potential for damage to infrastructure and the built environment, while social vulnerability considers the characteristics of populations that influence their capacity to anticipate, cope with, and recover from hazards. Economic vulnerability assesses the potential for losses and disruptions to economic activities, and environmental vulnerability examines the sensitivity of natural systems to human-induced stresses. Understanding these interconnected dimensions is crucial for developing comprehensive strategies to reduce vulnerability and enhance urban resilience.

To effectively assess and address vulnerability in urban planning, it is necessary to adopt a systems approach that considers the complex interactions and feedback between different components of urban systems [18]. This approach should also incorporate the knowledge and perspectives of diverse stakeholders, including local communities, policymakers, and experts from different disciplines [22]. However, traditional vulnerability assessment methods often struggle to capture the dynamic and multidimensional nature of vulnerability, highlighting the need for more integrated and adaptive approaches [16].

Vulnerability and Geodesign. The integration of geodesign principles and methods into vulnerability assessments offers a promising avenue for addressing the limitations of traditional approaches. Geodesign's emphasis on interdisciplinary collaboration, stakeholder engagement, and spatial analysis [1] aligns well with the need for a systems approach to vulnerability assessment [18]. By incorporating geospatial data, environmental factors, and stakeholder input, geodesign can help planners and decision-makers better understand the spatial distribution of vulnerability and identify targeted interventions [3].

Despite the existence of studies exploring the potential for geodesign in resilience planning [3,23] and climate change adaptation [24], more systematic evaluations are needed to understand how geodesign can be integrated into vulnerability assessments effectively across different contexts and scales.

3. Materials and Methods

This study employs a comparative case study approach to evaluate the application of the geodesign framework in conducting vulnerability assessments with a specific focus on flooding. Comparative case studies are particularly useful for understanding complex phenomena in real-world contexts and for identifying patterns, similarities, and differences across different cases [25]. The approach can facilitate a deeper understanding of underlying issues by providing contextual diversity [26]. The methodological processes of comparative analysis include case study methods, social-science analytical methods, and methods using various scales of analysis [27]. Comparative methods have been found useful for gaining insight into the causal mechanisms of both urban and landscape-scaled phenomena [28]. According to Lange [27], the use of comparative case-study methods can help balance "ideographic and nomothetic explanations of empirical phenomena". The approach has been used to compare nations [29], cities [30], and smaller landscapes [31]. Marsal-Llacuna [32] proposes the use of real-time data to construct smart-city indices, and Frey [33] uses a comparative framework to study sustainable urban form. Solecki et al. [34] use comparisons for urban climate change mitigation and adaptation analysis.

Comparative methods are not without limitation, however. Limitations identified in the literature include a lack of clear research questions [35], context-specificity [36], cultural biases [37], and limited applicability or understandability of the findings [38]. According to Timm [39], one major shortcoming of comparative methods is "a tendency to offer parallel descriptions of different places without a clear research question, hypothesis, or coherent system of comparison".

Two case studies are selected for this comparative research that represent two different contextual applications of the geodesign process in flood-prone contexts in the US. They range from a data-rich urban setting (League City, TX, USA) to a data-scarce indigenous community (Tohono O'odham Nation, Arizona, USA). This diversity allows for a more comprehensive evaluation of the adaptability and transferability of the framework across scales, geographies, and social contexts.

The decision to employ only two case studies is justified by considerations of resource constraints, the need for focused investigation, theoretical relevance, methodological rigor, and practical feasibility. Conducting an urban analysis entails comprehensive data collection, synthesis, and interpretation, which can be resource-intensive in terms of time, finances, and manpower. By focusing on a limited number of case studies, we can allocate resources effectively, ensuring in-depth investigation and robust analysis. Selecting a smaller number of case studies also allows for a more concentrated examination of specific phenomena. Carefully selecting our two case studies to represent diverse urban typologies, socio-economic conditions, and policy contexts can enhance the generalizability and applicability of our findings to broader urban contexts [40]. Furthermore, limiting the number of case studies facilitates comparability and consistency in data collection, analysis, and reporting (see Table 1).

Table 1. Evaluative Indicators and criteria.

Indicator	Criteria and Description
Spatial Analysis	**Integration of relevant data layers:** Incorporating diverse spatial data layers allows for a holistic analysis of physical, environmental, and social vulnerability factors. **Use of spatial analysis techniques:** Enables extraction of meaningful insights, identification of high-risk areas, and assessment of potential impacts. **Generation of spatial outputs:** Clear and informative spatial outputs communicate findings and support targeted mitigation strategies. **Spatial resolution:** High-resolution data improves reliability, enables the identification of localized vulnerabilities, and supports fine-grained decision-making.
Stakeholder Engagement	**Types of stakeholders:** Engaging diverse stakeholders captures unique perspectives, experiences, and needs, particularly of vulnerable or marginalized groups. **Engagement methods:** Multiple participatory methods cater to different preferences, uncover local knowledge, ensure accessibility, and enable broad participation. **Incorporation of stakeholder input:** Ensures findings are relevant, legitimate, and actionable.
Interdisciplinary Integration	**Representation of relevant disciplines:** Captures diverse drivers, impacts, and potential solutions related to vulnerability. **Application of integrative methods and tools:** Enables synthesis of knowledge and insights across disciplines.
Feedback Loops and Iterative Processes	**Presence of monitoring and evaluation mechanisms** : Allow tracking of performance, identification of gaps, and assessment of effectiveness in informing decisions. **Evidence of iterative refinements:** Demonstrate responsiveness to feedback, adaptability to new information, and the ability to remain relevant over time.
Resilience and Adaptability	**Incorporation of resilience indicators:** Measures assessing capacity to withstand, recover, and adapt to stresses and shocks. **Use of adaptive design principles:** Ensures proposed strategies are robust and adaptable to changing conditions. **Consideration of long-term climate change impacts:** Identifies potential future vulnerabilities and informs robust adaptation strategies.

Case Study Comparative Analysis. This study employs 2 contrasting cases to assess the adaptability and transferability of the geodesign framework across diverse contexts, scales, geographies, and social settings using flood vulnerability as a common factor.

The first case study, set in a data-rich urban context, allows for the evaluation of the framework's effectiveness in spatial analysis, stakeholder engagement, interdisciplinary integration, and iterative processes. In contrast, the second case study, situated in a data-scarce indigenous community, enables the assessment of the framework's adaptability, ability to integrate diverse data sources, and capacity to facilitate community-driven planning processes. The contrasting nature of these two cases provides insights into the strengths, weaknesses, and best practices for applying the geodesign framework in flood vulnerability assessments, contributing to a holistic understanding of its potential and limitations.

Evaluation Framework. From the literature, we identify five key variable categories to base our evaluation on: spatial analysis, stakeholder engagement, interdisciplinary integration, feedback and iterative processes, and resilience and adaptability. Spatial analysis evaluates the integration, sophistication, and relevance of spatial data and analytical methods. Stakeholder engagement refers to the extent and quality of participation by local communities, policymakers, designers, planners, and scientists in the assessment process. This variable is important for ensuring that the assessment incorporates a wide range of perspectives and knowledge bases [22]. Interdisciplinary integration is the extent to which the framework incorporates insights and methodologies from geography, urban planning, architecture, and environmental science into the analysis, design, and planning processes [41]. Iterative processes and feedback loops refer to the existence of mechanisms for continuous refinement of design decisions in response to new information, changing conditions, and ongoing monitoring of the system [1,42]. Finally, resilience and adaptability describe the framework's ability to enhance the resilience and adaptability of communities, considering both immediate and long-term impacts [23,41].

We use this evaluative framework to study our 2 cases to understand the potential contribution of geodesign to urban resilience and flood risk management, as well as to assess the effectiveness of geodesign approaches in vulnerability assessment.

We use a systematic 'document analysis' methodological approach for our comparative case study research to generate insights, identify patterns, and construct narratives within and across case studies [43]. Textual materials provide valuable insights into the beliefs, values, norms, practices, decisions, and actions of individuals, groups, organizations, and institutions within specific socio-cultural, political, economic, and historical contexts [44]. For each case, we extensively review and analyze a diverse set of textual resources, including published papers, reports, official documents, and other published materials (newspaper, magazine articles, etc.). Our analysis focuses on extracting information from each document source related to our selection of indicator variables. For example, we examine plans and reports to identify spatial datasets and outputs connected to our flood vulnerability indicator.

4. Results

Each resulting case study is presented using our noted 5 evaluation criteria (see Table 2).

4.1. Case Study 1: "Integrating a Resilience Scorecard and Landscape Performance Tools into a Geodesign Process"

This study demonstrates an application of geodesign principles in League City, Texas. League City is a part of the Houston metro area in Southeast Texas on Galveston Bay and is considered highly urbanized. The case is aimed at urban flood resilience and employs an approach that combines a resilience planning scorecard, geodesign tools, and a landscape performance calculator to assess potential design interventions for sea-level rise vulnerabilities.

4.1.1. Spatial Analysis

In the League City case study, spatial analysis played a crucial role in delineating the ecological and social vulnerabilities at multiple scales by integrating diverse datasets within a GIS framework. In this context, to accurately identify the vulnerable area, the study integrates comprehensive data within GIS, including environmental, social, urban planning, and economic datasets. Researchers on the project followed a specific methodology as outlined in a reported "Methods" section [23]. First, to project the flood vulnerability, researchers utilized the FEMA 100-year flood elevations on Digital Flood Insurance Rates Maps and data derived from the US Army Corps of Engineers sea-level rise calculator to project the 100-year flood plain in 2100 at the site. As a result, the researchers have mapped the FEMA 100-year floodplain for 2100, identifying 3 levels of flood risk in the regional projection and 4 distinct flooding scenarios at the site level.

To address the social dimensions of vulnerability, in "Vulnerability projection", the League City case study incorporated block-level socio-economic data into the spatial analysis, resulting in a comprehensive vulnerability map that classified the site as highly vulnerable. In addressing the social dimensions of vulnerability, the study integrated block-level social data into the suitability analysis. This included metrics such as percentage of individuals in poverty, per capita income in 2010, percentage of persons 65 years of age or older, percentage of persons 17 years of age or younger, percentage of minority, less than High School Diploma, High School Diploma, or Higher, and improvement values. Consequently, by integrating social and economic data into the geographic information system (GIS) at a regional scale, the researchers produced a final vulnerability map, depicted as "Figure 5 [23]. Hazard vulnerability outputs by census block in League City, TX, USA". Through this comprehensive analysis, the site was classified as having a high level of vulnerability.

4.1.2. Stakeholder Engagement

The League City case study employed participatory methods to effectively engage stakeholders, integrating local insights and community feedback into the planning process to align the master plan with site-specific needs and aspirations. Under the "Master planning strategy-Community engagement and feedback" section, the study explains that the research and design efforts addressing those issues in the neighborhood adopted a participatory approach. This collaborative process involved local community members and senior planners from League City and included, "green infrastructure, open space planning, and community design scenarios developed through several engagement sessions assisted by community input". This highlights that in this study, participatory methods were utilized to elicit community feedback on key issues such as green infrastructure, open space planning, and community design scenarios to enhance flood resilience. The research and design efforts within this neighborhood were conducted through a collaborative, participatory approach, engaging local community members and senior planners of League City. This approach facilitated the acquisition of local insights, which were subsequently integrated into the planning process. Evidence of this integration was manifest in the subsequent meetings between community members and the design teams. These sessions focused on providing feedback and critiquing the proposed design scenarios, leading to the iterative refinement of the community's master plan, therefore aligning it more closely with the site-specific needs and aspirations identified through stakeholder engagement.

4.1.3. Interdisciplinary Integration

The League City case study demonstrates the value of interdisciplinary integration in comprehensively evaluating flood vulnerability. It employs a multifaceted approach that combines resilience planning scorecards, Geodesign tools, and landscape performance calculators. This approach showcases how integrating methods and tools from different disciplines can provide a more holistic understanding of flood vulnerability and inform effective mitigation strategies. Extracting from their conclusion, the resilience planning scorecard methodology, combined with policy analysis techniques, provided insights into the evaluation of existing local plans, regulations, and policies, determining their efficacy in mitigating vulnerabilities to hazards like flooding, sea-level rise, and other climate-related risks. Geodesign tools facilitated the integration of datasets from various fields, including social science (via the American Community Survey), geology, and environmental science (flood projections), enhancing the understanding of the city and site in both human and environmental aspects. Landscape performance calculators were employed as impact models to evaluate the flood attenuation capabilities of the site's master plan, providing a quantitative measure of the landscape's potential to mitigate flood impacts.

4.1.4. Feedback Loops and Iterative Processes

The participatory involvement featured feedback loops and iterative processes, which were initiated four times over 8 months. As evident from the quote, "a third and fourth meeting involved a feedback loop between community members and the design team in which a series of design scenarios were presented and critiqued by neighborhood members". A total of 4 meetings were held, and the third and fourth meetings involved feedback from community members and the design team. During these meetings, the design team presented a series of design scenarios, which were then critiqued by neighborhood stakeholders and subsequently revised to reflect that feedback.

4.1.5. Resilience and Adaptability

The League City case study showcases the development of a data-driven, resilient community plan that incorporates a comprehensive array of flood attenuation strategies, ensuring the city's capacity to withstand and recover from projected extreme conditions up to 2100. In this study, a data-driven community plan was developed for the city, designed to withstand the extreme conditions projected for 2100. The master plan incorporates a comprehensive array of flood attenuation mechanisms, drawing from both structural and non-structural strategies informed by resilient community design practices globally. These strategies are structured to be executed in three phases: retreat from flood-prone areas, flood mitigation, and flood control. Each phase is rigorously evaluated through cost-benefit analysis, employing landscape performance calculators to project their efficacy up to the year 2100, therefore ensuring a strategic approach to building resilience.

The League City case study demonstrates the potential of geodesign as a comprehensive, data-driven approach for analyzing and mitigating landscape vulnerabilities, particularly in the context of flood resilience planning. Through the integration of spatial analysis, stakeholder input, and resilience planning, geodesign facilitates a comprehensive and adaptive framework for managing and mitigating landscape vulnerabilities, highlighting its vital role in sustainable and resilient community planning.

4.2. Case Study 2: Resilience-Based Adaptation in Data-Scarce Areas: Flood Risk Assessment Using Geodesign in the Tohono O'odham Nation

The Sif Oidak District is a part of the greater Phoenix metropolitan statistical area in Arizona. Its case study showcases the application of Geodesign in developing a community-based, flood-resilient land-use plan for a data-limited Native American community facing significant flood risks. This community faces significant flood risks due to a combination of factors, including shallow soils, sparse vegetation, heavy rainfall, significant elevation changes, and traditional living patterns close to water sources. The study demonstrates the integration of qualitative, statistical, and spatial analysis with local insights to develop a strategy aimed at community resilience against flooding.

4.2.1. Spatial Analysis

In contrast to the League City case, the Sif Oidak District faced a scarcity of data for assessing flood vulnerability, lacking resources such as FEMA flood maps, flood zones, or county-level risk data. To address this challenge, the researchers employed a different approach, as detailed in the "Spatial Analysis of Flood Risk" section under "Research Design" in their study. They utilized remote sensing data in conjunction with a flood risk model to generate a flood risk map for the Sif Oidak District. The determination of flood risk values was based on a 10-m digital elevation model, which provided insights into slope, flow accumulation, and the drainage network. This alternative methodology allowed the researchers to overcome the data limitations and assess flood risk in the study area. Additionally, rainfall data for the basin was considered, while land-use cover and soil data were sourced from the U.S. Geological Survey and the United States Department of Agriculture (USDA) Natural Resource Conservation Service, respectively. The Sif Oidak District has been delineated into five distinct risk levels for flood vulnerability assessment: very low, low, medium, high, and very high risk.

4.2.2. Stakeholder Engagement

Given the scarcity of data, engaging stakeholders is crucial for identifying flood-prone areas in this project. A "Working with the Tohono O'odham Nation" section in their report documents the project's collaboration with a diverse group of stakeholders. "Local community members, district council members, and officials from the Tohono

O'odham Nation's planning department" were involved in the process. For example, community members completed surveys to gather information on past flooding events and perceptions of flood risk and participated in the Geodesign workshop to create geospatial data mapping of known flood areas from the most recent and historical flood events. Through the use of mixed qualitative and quantitative methods, such as visioning sessions and Geodesign workshops, these stakeholders have actively participated, contributing their local knowledge and preferences to the planning process.

The comprehensive flood vulnerability assessment developed in the Sif Oidak District case study, which incorporates local input and knowledge, demonstrates the significant value added by stakeholder engagement in enhancing flood information and identifying risk reduction strategies. As a result, a thorough flood vulnerability assessment, "Mixed-Methods Flood Risk Map", has been developed. It not only spatially identifies flood zones but also comprehends the frequency and locations of flood events within the community. Incorporating local input, it assesses the severity of damage and the frequency of displacement among community members, evaluates their perceptions of flood risk for both their community and the district, and identifies viable strategies to reduce flood risks. This study has underscored the significant enhancement of flood information in the district through stakeholder engagement.

4.2.3. Interdisciplinary Integration

The Sif Oidak District case study demonstrates the value of interdisciplinary integration in analyzing landscape vulnerabilities, although the lack of social vulnerability considerations highlights an area for improvement in the Geodesign approach. In the "Working with the Tohono O'odham Nation"—"Geodesign workshop" section, the report outlines the integration of geodesign through various methods. These include community-based participatory GIS, a pre-workshop visioning session, a geodesign workshop integrated with GIS practice, and surveys. This multifaceted approach allowed for the incorporation of geodesign principles and community input throughout the process.

4.2.4. Feedback Loops and Iterative Processes

The Sif Oidak District case study illustrates the importance of feedback loops and iterative processes in adapting to evolving conditions and community input, although the lack of future flood risk projections raises questions about the long-term reliability of the developed flood risk map. In this instance, a continuous feedback loop was implemented during community engagement to gather perceptions of flood risks before and after the workshops, allowing for adjustments in project strategies based on this feedback. As part of the Geodesign process, the community identified flood areas, weakened and aging berms, and land uses that could withstand flood damage, as well as areas of severe and minimal flooding. Consequently, the Sif Oidak District community was successful in developing a map depicting historical or known flood risk areas.

4.2.5. Resilience and Adaptability

The study demonstrates the practical value of flood risk assessments in enhancing resilience and adaptability despite its limited approach to incorporating resilience indicators. By focusing primarily on flood risk, the study may have overlooked other critical dimensions of resilience, missing opportunities to provide a more holistic understanding of the community's capacity to cope with and adapt to stresses and shocks. Nonetheless, the case study showcases how flood risk assessments can inform land-use planning and guide development decisions to enhance community resilience. Leveraging local knowledge and scientific data, community members and researchers formulated flood-resilient land-use designations encompassing ten different land uses. The study's adaptability is demonstrated by the initiation of housing developments in low-risk areas identified through community insights and spatial flood risk analysis. However, future studies could benefit from incorporating a broader range of social, economic, and environmental indicators to fully capitalize on the potential of resilience indicators in vulnerability assessments. By taking a more comprehensive approach to assessing resilience, vulnerability assessments can provide a more complete picture of a community's capacity to withstand and recover from stresses and shocks and inform the development of targeted strategies to build resilience over time.

While the Sif Oidak District case study showcases the successful integration of community input and scientific research through the geodesign approach, it also reveals limitations in addressing future flood risks and incorporating iterative processes for land-use planning adaptations. This collaborative effort has successfully identified areas susceptible to severe and minimal flooding, deteriorating berms, and locations ideal for flood-resilient land use, thus promoting resilience in data-scare areas. The limitation of this study lies in the absence of iterative processes for future flood projections and the lack of feedback concerning proposed land-use planning adaptations based on future flood scenarios.

Table 2. Case Study Comparison.

Indicator	Case Study 1	Case Study 2
Spatial Analysis	**Integration of relevant data layers:** The study integrated diverse datasets within a GIS framework, including FEMA flood maps, sea-level rise projections, and socio-economic data. **Use of the spatial analysis techniques:** Overlay analysis and vulnerability mapping	**Integration of relevant data layers:** The study integrated a 10-m digital elevation model, rainfall data, land-use cover, and soil data to assess flood risk. **Use of the spatial analysis techniques:** Flood risk modeling and GIS toolkits
	Generation of spatial outputs: Projection of FEMA 100-year flood plain; regional hazard vulnerability map; flood vulnerability maps, delineating areas of low, medium, and high vulnerability.	**Generation of spatial outputs:** A flood risk map delineating five distinct risk levels (very low, low, medium, high, and very high);
	Spatial resolution: Census block level for socio-economic vulnerability analysis, and municipal and local scales for future flood-prone area projection.	**Spatial resolution:** A 10-m digital elevation model for flood risk modeling.
Stakeholder Engagement	**Types of stakeholders:** Local community members and senior planners of League City.	**Types of stakeholders:** Local community members, district council members, and officials from the Tohono O'odham Nation's planning department.
	Engagement methods: A collaborative, participatory approach, facilitating feedback and critique of proposed design scenarios.	**Engagement methods:** A mix of qualitative and quantitative methods, including surveys, visioning sessions, and Geodesign workshops, indicating a high level of participation and collaboration.
	Incorporation of stakeholder input: The planning process involved refining the community's master plan based on "responses from the community to the design team", which were "utilized to condense the scenarios into one unified revised master plan".	**Incorporation of stakeholder input:** The mapping of *"historical flood risk or known flood risk areas identified by Sif Oidak District community members"* and the development of a flood vulnerability assessment through the *"Sif Oidak Geodesign workshop to develop a community-based land-use plan"*
Interdisciplinary Integration	**Representation of relevant disciplines:** The study combined resilience planning scorecards, geodesign tools, and landscape performance calculators.	**Representation of relevant disciplines:** The study employed Geodesign as a framework for integrating different methods and perspectives.

Table 2. *Cont.*

Indicator	Case Study 1	Case Study 2
Interdisciplinary Integration	**Application of integrative methods and tools:** The study employed a sequential approach, utilizing resilience scorecards for policy analysis, geodesign tools for data integration, and landscape performance calculators for impact assessment.	**Application of integrative methods and tools:** The study combined participatory GIS, visioning sessions, and Geodesign workshops to facilitate interdisciplinary integration.
Feedback Loops and Iterative Processes	**Presence of monitoring and evaluation mechanisms**: As evident from the quote (from results), this demonstrates the presence of feedback mechanisms during stakeholder engagement and the revision of the master plan.	**Presence of monitoring and evaluation mechanisms**: The study implemented a continuous feedback loop during community engagement to gather perceptions of flood risks before and after the workshops.
	Evidence of iterative refinements: The study showed evidence of iterative refinements through the creation and refinery of the master plan. However, it reveals a shortcoming in comprehensively testing the master plan against various future flood scenarios, indicating a need for more robust iterative processes.	**Evidence of iterative refinements:** The study showed evidence of iterative refinements through the identification of flood areas, weakened berms, and flood-resilient land uses based on community input. However, it lacked projections of future flood scenarios, potentially limiting its ability to handle emergent issues related to long-term flood risks.
Resilience and Adaptability	**Incorporation of resilience indicators:** The study developed a data-driven, resilient community plan incorporating a comprehensive array of flood attenuation strategies.	**Incorporation of resilience indicators:** The study considered flood risk as a key indicator for assessing resilience.
	Use of adaptive design principles: The master plan employs a phased approach, including retreat, flood mitigation, and flood control strategies, ensuring adaptability to future conditions.	**Use of adaptive design principles:** The study developed flood-resilient land-use designations based on community insights and spatial flood risk analysis.
	Consideration of long-term climate change impacts and uncertainties: The study evaluates the master plan's performance up to the year 2100, considering projected extreme conditions and sea-level rise.	**Consideration of long-term climate change impacts:** The study did not explicitly address long-term climate change impacts or future flood risk projections.

5. Discussion

Our case studies demonstrate the role of geodesign in assessing flood vulnerability. The findings reveal that geodesign serves as a useful tool in both data-rich and data-scarce contexts. In each of our five indicators—spatial analysis, stakeholder engagement, interdisciplinary integration, feedback loops and iterative processes, and resilience and adaptability—geodesign processes were proven to be vital components in assessing and addressing flood vulnerability.

The case studies presented explore the synergies between geodesign and vulnerability analysis to evaluate the effectiveness of integrating geodesign principles into vulnerability assessments across diverse contexts. The League City case study demonstrates the application of geodesign in a data-rich urban environment, while the Sif Oidak District case study describes a more rural, data-scarce environmental setting. Both case studies highlighted the critical roles of spatial analysis, stakeholder engagement, and interdisciplinary integration in enhancing flood resilience planning. However, they also revealed limitations in addressing social vulnerability, future flood risks, and the need for more robust feedback loops and iterative processes. In this discussion, we synthesize the key findings from these case studies and examine their broader implications for the integration of geodesign and vulnerability analysis in urban and environmental planning.

The comparative analysis of the two case studies emphasizes the importance of utilizing the geodesign approach and integrating comprehensive spatial analysis, stakeholder engagement, and interdisciplinary methods in flood vulnerability resilience planning. In League City, Texas, geodesign principles have been applied comprehensively to assess urban flood resilience in a data-rich environment. On the other hand, the case of Tohono O'odham Nation's Sif Oidak District illustrates the unique challenges and innovative solutions that emerge when implementing geodesign strategies in data-limited environments. As a result of this comparison, it is evident that geodesign is adaptable and useful for enhancing flood resilience planning in various contexts.

Spatial Analysis: Both case studies highlight the indispensable role of spatial analysis in flood risk management. While League City leveraged detailed FEMA data and sea-level rise projections, Sif Oidak employed remote sensing and local knowledge to compensate for data scarcity. Spatial analysis techniques can adapt to varying data availability environments through these approaches. The League City case study demonstrates the effective delineation of ecological and social vulnerabilities at the municipal scale, although the analysis faced limitations in capturing fine-grained social vulnerability data for site-specific design and planning. They have pinpointed the flood-prone area with projections of the sea-level rise. Through the flood projections, they also critically evaluated the existing policy frameworks, assessing their efficacy in mitigating flood risks. Moreover, the articulated flood scenarios at the site-specific level provided essential insights, facilitating the refinement of land-use planning, community design, and landscape architecture initiatives. Nonetheless, the analysis faced challenges in addressing social vulnerability at the municipal scale. Despite utilizing block-scale census data, the resolution was inadequate for capturing the nuanced details of the site and its immediate surroundings, partly due to the data being aggregated into three broad categories. As a result, the study uniformly classified the entire site as having high social vulnerability, while the surrounding areas were categorized as medium to high vulnerability. The broad categorization of vulnerability constrained the data's applicability for detailed design purposes, impeding the capacity for conducting accurate, site-specific analyses. This limitation is significant in the context of landscape architecture and urban planning, where detailed, nuanced information is crucial for informed decision-making and effective design implementation. The Sif Oidak District case study highlights the critical role of spatial analysis in identifying vulnerable areas by integrating remote sensing datasets and GIS toolkits, particularly in data-scarce contexts. Its strength lies in the integration of remote sensing datasets and GIS toolkits. It is a vital step in identifying vulnerable areas.

Stakeholder Engagement: The engagement of local communities and stakeholders proved vital in both contexts, enhancing the relevancy and acceptance of the planning processes. In League City, stakeholder feedback refined urban planning initiatives, while in the Sif Oidak District, it directly influenced the development of a flood risk map and land-use strategies. Nevertheless, both cases revealed gaps in documenting the extent to which stakeholder input impacted the final outcomes, pointing to a need for more transparent and detailed engagement records. While stakeholder engagement was crucial in shaping the League City master plan, the lack of detailed documentation of community members' specific suggestions limits the ability to fully assess the impact of their contributions on the final design and planning outcomes. Its process ensured that the plan was reflective of the community's needs and aspirations, therefore enhancing its relevance and effectiveness in addressing the site-specific challenges and opportunities. However, in this context, there is a limitation in the detailed documentation of community members' specific suggestions, therefore restricting clarity on how their contributions tangibly shaped the planning outcomes. This gap hinders the ability to fully assess the impact of stakeholder engagement on the final design and planning process.

Interdisciplinary Integration: In addressing the complex nature of flood risks, these studies demonstrate the advantages of integrating interdisciplinary research. To conduct a comprehensive flood vulnerability assessment, it is necessary to combine various disciplines due to the complexity of natural and human systems. In the cases examined, the synthesis of design, Geographic Information Systems (GIS), geography, and urban planning was instrumental in delineating flood zones and formulating flood resilience strategies. Specifically, in League City, the adoption of additional methods, such as resilience scorecards and landscape performance tools, further enriched the analytical framework. In contrast, the Sif Oidak District benefitted from the incorporation of indigenous knowledge, which played a crucial role in comprehending areas susceptible to flooding. These examples illuminate the efficacy of interdisciplinary approaches in crafting holistic and sustainable solutions to flood risks, demonstrating their capacity to enhance resilience and adaptability in the face of environmental challenges. The sequential implementation of resilience planning scorecards, Geodesign tools, and landscape performance calculators in the League City case study facilitated a systematic and focused analysis of the city's vulnerabilities and the effectiveness of proposed mitigation strategies. This process allowed for a systematic examination of League City's susceptibility to flooding and other climate-related risks, followed by an assessment of the effectiveness of planning and mitigation strategies in addressing these identified risks. The sequential approach facilitated a focused and structured analysis, ensuring that each phase built upon the insights gained from the previous one, thus creating a comprehensive understanding of both the challenges faced and the potential effectiveness of the proposed solutions.

Although the League City case study initially considered social vulnerability, the master plan and subsequent evaluations primarily focused on flood mitigation, potentially overlooking the importance of social resilience and community needs. Despite the initial comprehensive considerations in policymaking and social vulnerability assessment, the context reveals that the master plan and its subsequent evaluations predominantly focused on flood mitigation, inadvertently sidelining social vulnerability issues. This suggests a potential imbalance in the approach, where the emphasis on physical and environmental resilience may have overshadowed the equally critical aspects of social resilience and community needs. To rectify this, it would be important for future revisions of the master plan to integrate social vulnerability considerations more thoroughly, ensuring that the strategies developed are holistic and address both environmental and social dimensions of resilience.

Similarly, the participatory GIS and Geodesign workshop are highlights of the Sif Oidak District case study, as they improve stakeholder engagement and expand access to geospatial data. Although this interdisciplinary approach helps with spatially identifying flood areas, there is no consideration of social vulnerability to floods. Social vulnerability is a critical aspect, as understanding the social dimensions of vulnerability, including factors like socio-economic status, age, disability, and access to resources, is essential for effective flood risk management. Addressing social vulnerability can ensure that flood management strategies are equitable and adequately protect the most at-risk populations. Integrating social vulnerability assessments into the GIS and Geodesign frameworks would be beneficial in enhancing this approach. This could involve collecting and analyzing data on the social, economic, and demographic characteristics of communities in flood-prone areas to identify vulnerable groups. By doing so, the strategy would be not only technically robust in identifying where floods might occur but also socially sensitive in understanding who is most at risk and why, allowing for more targeted and effective mitigation and response efforts.

Feedback. The significance of feedback loops and iterative processes in maintaining the effectiveness of flood resilience strategies over time is a key insight that emerges from the case studies. While current approaches have achieved improvements in flood resilience, the studies emphasize the need for more detailed scenarios and robust feedback mechanisms to facilitate systematic refinement of mitigation strategies in response to evolving flood scenarios. These findings underscore the importance of continuous evaluation and adaptation in ensuring the long-term relevance and effectiveness of flood risk plans. The League City case study demonstrates the importance of feedback loops and iterative processes in ensuring the adaptability and resilience of the master plan but also reveals a shortcoming in comprehensively testing the plan against various future flood scenarios. While feedback mechanisms were evident during stakeholder engagement and the revision of the master plan, the study failed to thoroughly evaluate the master plan's performance under projected flood scenarios, such as the floodplain projections for 2025, 2050, and 2100 with sea-level rise, as well as the 100 year floodplain. This gap indicates a need for a more robust application of feedback loops and iterative processes to ensure that the master plan remains resilient and effective under a range of potential environmental changes and challenges, enhancing its ability to adapt to future conditions.

However, the Sif Oidak District case study demonstrates the lack of focus on feedback loops and iterative processes concerning future flood risks since no projections of future flood scenarios are provided. This results in questions regarding the feasibility and reliability of the flood risk map for future use in the district.

Resilience and Adaptability. The case studies highlight the importance of incorporating resilience and adaptability into flood risk management strategies. League City's case study demonstrates the value of a strategic, long-term approach to building resilience. Similarly, the Sif Oidak District case study showcases how flood risk assessments can enhance resilience and adaptability by informing flood-resilient land-use designations and guiding practical applications. However, it also reveals limitations in addressing long-term resilience due to the lack of iterative processes for future flood projections and feedback on proposed land-use planning adaptations. To further enhance resilience and adaptability, future studies should consider integrating nature-based solutions and ecosystem-based approaches that recognize the critical role of ecosystems in providing essential services. These approaches can help identify strategies that work with nature, harnessing the resilience and adaptability of ecosystems to support human systems.

This work contributes to our understanding of vulnerability as it relates to landscape design processes. More specifically, it furthers knowledge of the integration of geodesign principles in flood risk and vulnerability assessments. Contributions include highlighting the role of spatial analysis in delineating ecological and social vulnerabilities at various scales, emphasizing the importance of stakeholder engagement in design initiatives, demonstrating the value of interdisciplinary integration in developing comprehensive flood resilience solutions, underscoring the significance of feedback loops and iterative processes in maintaining long-term effectiveness, and showcasing the potential of geodesign in informing flood-resilient land-use designations and guiding practical applications to enhance community resilience. Based on these findings, we have several (7) key planning and design recommendations. First, it is essential to invest in robust spatial data infrastructure including environmental, social, and economic datasets, to support comprehensive vulnerability assessments. In data-scarce contexts, exploring alternative data sources, such as remote sensing and local knowledge, can help overcome data limitations. Second, fostering active stakeholder engagement by involving a diverse set of stakeholders, including local community members, regional experts, and local decision-makers throughout the planning and design process is crucial. Developing transparent and detailed engagement records can help document the impact of stakeholder input on final outcomes. This helps justify the results and ease the transition from data to design decisions. Third, embrace interdisciplinary collaboration. Assembling multidisciplinary teams comprising experts from design, GIS, geography, urban planning, and other relevant fields can lead to the development of holistic, robust, defensible, and sustainable flood risk management strategies. Fourth, implementing iterative planning and design processes (like geodesign processes) that help establish mechanisms for continuous monitoring, evaluation, and refinement of flood resilience strategies based on evolving flood scenarios and stakeholder feedback is essential for long-term success. Fifth, integrate resilience and adaptability indicators that incorporate a broad range of social, economic, and environmental measures that can help assess community resilience and inform the development of targeted strategies for building long-term resilience. Sixth, explore nature-based solutions that integrate ecosystem-based approaches and that recognize the critical role of ecosystems in providing essential services for enhancing resilience. Finally, promote knowledge sharing and capacity building by encouraging the exchange of best practices and lessons learned to foster continuous improvement and innovation in design for flood risk management. By adopting these recommendations, planning professionals and authorities can harness the potential of geodesign to develop more effective, resilient, and adaptable flood risk management strategies that address the complex challenges posed by changing environmental conditions and social vulnerabilities.

6. Conclusions

This study questions the potential of integrating geodesign and vulnerability analysis to enhance resilience and sustainability in urban and environmental planning. The case studies of League City, Texas, and the Tohono O'odham Nation in Arizona provide insights into the application of the interdisciplinary approach across diverse contexts, highlighting the critical roles of spatial analysis, stakeholder engagement, and interdisciplinary integration in flood resilience planning and design.

Our investigation suggests that the integration of geodesign with vulnerability analysis can provide useful information for improving sustainable and resilient flood-based outcomes in terms of landscape design and urban and environmental planning. The interdisciplinary nature, engaged processes, and advanced spatial analytics in geodesign provide important components for identifying and addressing vulnerable places. The League City, Texas, and Tohono O'odham Nation case studies support its efficacy in improving flood resilience.

In League City, the integration of resilience planning scorecards, Geodesign tools, and landscape performance calculators allowed for a detailed assessment of flood vulnerabilities and design interventions. Conversely, the Tohono O'odham Nation study demonstrates the utility of geodesign in data-scarce environments, where community-based participatory methods were crucial in developing a flood resilience plan.

Both case studies underscore the importance of spatial analysis, stakeholder engagement, and interdisciplinary integration in crafting effective flood resilience strategies. However, they also revealed the need for more robust feedback and iterative processes to ensure the adaptability and long-term effectiveness of the resilience plans. The studies highlighted that while geodesign can significantly enhance the technical assessment of flood risks, integrating social vulnerability considerations remains essential for developing equitable and comprehensive strategies.

Future research should aim to fill the gap between technical assessments and social considerations. In addition, follow-up work should evaluate a more stringent and comprehensive scoring framework for determining whether design outcomes or plans truly reflect community needs and aspirations.

Author Contributions: Conceptualization, B.P. and B.D.; methodology, B.P. and B.D.; formal analysis, B.P.; investigation, B.P.; resources, B.P.; writing—original draft preparation, B.P.; writing—review and editing, B.P. and B.D.; visualization, B.P.; project administration, B.P. All authors have read and agreed to the published version of the manuscript.

Funding: This research received no external funding.

Data Availability Statement: Data are contained within the article.

Conflicts of Interest: The authors declare no conflicts of interest.

References

1. Steinitz, C. A Framework for Geodesign: Changing Geography by Design. *J. Landsc. Archit.* **2012**, *7*, 87.
2. McElvaney, S. *Geodesign: Case Studies in Regional and Urban Planning*; Environmental Systems Research Institute: Redlands, CA, USA, 2012.
3. Gu, Y.; Deal, B. Geodesign, resilience and planning support systems: The integration of process and technology. In *Handbook of Planning Support Science*; Edward Elgar Publishing: Gloucestershire, UK, 2020; pp. 110–131.
4. Flaxman, M. *Geodesign: Fundamental Principles and Routes Forward*; GeoDesign Summit ESRI: Redlands, CA, USA, 2010.
5. Dias, E.; Linde, M.; Scholten, H.J. Geodesign: Integrating geographical sciences and creative design in envisioning a 'New Urban Europe'. *Int. J. Glob. Environ. Issues* **2015**, *14*, 164–174. [CrossRef]
6. Slotterback, C.S.; Runck, B.; Pitt, D.G.; Kne, L.; Jordan, N.R.; Mulla, D.J.; Zerger, C.; Reichenbach, M. Collaborative Geodesign to advance multifunctional landscapes. *Landsc. Urban Plan.* **2016**, *156*, 71–80. [CrossRef]
7. Li, Y.; Kim, Y. Analysis of effects of sponge city projects applying the geodesign framework. *Land* **2022**, *11*, 455. [CrossRef]
8. Rolf, W.; Peters, D.G. Algorithmic Landscapes Meet Geodesign for Effective Green Infrastructure Planning: Ideas and Perspectives. *J. Digit. Landsc. Archit.* **2020**, *5*, 468–476.
9. Cutter, S.L. Vulnerability to environmental hazards. *Prog. Hum. Geogr.* **1996**, *20*, 529–539. [CrossRef]
10. Huang, L.; Xiang, W.; Wu, J.; Traxler, C.; Huang, J. Integrating GeoDesign with landscape sustainability science. *Sustainability* **2019**, *11*, 833. [CrossRef]
11. Turner, B.L.; Kasperson, R.E.; Matson, P.A.; McCarthy, J.J.; Corell, R.W.; Christensen, L.; Eckley, N.; Kasperson, J.X.; Luers, A.; Martello, M.L.; et al. A framework for vulnerability analysis in sustainability science. *Proc. Natl. Acad. Sci. USA* **2003**, *100*, 8074–8079. [CrossRef] [PubMed]
12. Birkmann, J. Measuring vulnerability to promote disaster-resilient societies: Conceptual frameworks and definitions. In *Measuring Vulnerability to Natural Hazards: Towards Disaster Resilient Societies*; The Energy and Resources Institute (TERI): New Delhi, India, 2006; Volume 1, pp. 3–7.
13. Holling, C.S. Resilience and stability of ecological systems. *Annu. Rev. Ecol. Syst.* **1973**, *4*, 1–23. [CrossRef]
14. Ludwig, D.; Walker, B.; Holling, C.S. Sustainability, stability, and resilience. *Conserv. Ecol.* **1997**, *1*, 21. [CrossRef]
15. Luthar, S.S. Resilience in development: A synthesis of research across five decades. In *Developmental Psychopathology: Volume Three: Risk, Disorder, and Adaptation*; Wiley: Hoboken, NJ, USA, 2015; Volume 3, pp. 739–795.
16. Cutter, S.L.; Boruff, B.J.; Shirley, W.L. Social vulnerability to environmental hazards. In *Hazards Vulnerability and Environmental Justice*; Routledge: London, UK, 2012; pp. 143–160.
17. Gu, Y.; Deal, B.; Larsen, L. Geodesign processes and ecological systems thinking in a coupled human-environment context: An integrated framework for landscape architecture. *Sustainability* **2018**, *10*, 3306. [CrossRef]
18. Berkes, F.; Colding, J.; Folke, C. *Navigating Social-Ecological Systems: Building Resilience for Complexity and Change*; Cambridge University Press: Cambridge, UK, 2008.
19. Campagna, M. Metaplanning: About designing the Geodesign process. *Landsc. Urban Plan.* **2016**, *156*, 118–128. [CrossRef]
20. Adger, W.N. Vulnerability. *Glob. Environ. Change* **2006**, *16*, 268–281. [CrossRef]
21. Pelling, M. *The Vulnerability of Cities: Natural Disasters and Social Resilience*; Routledge: London, UK, 2012.
22. De Chazal, J.; Quétier, F.; Lavorel, S.; Van Doorn, A. Including multiple differing stakeholder values into vulnerability assessments of socio-ecological systems. *Glob. Environ. Change* **2008**, *18*, 508–520. [CrossRef]
23. Newman, G.; Malecha, M.; Yu, S.; Qiao, Z.; Horney, J.A.; Lee, J.; Kim, Y.J.; Lee, R.J.; Berke, P. Integrating a resilience scorecard and landscape performance tools into a Geodesign process. *Landsc. Res.* **2020**, *45*, 63–80. [CrossRef]
24. Eikelboom, T.; Janssen, R. Collaborative use of geodesign tools to support decision-making on adaptation to climate change. *Mitig. Adapt. Strateg. Glob. Change* **2017**, *22*, 247–266. [CrossRef] [PubMed]
25. Yin, R.K. *Case Study Research: Design and Methods (Applied Social Research Methods)*; Sage Publications: Thousand Oaks, CA, USA, 2014.
26. Perrow, C. A framework for the comparative analysis of organizations. *Am. Sociol. Rev.* **1967**, *32*, 194–208. [CrossRef]
27. Lange, M. *Comparative-Historical Methods*; SAGE Publications Ltd.: Southend Oaks, CA, USA, 2012.
28. Abu-Lughod, L. Locating ethnography. *Ethnography* **2000**, *1*, 261–267. [CrossRef]
29. Burby, R.J.; Beatley, T.; Berke, P.R.; Deyle, R.E.; French, S.P.; Godschalk, D.R.; Kaiser, E.J.; Kartez, J.D.; May, P.J.; Olshansky, R.; et al. Unleashing the power of planning to create disaster-resistant communities. *J. Am. Plan. Assoc.* **1999**, *65*, 247–258. [CrossRef]
30. Portney, K.E. *Taking Sustainable Cities Seriously: Economic Development, the Environment, and Quality of Life in American Cities*; MIT Press: Cambridge, MA, USA, 2013.
31. Rose, R.; Mackenzie, W.J.M. Comparing forms of comparative analysis. *Political Stud.* **1991**, *39*, 446–462. [CrossRef]
32. Marsal-Llacuna, M.L. Measuring the Standardized Definition of "smart city": A Proposal on Global Metrics to Set the Terms of Reference for Urban "smartness". In Proceedings of the Computational Science and Its Applications—ICCSA 2015: 15th International Conference, Banff, AB, Canada, 22–25 June 2015; Proceedings, Part II 15; Springer: Berlin/Heidelberg, Germany, 2015; pp. 593–611.
33. Frey, H. *Designing the City: Towards a More Sustainable Urban Form*; Taylor & Francis: New York, NY, USA, 2003.

34. Solecki, W.; Seto, K.C.; Balk, D.; Bigio, A.; Boone, C.G.; Creutzig, F.; Fragkias, M.; Lwasa, S.; Marcotullio, P.; Romero-Lankao, P.; et al. A conceptual framework for an urban areas typology to integrate climate change mitigation and adaptation. *Urban Clim.* **2015**, *14*, 116–137. [CrossRef]
35. Azarian, R. Potentials and limitations of comparative method in social science. *Int. J. Humanit. Soc. Sci.* **2011**, *1*, 113–125.
36. Losos, J.B. Uncertainty in the reconstruction of ancestral character states and limitations on the use of phylogenetic comparative methods. *Anim. Behav.* **1999**, *58*, 1319–1324. [CrossRef] [PubMed]
37. Boas, F. The limitations of the comparative method of anthropology. *Science* **1896**, *4*, 901–908. [CrossRef]
38. Collier, D. The comparative method. In *Political Science: The State of Discipline II*; Finifter, A.W., Ed.; American Political Science Association: Washington, DC, USA, 1993.
39. Timm, S.N. Comparative Planning in Practice: Identifying the Cultural and Behavioral Barriers to Learning from International Sustainability Solutions. Ph.D. Thesis, University of Illinois at Urbana-Champaign, Champaign, IL, USA, 2016.
40. Miles, M.B.; Huberman, A.M. *Qualitative Data Analysis: An Expanded Sourcebook*; Sage: New York, NY, USA, 1994.
41. Li, W.; Milburn, L.A. The evolution of geodesign as a design and planning tool. *Landsc. Urban Plan.* **2016**, *156*, 5–8. [CrossRef]
42. Foster, K. Geodesign parsed: Placing it within the rubric of recognized design theories. *Landsc. Urban Plan.* **2016**, *156*, 92–100. [CrossRef]
43. Bollens, S.A. *Comparative Research on Contested Cities: Lenses and Scaffoldings*; Crisis States Research Centre London: London, UK, 2007.
44. Wong, K.Y.; Casey, R.G.; Wahl, F.M. Document analysis system. *IBM J. Res. Dev.* **1982**, *26*, 647–656. [CrossRef]

MDPI AG
Grosspeteranlage 5
4052 Basel
Switzerland
Tel.: +41 61 683 77 34

Land Editorial Office
E-mail: land@mdpi.com
www.mdpi.com/journal/land

www.ingramcontent.com/pod-product-compliance
Lightning Source LLC
Chambersburg PA
CBHW041946260926
51850CB00018B/972
* 9 7 8 3 7 2 5 8 6 3 4 8 8 *